CHAOS AND ORDER

CHAOS

New Practices of Inquiry

A Series Edited by Donald N. McCloskey and John S. Nelson

AND ORDER

Complex Dynamics in Literature and Science

Edited by
N. Katherine Hayles

The University of Chicago Press
Chicago and London

The University of Chicago Press, Chicago 60637
The University of Chicago Press, Ltd., London

Printed in the United States of America

00 99 98 97 96 95 94 93 92 5 4 3 2

Library of Congress Cataloging-in-Publication Data

Chaos and order : complex dynamics in literature and science / edited by N. Katherine Hayles.
p. cm.—(New Practices of inquiry)
Includes bibliographical references and index.
ISBN 0-226-32143-6 (cloth).—ISBN 0-226-32144-4 (paper).
1. Literature, Modern—20th century—History and criticism. 2. Chaotic behavior in systems in literature. 3. Literature and science. I. Hayles, N. Katherine. II. Series.
PN771.C44 1991
809—dc20 90-20264

⊗ The paper used in this publication meets the minimum requirements of the American National Standard for Information Sciences—Permanence of Paper for Printed Library Materials, ANSI Z39.48-1984.

Contents

Acknowledgments

This volume has benefited from many hands. It could not have seen the light of day without the cooperation and advice of the contributors. Jules van Lieshout provided invaluable assistance in tracking down errant references, insuring consistency, and preparing the manuscript. Sofia Lesinska also worked on the manuscript, and it is the better for her efforts. University House at the University of Iowa provided support through the use of its facilities; special thanks are due to Jay Semel and Lorna Olson. Douglas Mitchell gave encouragement and advice, and Elizabeth Churchwell helped to prepare the index.

1

Introduction: Complex Dynamics in Literature and Science

N. Katherine Hayles

The law of chaos is the law of ideas, of improvisations and seasons of belief.
Wallace Stevens, "Extracts from Addresses to the Academy of Fine Ideas"

I

This volume focuses on a question that is attracting increasing attention within the sciences and humanities: what is the relation between order and disorder? Traditionally, of course, they have been regarded as opposites. Order was that which could be classified, analyzed, encompassed within rational discourse; disorder was allied with chaos and by definition could not be expressed except through statistical generalizations. The last twenty years have seen a radical reevaluation of this view. In both contemporary literature and science, chaos has been conceptualized as extremely complex information rather than an absence of order. As a result, textuality is conceived in new ways within critical theory and literature, and new kinds of phenomena are coming to the fore within an emerging field known as the science of chaos.

A science of chaos may seem to be a contradiction in terms. In the scientific sense, however, chaos means something different than it does in common usage. At the center of chaos theory is the discovery that hidden within the unpredictability of chaotic systems are deep structures of order. "Chaos," in this usage, denotes not true randomness but the orderly disorder characteristic of these systems. The science of chaos seeks to understand behavior so complex that it defeats the usual methods of formalizing a system through mathematics. Hence the science of chaos has also been called the science of complexity—or more precisely the sciences of complexity, for fields as diverse as meteorology, irreversible thermodynamics, epidemiology, and nonlinear dynamics are included within the rubric. The kinds of systems to which chaos models have been successfully applied range from dripping faucets to measles epidemics, schizophrenic eye movements to fluctuations in fish populations.[1] Recently there has been considerable interest in using chaos theory to understand the stock market.[2] Although it is too soon to say where the discoveries associated with complex systems will end, it is already apparent that chaos theory is part of a paradigm shift of remarkable scope and significance.

Among the controversial issues within the science of chaos is the word "chaos" itself. No sooner did the term become widely associated with nonlinear dynamics than practitioners in the field began to shy away from it, regarding it as an imprecise, even sensationalized, word that was unnecessarily confusing. As the term gained notoriety, chiefly through James Gleick's popular book *Chaos: Making a New Science*, it lost credibility within the scientific community. To many, the word has now become so thoroughly deprofessionalized that its use is regarded as a signal that one is in the presence of a dilettante rather than an expert. Nevertheless, it will be retained in this volume precisely because of the ambiguous meanings that inhere within it. Marked by scientific denotations as well as historical and mythic interpretations, it serves as a crossroads, a juncture where various strata and trends within the culture come together.

The cultural attitudes that mark the word "chaos" can be read in its etymology. The word derives from a Greek verb-stem, KHA, meaning "to yawn, to gape"; from this comes the meaning given by the *Oxford English Dictionary*, "a gaping void, yawning gulf, chasm, or abyss." Creation myths in the West, from the Babylonian epic *Enuma Elish* to Milton's *Paradise Lost*, depict chaos as a negative state, a disordered void which must be conquered for creation to occur. Eugene Eoyang ("Chaos Misread"), writing about *Enuma Elish*, recalls the dramatic moment in the epic when the demoness Tiamat is slain by the young warrior Marduk. He explains that Tiamat stands for the entropic tendency toward dissolution, the force of decay that would return everything to darkness. Her death signals the beginning of a new era. Marduk, exultant over Tiamat's carcass, exclaims that he will create a new "savage," and that "'man' shall be his name" (quoted in Eoyang 272). The emphasis on man as savage implies that chaos is the aboriginal foe which must be vanquished before the civilizing process can begin. In this semiotic, chaos is opposed to civilized values as well as to the initiating act of creation.

Eoyang contrasts these assumptions with those encoded into the ancient Taoist creation myth recounted in the seventh chapter of the *Zhuangzi*. Shu (Brief) and Hu (Sudden) go to visit Hun-dun (Chaos), who graciously offers them his hospitality. Observing that Hun-dun lacks the seven openings through which men see, hear, eat, and breathe, Shu and Hu determine to create them. Each day they bore a new hole. On the seventh day, Hun-dun dies. Here the destruction of chaos, far from marking the beginning of civilization, bespeaks a provinciality unable to accept an other different than the self. In contrast to the triumphal climax of the Western epic, the Taoist story ends with an ironic twist.

The etymology of "Hun-dun" provides a similar contrast. Eoyang explains that the word belongs to a category of rhyming compounds whose sounds enact their meanings. The onomatopoetic quality that Hun-dun suggests is whirling water, flowing turbulence, swirling action. According to Eoyang, the word is not so much a nominative as an attributive adjective, a quality rather than an object (275). He quotes a scholar from the Jin Dynasty, Du Yü, who wrote that Hun-dun was "something whose appearance cannot be seen through" (quoted in Eoyang 275). Thus in the myth Hun-dun has no openings, and when his visitors insist on providing him with some, he dies. There is the delicate suggestion that Hun-dun stands outside the world of reified concepts Taoist irony tries to puncture. Chaos remains the necessary other, the opaque turbulence that challenges and complements the transparency of order.

That chaos has been negatively valued in the Western tradition may be partly due, Eoyang suggests ("Heuristics"), to the predominance of binary logic in the West. If order is good, chaos is bad because it is conceptualized as the opposite of order. By contrast, in the four-valued logic characteristic of Taoist thought, not-order is also a possibility, distinct from and valued differently than anti-order. The science of chaos draws Western assumptions about chaos into question by revealing possibilities that were suppressed when chaos was considered merely as order's opposite. It marks the validation within the Western tradition of a view of chaos that constructs it as not-order. In chaos theory chaos may either lead to order, as it does with self-organizing systems, or in yin/yang fashion it may have deep structures of order encoded within it. In either case, its relation to order is more complex than traditional Western oppositions have allowed.

This cultural history helps to explain why the science of chaos should seem revolutionary within the Western tradition. Many of the methods of nonlinear dynamics are not new, going back to the nineteenth century; and many of its central insights had been anticipated in the nineteenth century and before. Yet calling the object of inquiry "chaos" caused something to click that had not before. In 1975 when James

Yorke, with characteristic flair, chose the flashy title "Period Three Implies Chaos" for a seminal paper on deterministic disorder (Yorke and Li), he tapped into a network of presuppositions that may soon put the emerging science of chaos on a par with evolution, relativity, and quantum mechanics in its impact on the culture. The furor over chaos has made many practitioners in nonlinear dynamics feel it would be best to avoid the word altogether. They see themselves as solving practical and technical problems, not rewriting cultural history.

Yet cultural traditions are not so easily escaped. They are encoded not merely into words but also into practices, institutions, and material conditions. If on the one hand this means that chaos theory can scarcely avoid having implications for the culture beyond its technical achievements, on the other it implies that existing practices exert an inertial pull on new ideas. The embedding of chaos theory into existing disciplinary contexts makes it difficult to determine how revolutionary it is. The claim is sometimes heard, for example, that the science of chaos challenges traditional ideas of how science is done. But the science of chaos is not opposed to normal science. It *is* normal science. Its criteria for evaluating evidence, reproducing results, credentialling investigators, and so on, differs not at all from those of other physical sciences. Thus its insights have evolved within contexts that partially reinscribe the very assumptions these insights draw into question. The result is no simple revolution in which a new view replaces an old. Rather, change occurs through negotiations at multiple sites among those who generate data, interpret them, theorize about them, and extrapolate beyond them to broader cultural and philosophical significances.

This volume aims to investigate these negotiations. It treats chaos both as a subject of scientific inquiry and a crossroads where various paths within the culture converge. It explores how insights that work against the grain of Western culture mingle with other cultural currents, changing and being changed by the resistances they encounter. It articulates chaos theory together with developments in the human sciences and postmodern culture. Such an approach implies, of course, that the science of chaos is part of the culture, and that scientists, like everyone else, are affected by the culture in which they are immersed. Yet so strong is the ideology of scientific objectivity that practitioners and laymen alike often speak as if scientists were hermetically sealed within the laboratory, isolated from and immune to the thousands of experiences that constitute the fabric of everyday life.

This hermetic view of scientific inquiry has recently been challenged on a number of fronts. Among the challengers are feminist critiques that explore gender issues, sociological studies that investigate how science is socially constructed, and ethnographic analyses of scientific communities as tribes with their own vocabularies, rituals, and social practices.[3] These studies show that even if we leave aside such obvious and important influences on scientific enterprises as sources of money, governmental regulations, commercial possibilities, and popular opinion, there remains a host of other cultural factors less visible but no less constitutive of scientific inquiry.

The one most relevant here is language. In general, scientific discourse adopts as its ideal univocality—one word, one meaning. Closely related to this goal is the belief that a language exists, or can be forged, that is purely instrumental. Clearly and unambiguously, it will communicate to the world what the speaker or writer intends to say. Roland Barthes (*Rustle*) has ironically called this the belief that science can own a slave language, docile and obedient to its demands.[4] Anyone who has seriously studied how language works is aware, however, that it shapes even as it articulates thought. There is now an impressive body of work exploring how metaphors, narrative patterns, rhetorical structures, syntax, and semantic fields affect scientific discourse and thought. Prominent examples include Gillian Beer's *Darwin's Plots*, Donald McCloskey's *The Rhetoric of Economics*, Michael Arbib and Mary Hesse's *The Construction of Reality*, Charles Bazerman's *Shaping Written Knowledge*, and Bruno Latour's *Science in Action*. These studies, along with many others, demonstrate that language is not a passive instrument but an active engagement with a vital medium that has its own currents, resistances, subversions, enablings, pathways, blockages. As soon as discovery is communicated through language, it is also constituted by language.

That language is interactive rather than inert implies that chaos theory is influenced by the culture within which it arose. In my view, it is a mistake to assume that the science of chaos has initiated the attitudes that have made it an object of popular fascination. Rather, it is one site within the culture where the premises characteristic of postmodernism are inscribed. The postmodern context catalyzed the formation of the new science by providing a cultural and technological milieu in which the component parts came together and mutually reinforced each other until they were no longer isolated events but an emergent awareness of the constructive roles that disorder, nonlinearity, and noise play in complex systems. The science of chaos is new not in the sense of having no antecedents in the scientific tradition, but of having only recently coalesced sufficiently to articulate a vision of the world. It is no accident that this vision has deep affinities with other articulations that have emerged from the postmodern context.

What are these affinities? The material conditions under which the science evolved are relevant as well as linguistic and cultural factors. The microcomputer, for example, has been extremely important in the development of chaos theory, for it allows mathematics to be practiced as an experimental science. It has also affected how people have imaged themselves and their relation to the world. The two come together when someone sits down in front of a computer to model a dynamical nonlinear system. Because the computer permits interaction, the practitioner need not proceed through the traditional mathematical method of theorem-proof. Instead she can set up a recursive program that begins when she feeds initial values for the equations into the computer. Then she watches as the screen display generated by the recursion evolves into constantly changing, often unexpected patterns. As the display continues, she adjusts the parameters to achieve different effects. With her own responses in a feedback loop with the computer, she develops an intuitive feeling for how the display and parameters interact. She notices that small changes in initial values can lead to large changes in the display. She also sees that, although the displays are complex, there are underlying symmetries that impart a pleasing, sometimes a startlingly beautiful quality to them. And she is subliminally aware that her interaction with the display could be thought of as one complex system (the behavior described by a set of nonlinear differential equations) interfaced with another (the human neural system) through the medium of the computer.

What presuppositions does this situation embody and reinforce? Perhaps the most obvious is that the connective tissue holding the system together is the flow of information circulating through it. William Gibson, author of the cyberpunk novel *Neuromancer*, remarked in an interview that a teenager playing an arcade video game illustrates how an informational feedback loop connects human and machine.[5] Photons leave the screen, enter the teenager's eyes, and trigger neural responses that coordinate with hand movements, which in turn cause the electronic circuitry of the machine to produce more photons. When such experiences are everyday events, a context is created that makes information flow seem as real as the matter and energy that carries it—or more real.

This context affects how theoretically abstract ideas such as chaos are invested with meaning. An important turning point in the science of chaos occurred when complex systems were conceptualized as systems rich in information rather than poor in order. One could construct a history that would treat this development as if it emerged solely from the internal logic of the scientific tradition, for example by analyzing competing definitions of entropy and examining their relation to information theory.[6] But the context that made disorder appear as complex information is not confined to scientific inquiry alone. It is part of a cultural milieu that included World War II, which among other things was an object lesson in the importance of information; consolidation of power by multinational corporations and the accompanying sense that the world was growing at once more chaotic and more totalized; increasing economic interdependencies between nations, which brought home to nearly everyone that small causes could lead to large-scale effects; and rapid expansion of information technologies.[7] All of these factors, and more, contributed to the cultural matrix out of which the science of chaos grew.

It should not be surprising, then, to find other sites within the culture that also embody the presuppositions informing chaos theory. To distinguish between the science of chaos narrowly defined and this broader cultural phenomenon, I will appropriate a term suggested to me by Ihab Hassan and call the latter chaotics.[8] The term signifies certain attitudes toward chaos that are manifest at diverse sites within the culture, among them poststructuralism and the science of chaos. The question of how such isomorphisms arise is not easily answered. Let me say at the outset, however, that I do not assume they are the result of direct influence between one site and another. In particular, I am *not* arguing that the science of chaos is the originary site from which chaotics emanates into the culture. Rather, both the literary and scientific manifestations of chaotics are involved in feedback loops with the culture. They help to create the context that energizes the questions they ask; at the same time, they also ask questions energized by the context.

The impact of chaotics derives less from specific theories than from the general awareness it fosters of nonlinear processes and forms. To some extent, we see what we are taught to see. How does the world envisioned by chaotics differs from that of Newtonian mechanics? The Newtonian paradigm emphasizes predictability. Such a mindset is exemplified by Laplace's famous boast that, given the initial conditions and an intelligence large enough to perform the calculations, he could predict the state of the universe at any future moment. By contrast, chaotics celebrates unpredictability, seeing it as a source of new information. Whereas Newtonian mechanics envisions the universe through inertial reference frames that extend infinitely far in space and time, chaotics concentrates on complex irregular forms and conceptualizes them (in fractal geometry) through fractional dimensions that defeat tidy predictions and exact symmetries. Newtonian mechanics and the Euclidean geometry on which it is based are scale-invariant; whether the sides of an isoceles triangle are two centimeters or two kilometers long, the triangle still has the same properties. By contrast chaotics takes scale into account, recognizing that for complex systems and irregular forms, statements made about one scale level do not necessarily hold true at another. Moreover, it emphasizes that couplings between levels are complex and unpredictable. The Newtonian expectation is that small causes lead to small effects, but chaotics looks at systems where minute fluctuations are amplified into dramatic large-scale changes.

The world as chaotics envisions it, then, is rich in unpredictable evolutions, full of complex forms and turbulent flows, characterized by nonlinear relations between causes and effects, and fractured into multiple-length scales that make globalization precarious. The difference between the two paradigms is expressed by the icons often associated with them. Whereas the Newtonians focused on the clock as an appropriate image for the world, chaos theorists are apt to choose the waterfall. The clock is ordered, predictable, regular, and mechanically precise; the waterfall is turbulent, unpredictable, irregular, and infinitely varying in form. The change is not in how the world actually is—neither clocks nor waterfalls are anything new—but in how it is seen. The broadest implications of chaotics derive from this change in vision.

II

In the prescient lines quoted in the epigraph, Wallace Stevens refers to the "law of chaos" as the "law of ideas." Not long ago, the thought that chaos could demonstrate law-like behavior was an oxymoron. In the science of chaos, it has become a truism. One of the new science's remarkable discoveries is that complex systems follow predictable paths to randomness and trace recognizable patterns when they are mapped into time-series diagrams. Embodied within these regularities are ideas that have implications beyond the bounds of scientific inquiry. Two terms taken from the science of chaos will be useful in explaining them: strange attractors and recursive symmetries.

An attractor is any point of a system's cycle that seems to attract the system to it. The midpoint of a pendulum's path is an example. A pendulum, no longer pushed, spontaneously returns to this point. To illustrate a strange attractor, we can construct a device called a double planar pendulum.[9] Imagine a normal pendulum, with one end fixed and one end swinging free. When the amplitude of oscillation is small, its motion can be modeled using linear differential equations. Now suppose that we fasten a second pendulum at the swinging end of the first pendulum, so that the structure becomes double-jointed, as it were. Although the double pendulum follows Newtonian laws of motion as rigorously as a simple pendulum, the evolution of the double-jointed system cannot be predicted. Moreover, over a wide range of energies, its motion cannot adequately be described using linear equations. This complexity derives from the extreme sensitivity of the second pendulum's momentum and position to even very slight changes in the first pendulum. To know how the second pendulum will swing and therefore how the system as a whole will evolve, it would be necessary to know the starting conditions of the first pendulum *with infinite precision*. Since this is impossible, the system remains unpredictable despite its deterministic character.

Like many nonlinear systems, the double planar pendulum is capable of an astonishing array of complex behavior. These complexities become apparent when its behavior is mapped into phase space. To construct a phase space, we choose a single point in the system's path to observe—say, when the two pendulums form a straight line. Every time the double pendulum is in this configuration, its angular position (the angle it forms with a vertical line) and angular momentum are recorded. In effect, these data amount to taking a snapshot of the system every time it is in the specified configuration. The phase space diagram then shows how these snapshots change over time. It has as many dimensions as there are variables in the system.

When the double planar pendulum is mapped into phase space, its orbits do not wander indiscriminately but stay within a confined region. Within this region no two orbits ever exactly coincide. Several years ago, the mathematician Stephen Smale showed that orbits which act like this move as though the space had been stretched and folded over itself time and again, much as one does with pastry when making croissants—hence the picturesque name, the baker's transformation.[10] Imagine that the complex layering of dough in a croissant were infinitely thin. Points which started out very close to one another, as the folding and stretching continue, diverge unpredictably. Yet they continue to evolve within a confined region. This conveys the flavor (so to speak) of how a strange attractor behaves when mapped into phase space. Its strangeness is now apparent, for it combines pattern with unpredictability, confinement with orbits that never repeat themselves.

There is a certain irony in using a modified pendulum to illustrate a strange attractor, for the pendulum has long been used as a textbook example of Newtonian mechanics. The double pendulum demonstrates that chaotic systems need not be esoteric or rare. Indeed they are more common in nature than ordered systems. But they were not *perceived* to be so until a paradigm shift occurred that placed them at the center rather than the margin of inquiry. Once investigators began looking for strange attractors, they seemed to find them everywhere, from data on lynx fur returns to outbreaks of measles epidemics.[11] The pervasiveness of strange attractors was both exhilarating and puzzling—exhilarating because it suggested that the idea had a wide scope; puzzling because it implied that systems which seemed completely different from one another nevertheless had something in common. What characteristics could be shared by schizophrenic eye movements and fluctuations in cotton prices, dripping faucets and measles epidemics?

This question led to a deeper one, for it was apparent that the similarities could not reside in any identity of mechanisms within the systems. What does it mean to model a system when the model is concerned only with the information produced by the system and has no reference to how the system works? The question is significant, for it implies that something more than a technique is at issue. At stake is not just the revelation of patterns within the data, but the mode of conceptualizing necessary to bring the patterns into view.

The key to this way of looking is recursive symmetry. A figure or system displays recursive symmetry when the same general form is repeated across many different length scales, as though the form were being progressively enlarged or diminished. Railroad ties disappearing into the distance have this property; so does turbulent flow, with swirls inside swirls of the same form, inside of which are still smaller swirls. The importance of recursive symmetry to complex systems derives from the kind of perspective required to see the predictability that lies hidden within their unpredictable evolutions. Mitchell Feigenbaum was the first to realize that, although iterating a nonlinear function yielded unpredictable results, the rate at which the recursions occurred quickly approached a limit that proved to be a universal constant.[12] This constant expresses an orderliness amidst the unpredictability by showing that large-scale features relate to small-scale ones in a predictable way. In addition to being a prominent feature of nonlinear mappings, recursive symmetry is also important in explaining why complex dynamic systems are extremely sensitive to small fluctuations. The repetition of symmetrical configurations across multiple levels acts like a coupling mechanism that rapidly transmits changes from one scale level to another.

The affinities of the science of chaos with other postmodern theories can now be detailed more precisely. It provides a new way to think about order, conceptualizing it not as a totalized condition but as the replication of symmetries that also allows for asymmetries and unpredictabilities. In this it is akin to poststructuralism, where the structuralist penchant for replicating symmetries is modified by the postmodern turn toward fragmentation, rupture, and discontinuity. The science of chaos is like other postmodern theories also in recognizing the importance of scale. Once scale is seen as an important consideration, the relation of local sites to global systems is rendered problematic, for movement across different scale levels is no longer axiomatic. Virtually all of postmodern critical theory is informed by a similar suspicion of globalization, from Jacques Derrida ("Signature") to Fredric Jameson ("Postmodernism"), Jean-François Lyotard (*Postmodern Condition*) to Luce Irigaray (*This Sex*). Another convergence is the emphasis on iterative techniques and recursive looping. In deconstruction, as in the science of chaos, iteration and recursion are seen as ways to destabilize systems and make them yield unexpected conclusions.[13] Yet another parallel is implicit in the emphasis in chaos theory on nonlinearity, with a consequent recognition that small causes can lead to large effects. What is the preference for the marginal in deconstruction but a similar appreciation that seemingly trivial deviations can lead to large-scale effects? Finally, the science of chaos is like other postmodern theories in its recognition that unpredictability in complex systems is inevitable, because one can never specify the initial conditions accurately enough to prevent it. The attack on the idea of the origin in deconstruction leads to a similar conclusion; because the origin cannot be specified exactly, unpredictability is inevitable. The parallels, then, are extensive—so extensive that the most likely source for them is no single site but the cultural matrix as a whole.

Since the science of chaos shares with quantum mechanics an emphasis on unpredictability, it may be useful at this point to clarify what chaos theory adds to the quantum picture of the universe. As is well known, quantum mechanics implies that physical reality cannot be known with infinite precision but only to the limits specified by the uncertainty principle. The limitation expressed by the uncertainty principle has considerable philosophical importance. But since quantum fluctuations are extremely small and tend to cancel each other out, they are often considered not to affect macroscale events to any appreciable extent. (I remember a problem from my college physics class meant to demonstrate that quantum fluctuations do not prevent a baseball from following a predictable Newtonian arc when it is hit.) By contrast, the science of chaos is concerned with systems configured so as to bring even microscopic fluctuations quickly up to macroscopic expression. A counter-example to the baseball is the weather. Because the weather is a dynamic nonlinear system, small fluctuations can and do result in large-scale unpredictability. For systems like this, chaos theory provides the missing link that makes quantum fluctuations relevant to macroscopic experience.

Note, however, that the fluctuations which make a chaotic system unpredictable need not *necessarily* be conceptualized as deriving from the uncertainty principle. Any small change that can be amplified over time will do. In the case of the double planar pendulum, it could be a tiny puff of wind when someone opened a window or a slight vibration of the floor as someone walked across it. In this sense chaos theory is more general than quantum mechanics, for it highlights the importance of stochastic events at every level, from the molecular to the global.

III

Although the history of chaos theory has scarcely begun to be written, it has already become problematic. It has two main branches; and each seems determined to ignore the other. The first branch, represented within Gleick's book and discussed in the previous section, is concerned with the order hidden within chaotic systems. Important investigators in this branch include Mitchell Feigenbaum ("Universal Behavior"), Benoit Mandelbrot (*Fractal Geometry*), Robert Shaw ("Strange Attractors," and "Faucets"), and Kenneth Wilson ("Renormalization"). The second branch focuses on the order that arises out of chaotic systems. Important researchers here include Arthur Winfree, Ilya Prigogine (Nicolis and Prigogine [*Self-Organization*]; Prigogine and Stengers [*Order Out of Chaos*]), and René Thom (*Structural Stability*). A key concept in the second branch is self-organization. Like the first branch, the second has a vision of the world that extends its significance beyond technical results into the cultural realm. Addressing a long-standing dispute between the biological sciences and thermodynamics, it asks why complexity seems spontaneously to come into existence. It envisions a world that can renew itself rather than a universe that is constantly running down, as nineteenth-century thermodynamicists believed. Disorder in this view does not interfere with self-organizing processes. Instead disorder stimulates self-organization and, in a certain sense, enables it to take place.

Why should the split have occurred between the two branches? That it exists is apparent. After reading Gleick's book, one would not know that Ilya Prigogine's work has substantial connections with the science of chaos; and after reading Prigogine and Stenger's *Order Out of Chaos:*

Man's New Dialogue with Nature, one would not know that many of the figures lionized in Gleick's text had made significant contributions to the field. No doubt Prigogine's work is passed over in Gleick's book for the same reason it is sometimes ignored or downplayed within the scientific community—because its orientation is toward the philosophical implications of chaos theory rather than toward solving practical problems associated with nonlinear systems. Gleick accurately reflects the cautious reaction to Prigogine's work within the scientific community when he refers to *Order Out of Chaos* (in an endnote) as a "highly individual,. philosophical view" (339). Despite reservations among many scientists, the vision articulated in *Order Out of Chaos* has become an important component of the new paradigm. While the importance of Prigogine's scientific work may be debated, there can be little doubt that he and Stengers have written an influential book.[14]

At the heart of the Prigoginian vision is the constructive role that entropic disorder plays in creating order. Such a view rests on a reconceptualization of the second law of thermodynamics. The second law decrees that in a closed system, entropy (defined as a function of absolute temperature) always tends to increase. In practical terms, the second law means that in every real heat exchange, some heat is always lost to useful purposes. The Second Law thus embodies what Lord Kelvin (William Thomson), the distinguished nineteenth-century thermodynamicist, called a "universal tendency toward dissipation" (Thomson 514). Kelvin understood that if heat is constantly dissipated, the universe must eventually arrive at a point where no heat reservoirs will be left. The temperature would then stabilize at slightly above absolute zero (-273 degrees centigrade) and life of any kind would be impossible. Kelvin and many of his contemporaries believed that the universe was inexorably plunging toward this final point, the so-called "heat death" that was the finale toward which the second law pointed.

Prigogine and Stengers argue against this traditional view. They envision entropy as an engine driving the world toward increasing complexity rather than toward death. They calculate that in systems far from equilibrium, entropy production is so high that local decreases in entropy can take place without violating the second law. Under certain circumstances, this mechanism allows a system to engage in spontaneous self-organization. Self-organizing reactions had been known since the nineteenth century. But it was not until the mid-twentieth century that Prigogine and Stengers (among others) put them in a context that suggested the universe has the capacity to renew itself. Recently Prigogine and his collaborators have extended this vision to cosmology (Gunzig, Geheniau, and Prigogine, "Entropy and Cosmology"). They argue that before the Big Bang there was a quantum vacuum, and that fluctuations in it brought into existence the aboriginal matter of the universe. Thus the "order out of chaos" scenario is extended to cosmogonic proportions that explain why there is something rather than nothing. In this vision chaos is transformed, as in Hesiod's ancient account in the *Theogony*, into the progenitor of the universe, order's precursor and partner rather than its antagonist.

The mixed reaction to Prigogine's work within the scientific community contrasts with the enthusiastic welcome his work has had within the human sciences. There, the impact of the Prigoginian vision derives at least in part from its resonances with the emerging field of cultural studies. What in the present cultural moment has energized these ideas, made them seem significant and compelling? The reconceptualization of the void as a space of creation has deep affinities with the postmodern idea of a constructed reality. If reality is not natural and self-evident but constructed, it can obviously be deconstructed. Repeatedly in postmodern theory and literature, the constructed fabric of the world (or the text-as-world) is torn to reveal the void underneath. Prefigured in *The Education of Henry Adams*, this tearing of the narrative fabric becomes explicit in such works as Italo Calvino's *If on a Winter's Night a Traveler* and Stanislaw Lem's *The Cyberiad*. Out of the void comes a reconstituted world, as Lem's narrator says "honeycombed with nothingness," in which disorder and order, negation and creation, come together in a fruitful dialectic. This reconstitution makes clear that the world as humans experience it is a collaboration between reality and social construction. No longer simply what is there, reality is subject to constant revision, deconstruction, and reconstruction.

Prigogine's vision illuminates and validates the dialectic between order and disorder by finding analogous processes in physical systems. Moreover, it imparts an optimistic turn to such processes by positing them as sources of renewal for the universe. Thus Prigogine's idea of a creative void is reinforced by much else that is happening in postmodern culture. The postmodern milieu does not necessarily *cause* articulations such as Prigogine's theory or Lem's novel in any direct sense. Rather, it creates a context in which they become thinkable thoughts. Where opportunities exist to propose new solutions because old ideas have proven untenable, these are the kinds of constructions that present themselves as plausible and convincing.

Another example of the collaboration between postmodern theories and culture is the increased attention to the aleatory and stochastic elements in complex systems. As we have seen, the awareness that small fluctuations can lead to large-scale changes is a prominent feature of the science of chaos. Such perceptions are strongly reinforced in a world so bound together by communication networks and socioeconomic interdependencies that the entire globe resembles a complex system. A new leader comes to power in Iran; oil supplies diminish; the developed countries experience runaway inflation; the global economy plunges into a depression. Industrial pollutants are released into the atmosphere; along with carbon dioxide, also a by-product of technology, they create the greenhouse effect; the resulting climate changes wreak havoc with the global ecosystem. Cascading effects from initially small causes could, and have, been observed at any time. But whereas in earlier epochs they tended to be seen as anomalous or unusual, now they are recognized as paradigmatic of complex behavior.

For both branches of chaos theory, then, there are extensive parallels with other theories and events within the culture. What are we to make of these parallels? Do some cause others? Can some be used to interpret and understand others? No doubt because of the prestige accorded to science within the culture, scientific theories have often been used to validate cultural theses; social Darwinism is a case in point. The same tendency can already be observed with the science of chaos, for example in Prigogine and Stenger's call for a new philosophy of becoming in *Order Out of Chaos* and Lyotard's belief in *The Postmodern Condition* that "paralogy" can rescue us from totalitarianism . I believe that we should view such claims skeptically. Scientific results cannot be equated with social programs.

Yet many of the same presuppositions that inform social programs are often encoded into scientific theories, as various studies have shown (Haraway "Animal Sociology"; Stepan "Race and Gender"). Where to draw the line between legitimate and illegitimate extrapolations from chaos theory is not easily or lightly answered. If the science of chaos is identified as the source of the ideas, the perimeter is apt to be more narrowly drawn than if it is seen as one site among many participating in a new paradigm. Often the debate comes down to questions of language. Should terms appropriated from chaos theory be confined to their technical denotation, or is it valid to use them metaphorically or analogically? If they are used metaphorically, what do such arguments demonstrate? The problems associated with these questions can be illustrated with a set of terms that occur repeatedly in discussions of chaos theory: linearity and nonlinearity. The diverse meanings that cohere within these terms reveal the complexity of the issues raised by the relation of chaos theory and culture, the science of chaos and chaotics.

IV

Systems of interest in chaos theory are dynamic. They change over time. These changes are expressed through differential equations, which describe how the behavior of a variable—say, the angular position of a pendulum—changes with respect to time.[15] A crucial step in modeling a dynamic system is arriving at the differential equations that can describe it. In a linear differential equation, no higher powers of derivatives appear; that is, each term contains only a single derivative and no products of derivatives. From a practitioner's viewpoint, linear differential equations have properties that are highly desirable. They can additively map different variables, for example, and they have explicit solutions. Functions that can be described by linear differential equations include sine and cosine curves, exponential growths, and hyperbolic curves. Systems that can be modeled using linear equations include oscillating springs and damped vibrators. The motion of a swing that is given a push and allowed to oscillate back and forth until it comes to rest can be modeled using linear equations. So can the motion of a simple pendulum when the amplitude of the swing is small.

For more complex motions, nonlinear terms are necessary. Nonlinear terms are especially likely to appear when the behavior of a system is determined by two or more factors acting independently of one another. Nancy Cartwright, in her intriguing book *How the Laws of Physics Lie*, uses the example of her camilla plants, which she fertilized with manure that had been incompletely composted (50–53). Camilla plants like rich soil but cannot tolerate heat at their roots. The manure made the soil rich, but it also provided heat. These two interacting factors made the outcomes difficult to predict. Some plants died; some lived. Cartwright points out that it is not possible to add together these competing factors in any simple way. Some of the plants that lived, for example, may have died if they had not been fertilized; they thus lived not in spite of but because of the treatment she gave them. Many physical systems are acted upon by similarly diverse factors that point in different directions.

But here I am betrayed by my language and cultural tradition. For as soon as I begin to think of a system, I constitute it as a collection of factors. Cartwright argues that this view is not intrinsic to the situation but rather derives from the act of analysis. From the system's point of view, there is only the totality that is its environment. So strong is our belief in analysis, however, that we take the environment to be the artifact and the collection of factors to be the reality. Thus gases are said to obey the ideal gas laws, although "correction" terms always have to be added to account for deviations of actual gases from the "ideal" behavior they are supposed to follow. The laws of physics are not laws at all, Cartwright concludes. Instead they are simplifications that reflect how we divide up and analyze systems. We forget that reality is not these laws, just as we forget that most physical laws do not accurately describe the complexities of real situations.

Something of the same thing happened with linear and nonlinear differential equations. Because linear differential equations had additive properties and explicit solutions, it was tempting to regard them as describing normal behavior and nonlinear differential equations as dealing with unusual or aberrant cases. Until recently, most college physics courses relegated nonlinear equations to the final few weeks of term or to a "Special Topics" course. In fact, however, there are far more systems that require nonlinear equations than there are those that follow linear equations. Part of the change that the science of chaos has brought about is the recognition that nonlinear systems are all around us, in every puff of wind and swirl of water. That the emphasis is still on the *order* that can be found in these systems should not surprise us. Cultures do not change their fundamental orientations simply because new methods of analysis are employed.

Mingling with the technical denotations of linearity and nonlinearity as they are used in the science of chaos, then, are values that derive partly from the mathematical properties of the equations, partly from the disciplinary contexts in which these equations were developed and employed, and partly from the cultural tradition, dominant at least since Plato, that privileges ideal abstraction over empirical variation. (I observe in passing that I have analyzed the situation in terms of separate factors, whereas in reality they are one environment interacting continuously and holistically. One can speak only in the language one has inherited.) Although the mathematics itself may have been value-free, the context in which it was used and the rhetoric that developed around it was not.[16]

Within the last few years, the terms "linear" and "nonlinear" have been appropriated in cultural and literary discourses. Although they entered the humanities at multiple points, an important juncture was Michel Serres's seminal essay, "Lucretius: Science and Religion" (*Hermes* 98–124). Serres constructs his essay as a commentary on Lucretius's epic poem, *De rerum natura*. He claims that Lucretius's poem is scientific rather than mythic, in the sense that it accurately describes social and physical reality. In the poem Lucretius, following Democritus, imagines that atoms fell all in straight lines until a clinamen or swerve made interaction possible. Serres identifies the straight fall of the atoms with the reign of Mars; the swerving paths represent the order of Venus. The linearity of Martian order represents those aspects of Western culture that privilege war over love, order over creativity, abstraction over embodiment, aggression over sympathy, death over life. Serres suggests that these values have been predominant in the West through a long, bloody history of war and conflict. He celebrates the birth of Venus from the sea as signaling that another kind of reign is possible. The ocean from which she came, unruly in its turbulence and unknowable in its depths, hints that this reign will delight in the fecundity of disorder. Shunning linear reason and fragmented analysis, it will take as its emblems the vortex, the wave, the cloud, the waterfall.

Significantly, Prigogine and Stengers contributed a long afterword to the collection in which this essay appeared ("Postface: Dynamics from Leibniz to Lucretius," in Serres [*Hermes*] 135–58). They endorse and amplify Serres's suggestion that chaos represents not just hitherto unrecognized phenomena but an unjustly neglected set of values. Less impressionistic and poetic than Serres, they nevertheless arrive at similar conclusions. They add to the list of Martian/Venusian oppositions spatiality/temporality and being/becoming. Largely as a result of these two essays, "linear" became associated with a whole set of values in critical discourse, ranging from binary logic to abstraction and rationality. Similarly, "nonlinear" metonymically stood for another set of values, from multivalued logic to empirical variation and temporality.

Faced with this expansion of meaning, many scientists would no doubt feel that things had gotten out of hand. Certainly the polyvocality of the terms as they are used by Serres and others is very far from the univocality that is the goal of scientific discourse. Yet the distinction between legitimate and illegitimate usage is not so easily drawn. It would be difficult to argue, for example, that linearity and nonlinearity have no value connotations in the sciences. Morever, it is clear that these implicit values are not entirely unconnected with the reign of Mars that Serres describes. How are the distinctions to be made, if at all?

My purpose here is not to argue for a specific answer. Rather, I want to urge that the question cannot be dismissed out of hand. How one constructs an answer will depend on how one chooses to respond to a number of prior questions. Is the science of chaos the center from which these ideas spread into the culture, or is it one site among many? Does language convey ideas or help to constitute them? What presuppositions are implicit within an analysis that seeks to distinguish between denotation and connotation? Is scientific inquiry culturally conditioned? In the earlier part of this essay I staked out positions on many of these questions and tried to sketch, with broad strokes, how a case could be made. This is not the same, of course, as making the case. Exploring the complexities of these and related issues in light of specific instances is the business of the essays that follow.

They are divided into three sections. In the first, the essays ask what we are to make of implicit convergences between the science of chaos and postmodern critical theory and literature. How do such convergences come about, and what do they portend? In what sense are the correspondences between literature and science merely metaphorical, and in what sense do they go beyond metaphor? The essays in the second section are concerned with the past. They return to it from the vantage of the present, re-reading texts that were written before the new paradigm coalesced to excavate in them possibilities and potentialities that resonate with the postmodern moment. In the third section, the essays explore critical and literary works in which there is a tension between chaotics, with its new vision of disorder, and a sedimented history that constructs order in traditional ways. Whereas the essays in the first section emphasize a change in paradigms and those in the second a change in reading and interpretation, the essays in the final section explore the reactive currents that keep change from being homogeneous or total. As a group, the essays give form and substance to what chaotics can mean for literary and cultural studies.

V

In the past, studies in literature and science have tended to follow a characteristic pattern. First some scientific theory or result is explained; then parallels are drawn (or constructed) between it and literary texts; then the author says in effect Q.E.D., and the paper is finished. In my view, every time this formula is used it should be challenged: What do the parallels signify? How do you explain their existence? What mechanisms do you postulate to account for them? What keeps the selection of some theoretical features and some literary texts from being capricious? What are the presuppositions of the explanations you construct, and how do they connect with what what you are trying to explain? None of these questions is easy to answer. Nevertheless, if we are to arrive at a deeper understanding of the connections between literature and science (and hence implicitly of the underlying cultural dynamics), it is essential not to gloss over the hard issues. The essays in this volume do not all ask the same questions; still less do they arrive at the same answers. But none fails to grapple with some of the deeper problems. In exploring different modes of explanations and sometimes taking issue with each other, they begin the important work of understanding chaotics and, more fundamentally, of understanding the cultural constructions that authorize changing views of chaos and order.

One way to understand the connection between literature and science is to see science as a repository of tropes that can be used to illuminate literary texts. Quantum mechanics and relativity theory have often been used in this way; no doubt chaos theory will be as well. A bolder and perhaps more interesting move is to posit connections that go beyond metaphor. The essays in the first section make this move. For William Paulson in "Literature, Complexity, Interdisciplinarity," the connection comes from seeing the reading process as a complex system in action. Paulson observes that when we read a difficult literary work, for example a poem, there are typically parts of it that we do not understand, which we can process only as noise rather than information. As a result of the first reading our cognitive processes are slightly reorganized. When we read the poem again, more of it is processed as information because we now read at a higher level of complexity. Thus the reading process instantiates the symbiotic relationship between complexity and noise, for it is the presence of noise that forces the system to reorganize itself at a higher level of complexity.

Following Jurij Lotman in *The Structure of the Artistic Text*, Paulson argues that this increased complexity, far from being accidental, is in fact the desired outcome of an artistic text. He defines literature as communication crafted to maximize the positive role of noise. This view implies that the noisy or non-assimilated elements in a system are crucial to its continuing development. Thus literature can play an important role in contemporary culture, Paulson argues, precisely because it is marginal. Perceived as so much noise by a culture immersed in electronic media and increasingly illiterate, literature has a transformative potential it could not have if it were positioned at the center. A strength of Paulson's approach is the persuasive connection it forges between information theory and cognitive processing. Readers not familiar with information theory may not immediately realize the full scope of his achievement; but as one who has struggled with the technical literature, I can testify that it is no trivial feat to see how this mathematical theory can illuminate a literary text.

David Porush in "Fictions as Dissipative Structures: Prigogine and Postmodernism's Roadshow" takes issue with Paulson's assumption that literature is marginal to contemporary culture. While literature may not have the mass appeal of television and videos, Porush contends, it has as much or more influence in forming cultural values. Indeed, so centrally does he locate postmodern literature that he argues for a reversal of the usual flow of influence from science to literature. When it comes to the kind of complex, unpredictable behavior typical of nonlinear systems, literature has a longer history of dealing with it and is more suited to describe its complexities than science. To make his case, Porush argues that the new science is indifferent to a distinction crucial to earlier periods, namely the difference between artificial and natural systems. Drawing on Prigogine's analysis of dissipative (or entropy-producing) systems, Porush points out that Prigogine's theory has been applied to both natural and artificial phenomena, for example chemical reactions and traffic jams. To re-mark the territory, Porush groups together literary texts, the human mind, and scientific theories as "biosocial phenomena."

From this vantage he argues that a postmodern text such as William Marshall's *Roadshow*, which he describes as a mystery novel that hinges on seeing a traffic jam as a higher level of order rather than disorder, is as much a dissipative structure as the traffic jam it describes, or as the minds that produce mystery novels and traffic jams. For the mind, too, operates as a dissipative system, and some of its products are theories and novels about dissipative systems. Thus *Roadshow*, and much of postmodern literature, can be understood as one dissipative structure giving rise to another. In the free-flowing conduit that Porush conceives the new paradigm to be, connections between postmodern literature and science are more than metaphoric because the boundaries separating literal and figural usage have themselves been dissolved by the paradigm.

Peter Stoicheff in "The Chaos of Metafiction" seeks to define the characteristics that make contemporary fiction distinctively postmodern. The four attributes he identifies—nonlinearity, self-reflexivity, irreversibility, and self-organization—are also prominent in chaotic systems. Like other essays in this section, his argument pushes beyond a metaphoric connection to assert a deeper congruence. He finds it in metafiction's interrogation of the language that constitutes text and world. Whereas a mimetic text creates the illusion that it is transmitting information about the world, a metafictional text reveals the world's constructed nature. Refusing to arbitrate between levels of reality, metafiction generates multiple significations, much as a chaotic system generates infinite information. In the process, metafiction produces a different kind of reader than a mimetic text—and once this transformation has taken place, it is irreversible. For a reader conditioned by metafictional texts will find that he reads all texts, even mimetic ones, as if they were metafictions. Hence metafictional texts act like local vortices from which spread a general turbulence.

If real dynamic systems also act like this, does it mean that in a sense metafiction is mimetic after all? No, says Stoicheff. He attributes the convergence between postmodern literature and science to a "narrative of chaos" that is characteristic of the present moment. This narrative gives rise to both metafiction and to the science of chaos, just as a Euclidean master narrative gave rise to Euclidean geometry and the realistic novel. Since the claim comes at the essay's conclusion, the master narrative's origin, extent, and significance remain undeveloped. An important task for those who want to develop this line of inquiry will be to go beyond the assertion that such a narrative exists to a delineation of its workings.

In "The Emplotment of Chaos: Instability and Narrative Order," Kenneth Knoespel warns against thinking in such terms precisely on the grounds that it obscures the importance of local sites. Rather than look for metanarratives, Knoespel argues, we should attend to practices that make one site distinct from another. To illustrate, he chooses deconstruction and chaos theory. Both use iterative methods to analyze or liberate chaos in complex systems; both can (and have) been held up as theories embodying the new paradigm. But such comparisons obscure important differences between how these two theories work in practice. To demonstrate how attending to local sites can lead to interpretive insights, Knoespel concentrates on the use of examples in deconstruction and chaos theory. He treats examples as small local narratives that can perform different, even opposed, functions. He argues that tracking how examples are used at specific sites leads to a much more nuanced and precise mapping of different disciplinary inquiries than can be achieved by constituting such inquiries through paradigmatic shifts or metanarratives. Knoespel's essay forms an important contrast with the other chapters in this section, for it moves beyond metaphor by limiting rather than expanding the focus of inquiry.

VI

When a new paradigm asserts itself, it is not only the present that is changed; the past is also reinterpreted. The essays in the second section look at texts written before the new paradigm took hold, showing how it brings into focus aspects of these texts that may have been neglected or obscured in other interpretive matrices. Robert Markley in "Representing Order: Natural Philosophy, Mathematics, and Theology in the Newtonian Revolution" argues that the order so often associated with scientific progress is not an intrinsic property of reality but an unstable construct impinged upon by a variety of other discourses, including theology, history, and politics. He traces the tension between a theological view of nature as fallen, imperfect, corrupt, and an emerging vision of it in the seventeenth century as the embodiment of a perfect, transcendent order. The problem for scientific inquiry was not just asserting that order existed; it also had to be enacted in stable systems of representation. Newton believed that a transcendent order did exist, but that it receded into the distance whenever one tried to capture it within a notational system. Because Newton was aware of the limitations of any one system of representation, Markley argues, he engaged in a strategy of supplementation, adding theological and alchemical writings to his mathematical work. Far from solving the problem, this strategy introduced an additional difficulty, for it was then necessary to stabilize the relation between the various discourses. As the edifice grew more elaborate and consequently more unstable, Markley writes, it was increasingly apparent to Newton that "mathematical order was not a closed system of representation but a provisional inquiry that, by its very nature, can never be finalized."

Not so with those who followed in Newton's footsteps. Distinguishing between Newton's own work and "Newtonianism," Markley argues that the Newtonians suppressed inconsistencies and instabilities in Newton's texts, promoting instead a popularized idea of a thoroughly systematized mechanics. This mechanics had the double value of legitimating theological claims of a divinely ordered universe and of enabling practical applications that had immediate economic and political payoffs. Hence Newtonianism was not simply a scientific theory but an ideological project "that integrated scientific research into the political and economic operations of eighteenth-century British society." As Newtonianism gained ground, it ceased to be a legitimation of theological beliefs and subsumed the function of signifying a transcendent order. Thus Markley argues that "theology is not exorcised from the corpus of science but repressed within it." Despite the success of Newtonianism, the problem Newton wrestled with had not disappeared, for systems of notation continued to fall short of authoritative languages that could completely describe reality.

This led to what Markley calls the "theological imperative," the recognition that the languages of science had to be supplemented with invocations of a mysterious order that can be mystically apprehended but never entirely represented. In this sense, however far the new physics has departed from Newtonianism conceptually, it remains the inheritor of the Newtonian tradition. "We are still struggling within and against the confines of Newtonian notions of order," Markley concludes. His essay creates a powerful context in which to evaluate claims that the science of chaos has given us new ways to think about order, for it suggests that the lack of closure in fractal forms is *causally related* to accompanying claims of mystical beauty. Each necessitates the other, in an economy of supplementation that would have been as familar to Newton as to Derrida.

Sheila Emerson in "The Authorization of Form: Ruskin and the Science of Chaos" explores an insight Ruskin shares with the science of chaos—the realization that movement through time can be recorded as form. "No one has ever loathed chaotic deviation more eloquently than Ruskin," Emerson writes. The project of envisioning a new kind of order was for Ruskin a highly charged project on two counts. First was the possibility that the order he discerned was in fact disorder, evidence of his own willfulness and instability; second was the implication that he could perceive freshly only by rebelling against traditional ways of constituting, recognizing, and rendering objects. Ruskin responded to these threatening possibilities, Emerson argues, by displacing the authority for them from himself to a higher law. That spatial forms are representations of temporal flows was not merely his perception, Ruskin argued, but a fact of nature that Turner and other great painters had intuitively understood.

The argument turns self-reflexive when Ruskin recalls how he learned to read by recognizing words as pictorial forms, rather than as the compositions of syllables that his mother insisted they were. Seeing words as forms opened the way to seeing writing as a record of natural growth. Thus writing is not so much a signifier as a mapping of forces, having the "aspect" of the flows of energy and pressure that determine why something grows one way and not another. However much Ruskin deplored the "desultory" nature of his writing or decried his lack of capacity for composition, he also constituted his writing as a record of growth that could not be other than what it was, for its form was the visible inscription of forces dictated by nature. Thus his childhood resistance of his mother's tutelage, like his re-visioning of order, is constituted as a reinscription of the "forces he could not resist as a boy, including the force exerted by the look of print on the page."

The result is a complex denial of authority that paradoxically turns rebellion into obedience. "At the very moment when a pattern materializes out of his own mental and physical movements," Emerson concludes, "he devotes all his force to showing that the pattern certifies a force beyond his own." Her essay illustrates how problematic the assertion of a new order is when it occurs in a cultural context that does not reinforce it through prevailing paradigms. Lacking the sense that his vision was part of a ground swell of change, Ruskin found it necessary to engage in strategies that could negotiate the treacherous passage between the Scylla of rebellion and the Charybdis of disorder. If his insight that form was temporal flow was ahead of his time, his interpretation of that insight as a driving force beyond his control was the mark of his time.

Linda Hughes and Michael Lund in "Linear Stories and Circular Visions: The Decline of the Victorian Serial" relate shifting ideas of order to the demise of the serial as a vehicle for "high" literature. They point out that ideas of growth evident in nineteenth century biology and history were also at work in the Victorian novel, reinforcing the belief that continuing narration correlated with continuing development. They contrast these assumptions with the modernist aesthetic, particularly the supposition that an artistic work should be grasped as a preexistent whole. Just as linear progressions gave way in science to such field theories as relativity and quantum mechanics, so the serial form of publication gave way to modernist techniques that privilege the work as a totality. Yet change is never homogeneous or total. Choosing three serialized works as their examples—*The Woodlanders*, *Lord Jim*, and *The Dynasts*—they explore tensions between structures affected by serialization and an emerging aesthetic that drew them into question. In *Lord Jim*, for example, the presupposition that more narration would clarify motives and develop the plot is subverted as the novel circles around mysteries rather than moves linearly toward revelation. Yet there are also aspects of *Lord Jim* responsive to the epistemology of serial form. Thus the argument is not that one form of reading displaced another, but that serial and whole-volume reading strategies commingle in these works, imparting to them characteristic tensions and ambiguities.

Adalaide Morris in "Science and the Mythopoeic Mind: The Case of H.D." traces the American poet H.D's development from a poetry that was content to reinscribe the Newtonian paradigm to works that broke with the scientific conservatism of her male relatives. Morris marks the turning point with *HERmione*, a comic novel in which H.D. "blasts free of all her mentors." At the novel's center is H.D.'s autobiographical Heroine, Her Gart, who envisions a dynamic, vital world that eludes static classification as much as Her's name. Morris finds affinities with relativity theory and quantum mechanics in this work, both of which came into prominence during H.D.'s formative years. Direct influence cannot explain, however, the parallels Morris detects between chaos theory and H.D.'s late epic, *Helen in Egypt*, written in London during World War II. Morris argues that *Helen in Egypt* is organized according to recursive symmetries. Like chaos theory, it pulls "seemingly random or disorganized phenomena into dynamic relation by discovering patterns that repeat across scales or recur one inside the next." So far, it may sound as if Morris sees chaos theory as a useful metaphor through which to explain the structure of H.D.'s late work. H.D. herself, however, believed that these recursive patterns were not metaphors but "templates," models and instantiations of mythopoeic truths that cut across literature and science. She intuited that these structures implied that a minute cause could lead to incalculable effects. The intuition reinforced her belief that even so small an event as the stroke of a poet's pen could have efficacy in a world that seemed bent on destroying itself. Morris's reading of H.D.'s work shows how chaos theory can lead to a vision of the interrelation between literature and science very different than that issuing from a Newtonian paradigm.

VII

The changing meanings of chaos can be confusing as well as exhilarating. Even in scientific articles, the term continues to carry traditional associations that are at odds with its denotative usage. When chaos is expanded into chaotics, the jostling between new interpretations and historical constructions intensifies. Pronouncements of an apocalyptic break with the past abound; but change does not occur all at once, or in the same way at every site. The essays in the final section explore the complex dynamics that result when new concepts are embedded within stratified cultural formations. They fall into two groups. Thomas Weissert and Istvan Csicsery-Ronay focus on issues of representation and modeling; Eric White and Maria Assad are primarily concerned with political implications that have been associated with the new paradigm. The two concerns are related, for what the new paradigm means politically cannot be separated from how it is represented, any more than how it is represented can be purged of political implications.

Thomas Weissert in "Representation and Bifurcation: Borges's Garden of Chaos Dynamics" finds that Borges's story "Garden of the Forking Paths" anticipates many of the insights of nonlinear dynamics, especially bifurcation theory. To make the comparison, he distinguishes four levels of representation in nonlinear dynamics and in Borges's story. Aware that even the first, most immediate level is already a representation, Weissert compares the literal level of Borges's story to a laboratory model of a physical system. As the levels move upward in abstraction, they rely for their persuasiveness on cultural codings too complex to be formalized completely. Thus Weissert emphasizes that the scientist must use his intuition in constructing the nonlinear differential equations that describe the laboratory model, as well in creating the model itself. Whereas the scientific representation is believed to have some contact with reality throughout all of its levels, Borges's story contains information that draws the immediate level of reality into doubt (and hence all subsequent levels as well). Yet despite this deconstructive tendency, Weissert finds Borges's "Garden" to be more modernist than postmodern. Arguing that modernism in literature and science is characterized by global theories that affirm determinism, Weissert identifies several ways in which the "Garden" works to deny randomness. Just as the labyrinth only appears disorderly to those unfamiliar with it, so the "Garden" represents chaos as order not previously comprehended.

In working out the parallels between levels of representation in nonlinear dynamics and in Borges's story, Weissert arrives at an insight that I think deserves to be underscored. He points out that a phase space mapping is essentially a spatialization of a system's temporal flow. Thus complex evolutions through time are transformed into complex physical shapes that can be intuitively appreciated. I emphasize the *intuitive* aspect of this knowledge (as does Weissert) because the forms are so complex that they never resolve into completely ordered structures. No matter how fine the resolution, some chaotic or "fuzzy" areas always remain. Literary critics have long been aware that fiction could be considered in spatial as well as a temporal terms. The preexisting totalized shape of the novel enables Walter Benjamin ("The Storyteller" in *Illuminations*), for example, to argue that a novel is teleological in a way that a story is not. What happens to this way of theorizing fiction when a form of spatialization appears that cannot, even in theory, be grasped as a resolved whole? When, in other words, spatialization is no longer synonomous with totalization? Although I will not have space here to develop the implications of the question, perhaps even this brief discussion is enough to demonstrate that it could have important implications for narrative theory.

Istvan Csicsery-Ronay Jr. in "Modeling the Chaosphere: Stanislaw Lem's Alien Communications" takes up the problematics of representation in Stanislaw Lem's fiction. Lem is acutely aware of the problems posed by what he calls "carousel reasoning," in which conclusions are dictated in advance by premises, and premises by conclusions that make them seem logical starting points. Lem concurs with many contemporary philosophers of science that there are no brute facts. "Facts" are always already constructions arrived at by a given species, so that some degree of anthropomorphic projection is inevitable. In his fiction Lem accentuates this inevitability by bringing scientists into confrontation with alien intelligences. Unable to assume that they have anything in common with the objects of their investigations, the scientists-protagonists realize that they must operate within what Csicsery-Ronay calls the "chaosphere," a space where no representation can be certain to do anything more than reflect its creator's face, and where the carousel closes into a solipsistic circle. The only way out is through communication with the Other—which, because the Other is not human, can only take place through models. Thus modeling becomes a communicative necessity as well as a scientific endeavor.

Through three of Lem's novels—*Solaris, His Master's Voice*, and *Fiasco*—Csicsery-Ronay traces the potentialities of modeling as communication. They range from the optimistic possibility of an anthropomorphic creation who is enough like a human being to be able to love, to the frustrating instance of a communication so complete unto itself that it can never be understood or decoded, to a modeling process that finds itself in demonic communication not with the Other but with the Other's model. Focused on Lem's fiction, the argument has implications beyond it. Whereas Weissert's essay sought to forge links between literature and science by thinking of literary texts as models, Csicsery-Ronay's essay reminds us that analogies constructed through models can be as erroneous as the models themselves. If there is a commonality between postmodern literature and science, Lem's work suggests it lies in realizing that model-making, whether in science or in literature, can never entirely escape the carousel of self-reflexivity.

In "Negentropy, Noise, and Emancipatory Thought," Eric White interrogates the emancipatory role that has been ascribed to the new paradigm. The exact nature of this role differs, depending on who envisions it. For Prigogine and Stengers in *Order Out of Chaos*, it consists of a movement from being to becoming, essence to process; for Michel Serres in *La Naissance de la physique dans le texte de Lucrèce*, it is a repudiation of sterile repetition in favor of fecund unpredictability. White is skeptical of such claims. Taking Pynchon's *The Crying of Lot 49* as his tutor text, he suggests that revolutionary subversions of an old order can easily turn repressive when they take over and become the new order. Moreover, the belief that a revolutionary movement exists may itself be a paranoid delusion, proof that we are always already imprisoned within an order we cannot escape. White detects in Serres's later work a growing recognition that co-optation is not only possible but probable.

White locates the turning point at the end of *The Parasite*, when Serres renounces Venus, ocean-born goddess of turbulence, for the ocean itself. The renunciation signals Serres's recognition that when chaos leads to order, as soon as this new order comes into being, it too strives toward what White calls "maximum scope and power." The players may have changed, but the game remains the same. Although White stops short of concluding that the revolutionary potential of the new paradigm will be defused by its co-optation into traditional power structures, his argument implies that there can be no simple or clear-cut break between old and new. The "activity of making sense by way of recourse to chaos, noise, and chance circumstance is an interminable task because power, like desire, is protean and omnipresent."

Maria Assad in "Michel Serres: In Search of a Tropography" delves further into the work of Serres to explore how far an allegiance to chaos can extend. She argues that *Genèse* marks a turning point in Serres's corpus. After the renunciation at the end of *The Parasite*, Serres concludes that all of his previous work demonstrates one thing: as Assad puts it, that the "suppression of chaos [is] the condition of possibility for any organizing principle to function historically." The challenge for him after this is to to move as close as possible to the "black box" that is chaos itself.

Dedicated to escaping the "hell" of dualistic thinking and the "tyranny of the One," he turns to chaos as an antidote to the order/disorder dichotomy. He prefers to talk about noise rather than disorder. Whereas disorder is defined by its relation to order, noise is multiple and, Assad asserts, "falls outside the binary opposition of affirmation/negation." But his project of interrogating the "multiple as such" is inherently paradoxical, for as soon as noise moves into the realm of language, it is always already not noise but language. He struggles to use language to go beyond it. For him, noise is not a remote or abstract condition. Already present in the sensory murmurings that precede language, it has the power to deconstruct the inside/outside dichotomy separating us as subjects from the world. He wants to use noise to construct objects not as real entities but as a "tropographic space" (Assad's phrase) where the multiple as such may be inscribed. Such a project cannot succeed, at least not in the verbal representations of language. As he attempts to move back through the strata society has constructed to the genesis of language, of thought, of representation itself, he always uncovers yet more layers. In a sense his text *creates* them, for his search for chaos is itself a representation of eluding representation. In contrast to the essays of the first section, which sought to make chaos more than a metaphor, Assad's essay shows that any chaos which can be written or spoken must be no more than a trope.

The contrast illustrates that much of the strength of this volume lies in the diversity of perspectives articulated within it. The disagreements, controversies, and divergences demonstrate that chaotics is only beginning to take shape as a field of inquiry. If there is a common theme running through this diversity, it is that science and literature are not above or apart from their culture but embedded within it. Assad's essay can thus stand as an appropriate conclusion for the whole, for it suggests that the answers to whether chaos is more than a metaphor, whether and what kind of chaos can be captured within representations, and how chaos has been historically inscribed or suppressed lies in the complex dynamics that connect writer, reader and text with the culture. Chaotics emerges from the belief that these questions deserve our attention.

Notes

1. Dripping faucets are described in Shaw ("Faucets"); fish populations in May; measles in William Schaffer and Kot; and schizophrenic eye movements in Huberman.

2. For an assessment of these attempts, see William Brock, "Chaos and Complexity in Economic and Financial Science," in von Furstenberg.

3. Feminist critiques include Harding and Keller. Social constructivist arguments include Latour and Woolgar, and Shapin and Schaffer. An example of an ethonographic study is Traweek.

4. Barthes's critique appears in "From Science to Literature," first printed in the *Times Literary Supplement* in 1976 and reprinted in Barthes (*Rustle*).

5. The interview is by Colin Greenland.

6. For an analysis of the relation of entropy to information theory and its relevance to chaos theory, see Hayles, "Self-Reflexive Metaphors in Maxwell's Demon and Shannon's Choice: Finding the Passages," in Peterfreund 209–38.

7. For a discussion of the importance of information to World War II, see Hayles ("Text").

8. Private communication; see Hassan.

9. I am indebted here to a film on the double planar pendulum from the Institut für den Wissenschaftenlichen Film, Göttingen. My colleague Dwight Nicolson also helped to clarify matters. A discussion of complications even in

a simple pendulum can be found in D'Humieres, Beasley, Huberman, and Libchaber.

10. Stephen Smale. See also Crutchfield, Farmer, Packard, and Shaw.

11. For lynx fur returns, see William M. Schaffer.

12. Feigenbaum ("Universal") tells of his discovery of universality theory.

13. For an analysis of the relation of chaos theory to deconstruction, see Hayles ("Shifting Ground").

14. For a more extensive evaluation of Prigogine's work and its reception by the scientific community, see "From Epilogue to Prologue: Chaos and the Arrow of Time," in Hayles (*Chaos Bound* 91–114).

15. Differentiation can occur with respect to other variables also; but for dynamical systems, the variable of interest is time.

16. Whether mathematics is in fact value-free has been challenged by Gabriel Stoltzenburg ("Can an Inquiry into the Foundations of Mathematics Tell Us Anything Interesting about Mind?"(in Miller and Lenneberg 221–69) and Rotman (*Signifying*, and "Towards").

Works Cited

Arbib, Michael, and Mary Hesse. *The Construction of Reality*. Cambridge: Cambridge University Press, 1983.

Barthes, Roland. *The Rustle of Language*. New York: Hill and Wang, 1986.

Bazerman, Charles. *Shaping Written Knowledge: The Genre and Activity of the Experimental Article in Science*. Madison: University of Wisconsin Press, 1988.

Beer, Gillian. *Darwin's Plots: Evolutionary Narratives in Darwin, George Eliot, and Nineteenth-Century Fiction*. London: Routledge & Kegan Paul, 1983.

Benjamin, Walter. *Illuminations*, trans. Harry Zohn. New York: Schocken, 1969.

Cartwright, Nancy. *How the Laws of Physics Lie*. New York: Oxford University Press, 1983.

Crutchfield, James P., J. Doyne Farmer, Norman H. Packard, and Robert S. Shaw. "Chaos." *Scientific American* 255 (1986): 46–57.

Derrida, Jacques. "Signature Event Context." *Glyph I: Johns Hopkins Textual Studies*, eds. Samuel Weber and Henry Sussman. Baltimore: Johns Hopkins University Press, 1977.

D'Humieres, D., M. R. Beasley, B. A. Huberman, and A. Libchaber. "Chaotic States and Routes to Chaos in the Forced Pendulum." *Physical Review* A26 (1982): 3483–96.

Eoyang, Eugene. "Chaos Misread: Or, There's Wonton in My Soup!" *Comparative Literature Studies* 26 (1989): 271–84.

———. "The Heuristics of Chaos: Seeing from the "I" of the Storm." Paper delivered at the Chaos Symposium, University of Tennessee, October, 1989.

Feigenbaum, Mitchell J. "Universal Behavior in Nonlinear Systems." *Los Alamos Science* 1 (1980): 4–27.

Gibson, William. *Neuromancer*. New York: Ace Books, 1984.

Gleick, James. *Chaos: Making a New Science*. New York: Viking, 1987.

Greenland, Colin. "A Nod to the Apocalypse: An Interview with William Gibson." *Foundation* 36 (1986): 5–9.

Gunzig, Edgard, Jules Geheniau and Ilya Prigogine. "Entropy and Cosmology." *Nature* 330 (1987): 621–24.

Haraway, Donna. "Animal Sociology and a Natural Economy of the Body Politic, I and II." *Signs* 4 (1978): 21–60.

Harding, Sandra. *The Science Question in Feminism*. Ithaca: Cornell University Press, 1986.

Hassan, Ihab. *The Postmodern Turn: Essays in Postmodern Theory and Culture*. Columbus: Ohio State University Press, 1987.

Hayles, N. Katherine. "Chaos as Orderly Disorder: Shifting Ground in Contemporary Literature and Science." *New Literary History* 20 (1989): 305–22.

———. *Chaos Bound: Orderly Disorder in Contemporary Literature and Science*. Ithaca: Cornell University Press, 1990.

———. "Text Out of Context: Postmodernism in an Information Society." *Discourse* 9 (1989): 24–36.

Huberman, Bernardo. "A Model for Dysfunctions in Smooth Pursuit Eye Movement," preprint, Xerox Palo Alto Research Center, Palo Alto, California.

Irigaray, Luce. *This Sex Which Is Not One*, trans. Catherine Porter and Carolyn Burke. Ithaca: Cornell University Press, 1985.

Jameson, Fredric. "Postmodernism, or the Cultural Logic of Late Capitalism." *New Left Review* 14 (1984): 653–92.

Keller, Evelyn Fox. *Reflections on Gender and Science*. New Haven: Yale University Press, 1985.

Latour, Bruno. *Science in Action*. Milton Keynes, England: Open University Press, 1987.

Latour, Bruno, and Steve Woolgar. *Laboratory Life: The Social Construction of Scientific Facts*. Beverly Hills: Sage Publications, 1979.

Lyotard, Jean-François. *The Postmodern Condition: A Report on Knowledge*, trans. Geoff Bennington and Brian Massumi. Minneapolis: University of Minnesota Press, 1984.

Mandelbrot, Benoit. *The Fractal Geometry of Nature*. New York: W. H. Freeman, 1983.

May, Robert. "Simple Mathematical Models with Very Complicated Dynamics." *Nature* 261 (1976): 459–67.

McCloskey, Donald. *The Rhetoric of Economics*. Madison: University of Wisconsin Press, 1985.

Miller, George A., and Elizabeth Lenneberg. *Psychology and Biology of Language and Thought: Essays in Honor of Eric Lenneberg*. New York: Academic Press, 1978.

Nicolis, G., and I. Prigogine. *Self-Organization in Nonequilibrium Systems: From Dissipative Structures to Order Through Fluctuations*. New York: Wiley, 1977.

Peterfreund, Stuart. *Literature and Science, Theory and Practice*. Boston: Northeastern University Press, 1990.

Prigogine, Ilya and Isabelle Stengers. *Order Out of Chaos: Man's New Dialogue with Nature*. New York: Bantam, 1984.

Rotman, Brian. *Signifying Nothing: The Semiotics of Zero*. London: Macmillan Press, 1987.

———. "Towards a Semiotics of Mathematics." *Semiotica* 72 (1988): 1–35.

Schaffer, William M. "Stretching and Folding in Lynx Fur Returns: Evidence for a Strange Attractor in Nature." *The American Naturalist* 124 (1984): 798–820.

———, and Mark Kot. "Nearly One-Dimensional Dynamics in an Epidemic." *Journal of Theoretical Biology* 112 (1985): 403–7.

Serres, Michel. *Hermes: Literature, Science, Philosophy*, ed. and trans. Josué Harari and David F. Bell. Baltimore: Johns Hopkins University Press, 1982.

———. *La Naissance de la physique dans le texte de Lucrèce*. Paris: Minuit, 1977.

Shapin, Steven, and Simon Schaffer. *Leviathan and the Air Pump: Hobbes, Boyle, and the Experimental Life*. Princeton: Princeton University Press, 1985.

Shaw, Robert S. *The Dripping Faucet as a Model Chaotic System*. Santa Cruz: Aerial, 1984.

———. 1981. "Strange Attractors, Chaotic Behavior, and Information Flow." *Zeitschrift für Naturforschung* 36A (1981): 79–112.

Smale, Stephen. *The Mathematics of Time: Essays in Dynamical Systems, Economic Processes, and Related Topics*. New York: Sprınger-Verlag, 1980.

———. "Differentiable Dynamical Systems." *Bulletin of the American Mathematical Society* (1967): 747–817.

Stepan, Nancy Leys. "Race and Gender: The Role of Analogy in Science." *Isis* 77 (1986): 261–77.

Thom, René. *Structural Stability and Morphogenesis: An Outline of a General Theory of Models*, trans. D. H. Fowler. Reading, MA: W. A. Benjamin, 1975.

Thomson, Sir William (Lord Kelvin). *Mathematical and Physical Papers*. Vol. 1. Cambridge: Cambridge University Press, 1881.

Traweek, Sharon. *Beamtimes and Lifetimes: The World of High Energy Physics*. Cambridge: Harvard University Press, 1988.

von Furstenberg, George M., ed. *Acting Under Uncertainty: Multidisciplinary Conceptions*. Boston: Klumer Academic Publishers, 1989.

Yorke, James, and Tien-Yien Li. "Period Three Implies Chaos." *American Mathematical Monthly* 82 (1975): 985–92.

Wilson, Kenneth G. "The Renormalization Group and Critical Phenomena." *Reviews of Modern Physics* 55 (1983): 583–600.

Winfree, Arthur. *The Geometry of Biological Time*. Biomathematics, vol. 8. Berlin and New York: Springer, 1980.

I

CHAOS: MORE THAN METAPHOR

2

Literature, Complexity, Interdisciplinarity

William Paulson

And one can wisely give to literature the task of revealing to us a part of man and of the world where science does not reach.

Jean Paulhan, *Les Fleurs de Tarbes ou la terreur dans les lettres*

We might say that in creative art man must experience himself—his total self—as a cybernetic model.

Gregory Bateson, "Conscious Purpose versus Nature"

In conclusion to his sibylline apology of rhetoric, Jean Paulhan, reviewing the paradoxes turned up by his implacable probing of extreme demands for purity in language and thought, names the method that has failed to seize either critical terror or rhetorical flowers. It is the method of detectives, philosophers, mechanics, and physicists: the method of Descartes. Paulhan quotes its rules from the autobiographical *Discourse on Method*, in which Descartes rejected the learning provided by texts and tradition and set forth principles for obtaining certain knowledge by reducing problems to simple, self-evident components and linking these in a linear chain of reasoning. What the Cartesian rules collectively imply, notes Paulhan, is "that our thought is in no case subjected to or confused with its objects, but independent" (244). The dualism of Cartesian metaphysics is of a piece with an epistemology that separates mind from its objects. Paulhan argues that in confronting the commonplaces of rhetoric, where thought and language fuse, neither the mind's separation from its objects nor the reduction to simplicity are possible. Thinking about the language of thought, thought is enmeshed in the web of its own complexity. The experience of literature, like the theories of modern physics, suggests to Paulhan that reason itself must change, give up its Cartesian vantage point and certitudes.[1]

With the *Discourse on Method*, Descartes joined his older contemporary Bacon as a founder of the kind of distinction between ways of knowing that now resides in the contrasts between the literary and scientific disciplines. The study of literature often seems a residue of the sort of education that these instituters of modern science were so quick to reject. Yet if today there can be a study of literature and science, if such a study can be theorized or practiced, this can only be because literature has a particular cultural status that relates it to science in particular ways. There must be something in "literature"—and in "science"—that makes talking about "literature and science" possible. In the pages that follow, I will argue that what most significantly unites literature and science in our age of noise and chaos is the notion of *complexity* and its implications for interdisciplinary understanding.

"The scientific revolution of the past fifty years," writes Anthony Wilden, "had its origins in a revolt against simplicity" (303). Modern science at its beginnings could get along well with the epistemological side of Cartesian dualism, with an absolute separation between the physical world to be known and the mind that does the knowing. The perspective of the scientific observer, uncontaminated by its objects, is comparable to that of a mind admitting only clear and distinct ideas. The Newtonian paradigm, by suggesting that wide ranges of phenomena could at least in theory be understood to be caused by the deterministic motion of bodies, held out the hope that certainty could be attained by the reduction of the complex to the simple. The most extreme conceptual figure of this project was the demon imagined by Pierre Simon Laplace. Capable of deriving all past and future events from complete knowledge of the position, mass, and velocity of all the fundamental particles of the universe, this demon—omniscient and unconnected to the dynamic system it observes—was an idealization of the scientific observer attempting to apply the paradigm of Newtonian mechanics.

Claims of universality for the Newtonian paradigm were shattered early in the nineteenth century by Fourier's formulation of a law for the propagation of heat that was independent of, and irreducible to, the mechanical laws of motion (Prigogine and Stengers 103–5). Twentieth-century discoveries, from relativity to quantum mechanics to dissipative structures, have put an end to at least the more metaphysical claims of objectivity for the scientific observer. No less important, science has changed in objects as well as theoretical models. Fields from biology to psychology to anthropology have joined the ranks of the sciences since the end of the Newtonian model's hegemony. We speak of the social sciences or the human sciences, and can no longer fix a boundary between these and what we still distinguish as the natural sciences. Where in the brain does neurophysiology stop, and psychology begin? Since this question has no clear answer, it becomes impossible to identify what has long been called "mind" with the subject rather than the object of science. And it becomes reasonable (though it is by no means the only intellectual option available) to suppose that "mind" is a name for a particular kind of natural system, entirely realized in physical phenomena no different, ontologically, from other physical phenomena that have long been the object of science. Yet the approach to the phenomenon of mind poses seemingly unprecedented problems of inherent complexity, because the mediation between physiochemical components and the mental effects that emerge from them must involve variety and diversity comparable to the resources of the mind that seeks to know the phenomenon.

The very possibility of a science of mind depends to a large extent on the prior development of scientific approaches to complexity. Prior to the "revolution against simplicity" embodied in information theory and cybernetics, science dealt above all with systems of matter and energy, in general recognizing no distinct role for information. As Wilden has noted, information becomes distinct and abstracted from matter-energy in systems (organic, social, cultural) of increasing complexity. Scientific explanations of these generally adaptive or goal-seeking systems cannot be obtained by reduction to components describable in terms of matter-energy, but instead must take into account the emergent function of information at the level of the whole. Classical causal explanation must be supplemented by what Bateson called "cybernetic explanation" (399–410).

Mathematical information theory—one of the foundations of the cybernetic approach—was developed to resolve problems in the transmission of signals. It begins by quantifying information: the information of a message can be measured as the number of binary bits required to encode it. Information is thus a measure of a quantity of possibilities out of which a single actual message is selected; it is, in other words, a measure of the uncertainty of a receiver that will be resolved by the reception of a given message. It can also be used as a measure of the organizational variety of a system or text.

Almost from the beginnings of information theory as a discipline, researchers began considering the possibility of using information as a parameter for measuring the organizational complexity or adaptive variety of systems that were more than simple communicative messages, most notably biological systems. More important than measurement, however, was the role of information as a concept for understanding biological organization, the preservation (and development) of beings characterized by ordered complexity in a universe governed by the second law of thermodynamics. Some early theorists of information in biology (the physicist Schrödinger in particular) assumed that the large quantities of information possessed by organisms must come from their environment. But if organisms are autonomous, how can what is outside them be information for them, how can it reliably be anything but perturbation or noise?

The concept of self-organization appeared around 1960, in part as an attempt to overcome this difficulty. Heinz von Foerster and later Henri Atlan developed the following line of thinking: organisms find not only information but also noise on their menu, and they make information from noise.[2] Out of the perturbations that threaten to destabilize organisms, to modify their structure and possibly undo their organization, they produce new and more complex forms of organization. They can do so, Atlan argues, because they are multilevel systems.

Self-organization from noise, which I will describe here following Atlan's formalization, provides a framework for understanding how organized variety, information, even meaning can arise from interaction with disorder. Suppose that in a system the totality of information contained in subsystem A is to be sent to subsystem B. If the transmission takes place without noise, B gets a copy of A and now has the same information as A. But if there is noise in the channel, the information received at B is diminished by a function known as the ambiguity of the message, the quantity of information coming *from* the channel that is independent of the information that entered the channel. From the viewpoint of B, defined as a pure and simple receiver of the message from A, information has been lost. But consider now the point of view of an *observer* of the combined subsystems A and B (and the channel). Such an observer might be a scientist studying the system, but it might also be a higher level within the the system itself, a subsystem related to (A→B) as B was to A. What is the quantity of information emitted by (A→B)? It is the information received by B *plus* the ambiguity: because the message from A to B has been altered, B has not just a copy of A, but contains information independent of A. Atlan calls the two effects of ambiguity "destructive ambiguity" (B's copy of A is a poor one) and "autonomy-ambiguity" (B is now something other than a copy of A). The system (A→B) contains potentially more information if B is an imperfect copy of A that if it is a perfect copy. The partial destruction of a transmitted message within one level of a hierarchy leads to increased variety in the message that this level in turn transmits to another part of the system (Atlan 39–90).

Of course, the creation of variety is only a necessary, not a sufficient, condition for the creation of meaning. Variety that was noise in one context *can* but does not necessarily become information in a new or reorganized context. As Atlan has recently written, "permitting chance to acquire a meaning *a posteriori* and in a given context of observation—that is ultimately how we can describe what self-organization is" ("Créativité" 169).

The process of self-organization from noise provides a suggestive model for the understanding of literary signification, a model that accounts for meaning by accepting, rather than resisting, the rhetorical dimension of language. Of course, this kind of application of a scientific model to literary texts will succeed only if there are features of both model and object that will make the former pertinent to the latter. In an apparent paradox, the current literature-science dialogue (to which my argument belongs) is made possible by the very properties of literature that long made it seem the antithesis of a scientific object.

The modern sense of the word "literature" was born in a movement of dissension from Cartesian-Newtonian reductionism. In the *Encyclopédie*, d'Alembert's "Discours préliminaire" surveys the state of different branches of knowledge and cultural activity in light of the seemingly definitive triumph of Newtonian science, and concludes that the place of imaginative creation has been inevitably reduced. But his collaborator, Diderot, had other ideas, thereby showing that what we now call complexity, which is inextricably bound up with our modern notion of literature, was already serving as a countermodel to physical reductionism. Diderot's speculations about the nature of the aesthetic object stand at a critical point in the history of those ideas and discourses that even today fix the place of literature and thus the conditions of possibility of the study of "literature and science." Writing on "the Beautiful" in the *Encyclopédie*, the great eighteenth-century dissenter from the ideal of simplicity suggested that the source of aesthetic pleasure was to be found in objects and perceptions quite untouched by the explanatory powers of the Newtonian world view. Beauty, he argued, depends on the idea and perception of relations. This was not in itself an original position, but Diderot further argued that the perception of relations is a fundamental human experience because we live in a culture of machines and devices, of made things:

> We are born with needs which oblige us to have recourse to different expedients. . . . Most of these expedients are tools, machines, or some other invention of this kind, but every machine supposes combination, arrangement of parts tending towards the same goal, etc. Here then . . . are ideas of order, arrangement, symmetry, mechanism, proportion, unity; all these ideas come from the senses and are artificial; and we have passed from the notion of a multitude of

artificial and natural beings—arranged, proportioned, combined, and made symmetrical—to the positive and abstract notion of order, arrangement, proportion, comparison, relations, symmetry, and to the negative notion of disproportion, disorder and chaos. (415–16)

Only secondarily do we find these same characteristics in natural beings; they originate with artificial, utilitarian objects.

Writing in the 1780s, Karl Phillip Moritz gave the notion that beauty consists in the harmony of a work's parts a romantic turn by associating it with the idea that the work of art is a totality that constitutes its own finality. The work is thus, like nature, a created being. For Moritz, internal organization appears when it is recognized that there is no external finality (3–9). With this kind of change in the notions of finality and imitation, the organism superseded the machine as a conceptual model of organization, and complexity replaced simplicity as a formal ideal. Schelling articulates this new organicism in the Introduction to his *Philosophy of Art*:

If it interests us to investigate as far as possible the structure, the internal dispositions, the relations and the intricacies of a plant or of an organic being in general, how much more should we be tempted to know these same intricacies and relations in that plant, considerably more organized and intertwined with itself, that is called the work of art. (358)

The interest in the complexity of harmonies and relations is thus one with the notion of the totality and autonomy of the work of art. Artistic autonomy implies that the authentic work of art participates in a dialectic of formal innovation, its rhetorical or stylistic devices becoming a matter of invention and deviation rather than inherited convention. The several formalisms of the twentieth century have attempted to investigate systematically the structures and procedures of this autonomy.

The Russian formalists emphasized such notions as the device and the poetic function of language so as to define as specifically as possible how literary texts differ both from communicative uses of language and from each other. Other formalist schools have taken the rhetorical figure as a fundamental unit of study, while others (much of the New Criticism, for instance) have remained closer to their Romantic origins by concentrating on what they call the work's "organic unity." In critical practice, all formalisms must confront the problem of relating the subordinate parts to the whole, the local to the global, the device or figure to the work as a totality of relations. Self-organization from noise can be taken to form the basis of such a formalism, one that uses concepts first articulated in the sciences of complexity.

Insofar as literary texts are both communicative and ambiguous, they are noisy channels. A rhetorical figure—a minimal unit by which literary language differs from a hypothetical language of pure communication—can be characterized as a departure from expectations and thus as a reduction of informational redundancy. It makes communication less reliable, but increases the informational variety of the system in which the communication takes place (see Eco 52–65; and Paulson 64–65, 87–91). Now if all rhetorical figures could be decoded unambiguously by logical or grammatical means, then any uncertainty in which they plunge the reader would be only temporary, and they would be entirely a matter of information, not noise. But deconstructive criticism has repeatedly shown that rhetoric is not in general reducible to grammar. Texts are thus communicatively imperfect, and this is what information theorists mean when they say that a channel is noisy. For de Man, Derrida, and other theorists of textual rhetoric, figures are not simply ornamental or persuasive, but pose fundamental problems of decidability, preventing texts from being fully decoded by unambiguous, grammar-like procedures.

Literary texts inevitably contain elements that are not immediately decodable and that therefore function for their readers as what information theory would call noise. With this in mind, we are in a position to extend a crucial conjecture advanced by Jurij Lotman in *The Structure of the Artistic Text* and argue that noise both within and outside the text can lead to the emergence of new levels of meaning neither predictable from linguistic and genre conventions nor subject to authorial mastery (Paulson 80–100). Here is what Lotman, one of the most important contributors to an information theoretic approach to literature, has to say about organization from noise:

> Art—and here it manifests its structural kinship to life—is capable of transforming noise into information. It complicates its own structure owing to its correlation with its environment (in all other systems the clash with the environment can only lead to the fade-out of information). (75)

This potential stems, for Lotman, from the organizational nature of works of art, itself a particular type of play between redundant order and informative surprise. The artistic text begins as an attempt to go beyond the usual system of a language—in which the word is a conventional sign—to a specifically artistic system such as that of poetry, in which sounds, rhythms, positional relations between elements will signify in new ways. The poetic text, in other words, demands of its reader that she create new codes, that she semanticize elements normally unsemanticized (Lotman 55, 59). New levels of constraint produce new kinds of variety and coding, new contexts in which aspects of language that in nonartistic communication would be extraneous to the message become elements that enter into secondary or tertiary signifying systems. Whereas in nonartistic communication there can be *extrasystemic* facts, which are simply ignored or discarded because they are not dealt with by the codes being used to interpret the message, in an artistic text there are only *polysystemic* facts, since whatever is extrasystemic at a given level, and thus destructive of regularity or predictability on that level, must be taken as a possible index of another level, another textual system with a new kind of coding.[3] The multiplication of codes, or rather the creation of new and specific codes within a given genre and a given text, is the essence of artistic communication and the emergence of meaning in artistic texts. "What is extra-systemic in life is represented as poly-systemic in art" (Lotman 72). What appears to be a perturbation in a given system turns out to be the intersection of a new system with the first. In becoming aware of such a relation, the reader in effect creates a new context in which the previously disruptive event or variety is reread. The principle of constructing a pattern out of what interrupts patterns is inherent in artistic communication.[4]

The qualities of a literary text, in this view, are *emergent*, i.e., "qualities not included in, and generally not predictable from, knowledge of the qualities of the systems in which they arise" (Wilden 170). One can speak of emergent qualities in many kinds of systems—inorganic, organic, sociocultural. The chemical properties of a molecule cannot in general be predicted from the physical qualities of its atoms, even though we suppose that the molecule is nothing but a particular set of atoms interacting in a given way. The same can be said for the properties of a living cell with respect to its chemical constituents, or the function of a brain with respect to the function of its neurons, or the distinctive features of a culture with respect to the socioeconomic structures within it (and which it may have in common with those of quite different cultures). A series of levels of emergent qualities define what Wilden calls a "dependent hierarchy," in that the existence of each level depends on the existence of the level from which it emerges: without atoms there can be no molecules, without macromolecules no living organisms, without organisms no society, without social relations of production and reproduction no culture. At the "independent," inorganic end of the hierarchy, constraints are extremely general (laws of physics and chemistry) and information has virtually no existence independent of matter-energy. As we move through the organic and social to the cultural, to the pole of maximum dependency, constraints become increasingly local, context-dependent, and complex. Information becomes increasingly abstracted from the matter-energy structures that bear it, and the increasingly autonomous interactions of information come to organize and direct matter-energy phenomena (Wilden 73–79, 167–76).

The idea of emergence stands in direct opposition to the reductionist project represented by the demon of Laplace. Already in 1903 Henri Poincaré suggested that an arbitrarily small uncertainty concerning the initial conditions of a system may lead to very large errors in the prediction of its subsequent states, so that reductionist prediction in any meaningful sense becomes impossible. Recent work by Ilya Prigogine and many others on the role of chaos in a variety of natural systems and mathematical models confirms and strengthens Poincaré's argument. In the words of a recent *Scientific American* article, "Chaos brings a new challenge to the reductionist view that a system can be understood by breaking it down and studying each piece. . . . The interaction of components on one scale can lead to complex global behavior on a larger scale that in general cannot be deduced from knowledge of the individual components" (Crutchfield et al. 56).

It is doubtless also the case that the cultural circulation of such concepts as complexity and emergence leads to their use to describe other kinds of systems, in which reductionism would indeed be ultimately possible but is rejected for reasons of economy and pertinence. In the pragmatics of knowing there must be different kinds of descriptions for different levels of phenomena, and often reductionism is simply not a practical or interesting option. The finitude of knowledge requires working with the discontinuities between levels, given that the complete description, for example, of a biological phenomenon in physico-chemical terms—should such a reductive description be theoretically possible, which is far from certain—would be inordinately cumbersome and complicated. Those who describe phenomena as complex or emergent thus renounce the Cartesian dream of maximal certainty by reduction to the simple, and assume the risk of choosing a pertinent level of description (Stengers, "Découvrir" 250–51).[5]

Emergence and complexity in this sense pertain to the categories of knowledge, to the discursive systems and disciplinary matrices by which the totality of the universe to be known is divided up and the several forms of knowledge produced. Different disciplines, in other words, correspond to different levels of description, or phenomenal domains. In the sciences, a discipline is a set of cognitive maps and decision procedures, presumably effective at reducing the unknown to the known at its level of pertinence. Because of the very success of the disciplines,

Atlan has noted, "the frontiers of knowledge are found not only, as is often believed, in the very small or the infinitely large, but in the articulations between levels of organization of the real that correspond to different fields of knowledge whose techniques and discourses do not overlap" ("L'Émergence" 129). To confront the passages between disciplines is to confront what Atlan calls *complexity*, large quantities of organizational information that appear to an observer as uncoded variety. In the absence of an operational explanation of the passage between levels, observers experience the system's complexity as disorder or noise, and must often negotiate with this presence of apparent randomness in constructing explanations.[6]

The problem posed by what Atlan calls interdisciplinary cognition is thus related to the attempt to construct meaning out of what initially seems to be noise: in his words, "*how to speak of that for which we do not yet have an adequate language?*" ("L'Émergence" 123). The attempt to bridge the discontinuity between an emergent level and its environment implies a process of self-organization from noise. As a schema of cognition, self-organization from noise describes situations in which knowledge is but partial, the ignorance of codes bound up with the presence of information. If the phenomena that we identify as emotions and ideas are truly emergent with respect to the physiological interactions of brain cells, then even complete knowledge of the brain at the neuronal level would not explain the mental level. The certainties of reductionism simply are not available. (Conversely, if a reductionist explanation of mind by neurons were possible, then mental properties would be emergent only in a temporary, illusory sense, insofar as for pragmatic reasons we might be forced to do without a complete physiological map.) To the observer of emergent phenomena, and thus to anyone who seeks to relate such phenomena to the less complex phenomena from which they emerge, unpredictable order is being produced because of a change in context. Attempting to mediate between levels of explanation, we face the task of explaining how what we experience as noise can become informative.

Viewed from this perspective, the defining feature of literature and its study would be their *interdisciplinarity from within*. I use the term *interdisciplinarity* here not in its usual sense, but in the context of Atlan's identification of a discipline with a level of operational description, and thus of interdisciplinarity with the forging of a language with which to speak of the passages between the phenomena that our several descriptive languages enable us to describe. This definition of a discipline fits the natural sciences well enough but runs counter to our empirical sense of what a discipline is in the social sciences and humanities. Fields such as literary studies, history, anthropology, and even economics deal with multiple levels of phenomena: by the definition adopted, many recognized disciplines would be "interdisciplinary from within" in an apparently trivial sense.

Yet there remains the crucial question of how these disciplines in fact handle the multiplicity of explanatory levels contained within them. In particular, do they become separated into subdisciplines, each devoted to the refinement of techniques of analysis for a particular level of phenomena, or are these subdisciplines united by a sense of "interdisciplinarity from within?" It sometimes appears, for example, that economic and social historians constitute a group quite distinct from political historians, and that members of each group argue that their own activity is the right and proper task of the discipline. Nonetheless, many if not most historians see these kinds of history not as alternatives but as components of a cognitive enterprise that must ultimately synthesize the two.

I have argued elsewhere that literary studies are less a discipline in the modern sense than a residue of a prescientific, predisciplinary form of knowledge that can best be evoked by referring to the broad sense of the term "literature" in the seventeenth and eighteenth centuries (Paulson 8–29). Yet many of the theoretical efforts characteristic of modern literary studies are in effect attempts to define a narrower disciplinary formation by privileging a particular level of phenomena and establishing methodological principles for its investigation. Schools of theorists work on procedures for describing different, virtually separate domains: rhetoric, narrative structure, intertextuality, the reading process, psychoanalytic structures, sociopolitical signification . . . the list could grow long. There is a risk that these critical schools, each believing itself to be providing disciplinary rigor, could come to have less and less to say to one another. The structural nature of literary texts, however, argues against any such exclusionary reductionism.

Literary works exhibit the complexity of emergent systems possessing singular, context-dependent constraints and forms. Although texts are made of language, the passage from linguistic structure to textual effect cannot be described with anything like the regularity or predictability to be found in, say, the grammatical description of sentences. The text is not fully determined by the linguistic features of which we know it to be made. We suppose, for example, that a poem presents itself to the reader as a complex system of relations. These include the usual relational structures of language considered as a system of signification; the relations between sounds established by rhyme, rhythm, meter, assonance; the thematic relations between signifieds; most important, the relations *between* these categories: between meter and syntax, rhythm and theme, and so forth. The reader brings to her assimilation of the poem a knowledge of the linguistic codes; if she is an experienced reader of poetry, she also brings some general sense of where and how to look for the further relations that make up the poem. But she does not begin with precise, operational knowledge of how all these different phenomena will interact and thereby contribute to the poem's effect, exactly how they will combine to produce what she will call the poem's meaning. In other words, the reader does not initially possess all of the codes needed to understand the poem, so that some of its variety is uncoded, or in other words is noise.

This situation arises not from the individual reader's incompetence but from the nature of literature, at least of romantic and postromantic literature. Under an aesthetic of formal innovation and uniqueness, the specific relations between elements of a text are to some degree unique to that text and so cannot have been learned anywhere else.[7] Literary and communicative utterances are thus distinguished: the latter are deemed effective if the reader possesses in advance as much as possible of the necessary codes, so that he can immediately receive *as information* as high a percentage as possible of the message's variety. The literary utterance, by contrast, is precommunicative, for whereas writer and reader share the natural language in which the text is written, the reader does not yet possess the specifically literary codes pertinent to the diversity of that text. Thus to the reader of a poem who starts out as a reader of prose, whatever "prosaic" information is transmitted by the poem will initially be accompanied by, and very likely obscured by, "poetic" diversity that at first, in the absence of an understanding of its articulation, can be experienced only as noise. The reader's construction of a meaning for the poem seems to proceed by a process of self-organization from noise: variety (and kinds of variety) not explainable in and of themselves become ingredients of a new level of explanation, a new context in which they may be informative rather than noisy.

The literary text cannot therefore be described at a single level, by a single discipline of a quasi-scientific kind; it can be successfully reduced neither to linguistic phenomena, nor to rhetorical figures, nor to global structures of poetic or narrative form, nor to psychological or ideological categories. Attempts to make such reductions, to account for the whole in terms of a single set of parts or a single procedure for organizing the parts, have never obtained general assent, precisely because they are always incomplete. Moreover, the strongest literary texts presumably attempt to "speak of that for which we do not yet have an adequate language," and such too is the task that confronts serious critics of these texts. The critical game is repetitive and sterile if played only in the domains of established codes, never venturing into the margins of noise and thus the emergence of new contexts of understanding.

Since the beginnings of romanticism, the aesthetic object and in particular the literary text have served as a cultural apprenticeship of complexity. In the modern academy, the pertinence of this kind of apprenticeship now manifests itself in the critical role of interdisciplinary study: the exchange of codes and information across the boundaries of their conventional academic contexts. Beyond the necessary efforts to combat fixation on a single level of description *within* the established disciplines, more and more scholars and researchers find that the frontiers of knowledge are now located in the spaces and eventual connections between aspects of the real that are studied in different departments, often with radically diverging traditions of discourse and method. If we now choose to investigate connections between literature and science, this is because we are convinced that the two are somehow related, parts of a system much larger than either, and yet separated by the very difference between the kinds of language that have arisen to describe each. And if I practice interdisciplinarity by importing terms and concepts such as those of information theory and self-organization, violating conventional boundaries by identifying textual ambiguity and rhetoricity with noise, I do so not to produce a Grand Synthesis but to disturb, enrich, and perhaps displace the study of literature by injecting into it some information sufficiently foreign as to function intially as "noise, the only possible source of new patterns" (Bateson 410). From the interference between disciplines can arise new forms of explanation, new articulations between levels of phenomena in a world of emergent complexity.

As a conclusion to these remarks, I wish to propose that the ideas discussed here may have their most important potential application not in the study of literature and science but rather in the study of literature and society. Literature and science, depending on how one defines the latter, are either very close together or very far apart in what I call, following Wilden, the dependent hierarchy of nature and culture. On the one hand, both literature and science are parts of the cultural superstructure; on the other hand, the objects of science are physical and organic systems, while the objects of literary studies are systems of greater complexity and more local constraint that emerge from a social environment, which itself emerges from an organic environment (see Wilden 74, 168). The interpretation of cultural texts in the context of means and social relations of production and reproduction involves the bridging of only one major hierarchical gap, the one between the study of society and the study of culture.

In *The Political Unconscious: Narrative as a Socially Symbolic Act*, Fredric Jameson presents and revises Althusser's concept of a mode of production as a structure uniting levels that range from the material and economic forces of production through political institutions and ideology to culture (35–39). He notes that the legitimate insistence on the semi-autonomy of these levels runs the risk of reinforcing the reification and isolation of the academic disciplines devoted to their individual study. He thereby identifies, as does Atlan, discipline and hierarchical level of description. The problem of mediation between cultural text and sociohistorical context is thus for Jameson a problem of interdisciplinarity that must be worked on by "a process of *transcoding*" that is local, emergent, and risk-taking rather than global, fixed, or certainty-producing: "the *invention* of a set of terms, the *strategic choice* of a *particular* code or language, such that the same terminology can be used to analyze and articulate two quite distinct types of objects or 'texts' or two very different structural levels of reality" (40; emphasis added in second quotation). Mediations through transcoding are thus constructed by an observer, just as for Atlan the observer of distinct levels constructs meaning out of the noise between them, but the process is validated as a mode of knowledge by the ultimate underlying unity of the systems that only disciplinary descriptions have cut into separate hierarchical levels.

Jameson's description of the cultural text—in particular the postmodern text and those features of the cultural text that are most strongly brought out by poststructuralism—emphasizes the "noisy" qualities of interference, disorder, and fragmentation: "the authentic function of the cultural text is staged as an *interference* between levels. . . . The current poststructural celebration of discontinuity and heterogeneity is therefore only an initial moment in an Althusserian exegesis, which then requires the fragments, the incommensurable levels, the heterogeneous impulses, of the text to be once again related" (56). Poststructuralism and the communicative disorder it celebrates thus apppear as a negative moment in the dialectical task of saving the interpretation of culture from disciplinary reification and transforming it into a mode of understanding the relatedness of cultural texts to the "great collective story" of human history.

One major reason why students of literature and society need to study cybernetics, emergence, order from fluctuation, and self-organization from noise is the potential pertinence of new kinds or concepts of causality—to be added, for example, to the mechanical, the expressive, and the structural discussed by Althusser, or to Jameson's "active" terms of relation between text and social subtext: "production, projection, compensation, repression, displacement" (44). What Bateson calls "cybernetic causality" is a dynamic form of structural causality, diachronic rather than synchronic, potentially homeorhetic rather than inevitably homeostatic. Self-organization from noise and order from fluctuation formalize the causal link between microscopic event and global consequences. Such figures of causality are essential ingredients of knowledge, and they are nearly always drawn from the concrete world of experience or else from the material world as elaborated and articulated for us by the sciences. And it does not suffice to recognize and then perhaps denounce these forms of causality as metaphors, for—as Fontenelle already knew in the eighteenth century—we cannot do without them, metaphors or not, even though we recognize that in given contexts some of them are less appropriate than others. As Althusser wrote, "the 'reprise' of a fundamental scientific discovery in philosophical reflection, and the production by philosophy of a *new form of rationality* . . . mark the great breaks in the history of the Theoretical" (185).

We need the forms of rationality opened up by the postmodern sciences of information and disorder in order to understand the texts and criticism of postmodernism in the only context that really matters: our material, organic, and social world.

Notes

1. Referring to Bachelard (*Le Nouvel Esprit scientifique*), Paulhan cites both relativity theory and the uncertainty relation of quantum mechanics as instances of changed reason. On the implications of these field models, see Hayles 31–59.

2. On the history of the concepts of self-organization, see Stengers ("Généalogies") and Atlan (*Cristal* 39–44).

3. Of course, misprints and other deformations that befall artistic texts are in some sense extra-systemic, but so strong are the autonomy conventions that even this external noise is often treated as poly-systemic fact: misprints, editorial mischief, even O-shaped cigar burns on a photocopied manuscript give rise to interpretation and commentary. See Parker 72–79.

4. Lotman himself restricts the term "noise" to external disturbances of the

artistic text, but his theorization of the role of external noise presupposes that internal features of the text already function in an analogous manner. If noise arising outside the text can be understood to produce new meaning, this can only be so because "irregularities" within the text are already being interpreted as part of multiple systems of coherence and signification.

5. They also renounce Cartesian dualism: for Atlan, the most important example of a system that must be treated as complex is the relation of brain components to thought, the contemporary locus of the mind-body problem (*A tort* 57–76).

6. Wilden sensibly restricts the term "complexity" to qualities based on *diversity*, "the combination of two or more kinds of variety" produced by more than one kind of constraint (172). Atlan does not explicitly make such a distinction, and sometimes appears to define complexity simply in terms of uncoded variety, but given the kind of system—biological—with which he is primarily concerned, it is clear that the organizational information linking two levels—information which appears as noise to the observer—would have to exhibit diversity and not just variety of a single kind, such as the random distribution of molecules in a gas of uniform temperature.

7. On the aesthetic necessity of innovation, and the replacement of a stable rhetoric of convention by a series of rhetorics of surprise, each accompanied by denunciations of its immediate predecessors as rhetorics of convention, see Paulhan 156–61.

Works Cited

Althusser, Louis. "Marx's Immense Theoretical Revolution." *Reading Capital*. Trans. Ben Brewster. London: New Left Books, 1970. 182–93.

Atlan, Henri. *A tort et à raison: Intercritique de la science et du mythe*. Paris: Seuil, 1986.

———. "Créativité biologique et auto-création du sens." *Cognition et complexité*. Cahiers du C.R.E.A. 9. Paris: Centre de recherche épistémologie et autonomie, 1986. 145–89.

———. *Entre le cristal et la fumée*. Paris: Seuil, 1979.

———. "L'Emergence du nouveau et du sens." *L'Auto-organisation: de la physique au politique*. Ed. P. Dumouchel and J.-P. Dupuy. Paris: Seuil, 1983. 115–30.

Bateson, Gregory. *Steps to an Ecology of Mind*. New York: Ballantine, 1972.

Campbell, Jeremy. *Grammatical Man: Information, Entropy, Language, and Life*. New York: Simon and Schuster, 1982.

Crutchfield, James P., J. Doyne Farmer, Norman H. Packard, and Robert S. Shaw. "Chaos." *Scientific American* 255, no.6 (1986): 46–57.

De Man, Paul. "The Resistance to Theory." *Yale French Studies* 63 (1982): 3–20.

Diderot, Denis. *Oeuvres esthétiques*. Ed. Paul Vernière. Paris: Garnier, 1968.

Eco, Umberto. *The Open Work*. Trans. Anna Cancogni. Cambridge: Harvard University Press, 1989.

Hayles, N. Katherine. *The Cosmic Web: Scientific Field Models and Literary Strategies in the Twentieth Century*. Ithaca: Cornell University Press, 1984.

Jameson, Fredric. *The Political Unconscious: Narrative as a Socially Symbolic Act*. Ithaca: Cornell University Press, 1981.

Lotman, Jurij. *The Structure of the Artistic Text*. Trans. G. Lenhoff and R. Vroon. Michigan Slavic Contributions, 7. Ann Arbor: University of Michigan, Department of Slavic Languages and Literatures, 1977.

Moritz, Karl Phillip. *Schriften zur Ästhetik und Poetik*. Ed. H. J. Schrimpf. Tübingen: Max Niemeyer Verlag, 1962.

Parker, Hershel. "Lost Authority: Non-sense, Skewed Meanings, and Intentionless Meanings." In *Against Theory*. Ed. W. J. T. Mitchell. Chicago: University Chicago Press, 1985. 72–79.

Paulhan, Jean. *Les Fleurs de Tarbes ou la terreur dans les lettres*. 1941. New ed. Paris: Gallimard "Idées," 1973.

Paulson, William. *The Noise of Culture: Literary Texts in a World of Information*. Ithaca: Cornell University Press, 1988.

Prigogine, Ilya, and Isabelle Stengers. *Order Out of Chaos: Man's New Dialogue with Nature*. New York: Bantam, 1984.

Schelling, Friedrich. *Sämmtliche Werke*. Part 1, vol. 5. Stuttgart: I. G. Cotta'scher Verlag, 1859.

Stengers, Isabelle. "Découvrir la complexité?" *Cognition et complexité*. Cahiers du C.R.E.A. 9. Paris: Centre de recherche épistémologie et autonomie, 1986. 223–54.

———. "Les Généalogies de l'auto-organisation." *Généalogies de l'auto-organisation*. Cahiers du C.R.E.A. 8. Paris: Centre de recherche épistémologie et autonomie, 1985. 7–104.

Wilden, Anthony. *The Rules Are No Game*. London: Routledge and Kegan Paul, 1987.

3

Fictions as Dissipative Structures: Prigogine's Theory and Postmodernism's Roadshow

David Porush

"It was awful, ghastly. It was man's worst modern nightmare: the ultimate expression of the ant running hopelessly, uselessly, and totally pointlessly around and round in ever decreasing circles. It was traffic."

William Marshall, *Roadshow*

"School's completely antique here. They make us read Lindsay's book. Shakespeare, it's called. Translated into the modern English by Abelard Lindsay."

"Is it that bad?" Lindsay said, tingling with déjà vu.

"You're lucky, old man. You don't have to read it. Not one word in there about spontaneous self-organization."

Bruce Sterling, *Schismatrix*

Trafficking in Apocalypse

Kerouac's *On the Road* is only one of dozens of novels and films in which heroes in cars move across the imaginative terrain of contemporary narrative, symbols for speed and spontaneity. This is not necessarily American, though no other country has so profound and extensive a romance with the automobile nor offers so varied and inviting a combination of roadways and landscape on which to play out modern versions of the romantic journey.

A free-flowing traffic system also is a sign of civilization's complexity. "The Night Driver," a short story by the late Italian postmodernist, Italo Calvino, portrays its hero driving furiously through the night, trying to get to his lover so he can correct a misunderstanding they had over the phone. Calvino clearly intends to show us that the driver has been transformed into a message himself—part of a larger flow of information—and that the car-road-mind nexus represents a communications system that rivals the telephone. In fact, the lover who decides to drive along the road to intercept his rival and reach his paramour has decidedly rejected the telephone as a means of communication.

John Hawkes' masterful narrative tour of the automobile quest-as-desire, *Travesty* (1976), traps a triangle of lovers in a car driven by the betrayed husband hurtling towards revelation and a deathly finale. We can multiply such examples of the speeding automobile as a literary trope for desire indefinitely.

But even these darker images of automobile-as-desire rely on freely flowing traffic. By contrast, there are very few works of narrative literature about traffic *jams*. Considering how much time urban folks spend stuck in their cars battling other cars for a few inches of purchase, it is surprising that this unfortunate fact of civilized life has not been granted greater space in literature along with other forms of pollution, for clearly, although we often associate automobile traffic with atmospheric pollution, that is only a side-effect compared to the temporal-spatial pollution we experience when stuck in traffic.

There are, however, a few noteworthy portraits of traffic. Jean Luc Godard's well-known cinema noir film, *Weekend* (1970) uses the traffic jam as a metaphor for the way civilization pollutes psychic life. A recent work by Don DeLillo, *White Noise*, shows citizens trapped in a traffic jam as they flee a hazardous chemical spill. But beyond these and a few other examples, the highbrow narrative imagery of traffic jams does not nearly rival the literature of the free-moving automobile. Where the moving car evokes the heroism and promise of gleaming chrome, the car caught in traffic evokes bent fenders. The headlong speed of a moving car (or sometimes the motorcycle, as in Robert M. Pirsig's 1980 novel-essay, *Zen and the Art of Motorcycle Maintenance*) carries us into sweeping expanses of landscape, back to nature (a nice paradox) and contemplation. The jammed car traps us claustrophobically with our own short tempers and frustrations, making us aware of our diminution by the conditions of urban life. Zipping in and out of lanes on a freeway or choosing a lonely back road, we assert our individuality. But stuck in traffic we become indistinguishable from thousands of others who are leveled by the same circumstance, rehearsing a kind of urban *crise de dédifférentiation*' as René Girard calls it—a crisis of undifferentiation; or *la fête qui tourne mal*—the sick ritual in which individuality is lost in mob behavior. When it moves, the automobile is a symbol for freedom. When snarled, it is a symbol for breakdown, alienation, despoliation, and decay, for the collapse of civilization and the contamination of sense and sensibility by the dull pedestrianisms, the pedestrian accidents of life.

Don DeLillo clearly indicates his understanding of this association between the traffic snarl and the whimper of slow modern apocalypse in a striking passage in *White Noise*:

Twenty minutes later we were in the car. . . . We were among the latecomers . . . and joined the traffic flow into the main route out of town, a sordid gauntlet of used cars, fast food, discount drugs and quad cinemas. As we waited our turn to edge onto the four-lane road we heard the amplified voice above and behind us calling out to darkened homes in a street of sycamores and tall hedges.

"Abandon all domiciles. Now, now. Toxic event, chemical cloud."

We made it out onto the road as snow began to fall. We had little to say to each other, our minds not yet adjusted to the actuality of things, the absurd fact of evacuation. Mainly we looked at people in other cars, trying to work out from their faces how frightened we should be. Traffic moved at a crawl but we thought the pace would pick up some miles down the road where there is a break in the barrier divide that would enable our westbound flow to utilize all four lanes. The two opposite lanes were empty, which meant police had already halted traffic coming this way. An encouraging sign. What people in an exodus fear most immediately is that those in positions of authority will long since have fled, leaving us in charge of our own chaos.[1]

"Chaos" and Prigogine's Postmodern Synthesis: From Being to Becoming

That last word in the passage above, chaos, is the point of departure for this essay. For strangely enough, from a mathematical or systems perspective, the traffic jam does not represent chaos or disorder or collapse or dedifferentiation, whatever its cost in havoc to human affairs. Rather, as Ilya Prigogine and his collaborators have shown, the traffic jam is one of a curious set of phenomena that represent *the spontaneous emergence of order out of disorder*—"a self-organizing system."[2] These phenomena, which Prigogine calls "dissipative structures," are ubiquitous in the biosphere or macroscopic world. Indeed, the very fact of the biosphere—with its seething complexity and diversity and apparent tendency to evolve in the direction of increased differentiation and complexity—is itself explained by Prigogine's theory of dissipative structures.

In order to appreciate the power of Prigogine's theory, it helps to view it in a historical context. While Sadi Carnot, Lord Kelvin, and others were developing the laws of thermodynamics which portrayed the universe winding down inexorably towards randomness and cold, Charles Darwin described a more heated aspect of the cosmos that evolved towards complexity and differentiation. This contradiction between the thermodynamic and Darwinian cosmologies, between entropy and evolution, has long been recognized. We can trace its source to the Greek fascination with the question of the divisibility of matter: is reality what we have before our senses or does it reside at the level of the atom, the smallest divisible part of nature? Lucretius, following Democritus, thought that the world was merely "atoms and the void, nothing more." Norbert Wiener described the entropy vs. evolution problem in 1948 in his popularization of cybernetics, *The Human Use of Human Beings*: biology is an "island of order in the universal entropic tide." But this island metaphor doesn't communicate the dynamic, highly unstable growth (or morphology) of such orderly systems. Instead, to shift the metaphor, a dissipative structure is more a *raft* which floats inexplicably but definitely upstream, *against the current*, gathering flotsam and organizing it into its flotilla with some sort of autonomous force or direction. Dissipative structures seem to have a mind of their own. They are self-organizing systems that locally contradict the second law of thermodynamics.

Although the contradiction between an evolving, self-organizing macroscopic world and the orderly, entropic dynamics of the microscopic world has long been self-evident, classical science has always resolved it in favor of the cosmology it found in the world of very small particles, at least until Heisenberg. Perhaps because the microscopic world could be more readily idealized while the macroscopic world intrinsically resisted simple modeling, modern science has typically presumed that reality flows from the microscopic. It is simply easier to ignore the messiness of irreversible reactions and focus instead on idealized vector diagrams modeling the dynamics of particle interactions. By contrast, literature has typically had little use for the microscopic except as it offers up some interesting metaphors.[3]

The fulcrum for the contradictory views of reality is the notion of time. In a classical conception of physical reality, time is reversible, even though this requires a highly idealized version of interactions among dynamic bodies, bringing physics to focus on the microscopic level where such things are theoretically possible. Certainly at the level of our senses, time is irreversible. It has been left primarily up to two disciplines to worry about the contradiction: philosophy and chemistry. It fell to chemistry to worry about the contradiction for two reasons: first, chemistry shuttles between the orderliness of physics and the complexity of biology; and second, even in the early nineteenth century it was recognized that most chemical reactions were irreversible. And philosophy worried because of the metaphysical consequences of such an unresolved conflict. As Prigogine the chemist himself notes in a philosophical vein:

> It was inevitable that the timeless conception of classical physics would clash with the metaphysical conceptions of the Western world. It is not by accident that the entire history of philosophy from Kant through Whitehead was either an attempt to eliminate this difficulty through the introduction of another reality (e.g. the noumenal world of Kant) or a new mode of description in which time and freedom, rather than determinism, would play a fundamental role. (*Becoming* xvii)

From the point of view of classical physics, then, the unruly reality of evolving life forms was somehow less real than the classically ordered reality of simple interactions among particles. Prigogine even notes that the French call classical dynamics *les mécaniques rationnelles*—"rational mechanics"—"implying that the laws of classical dynamics are the very laws of reason" (*Becoming* 20). Strict determinism and reversibility hold throughout. The probabilistic and irreversible are proscribed. From Newton until quantum mechanics, interactions among particles were traditionally viewed as reversible, like those of frictionless billiard balls. They escaped the dictum of entropy. Prigogine, adopting consciously the terminology of existential philosophy, calls this the science of *being*. Even in our century the search for a GUT—the Grand Unified Theory that would organize the four forces, strong and weak, under some simpler, more idealized whole—has taken precedence over attempts to describe the unruliness of reality, and also qualifies as a science of being.

While some philosophers like Henri Bergson were bothered by this wholesale dismissal of the world as we experience it, the contradiction wasn't resolved until this decade with the emergence of what Americans have come to know as the "new science of chaos." The word is a misnomer; it belongs in scare quotes. Though highly dramatic, to call the phenomena described by the mathematics of chaos *chaotic* is to stress not what is new but what has been discarded, for the revolution of the science of so-called chaos is precisely to show that systems that behave in what seemed like random or disordered fashion actually could be described by mathematics. Thus, to call "Chaos" the entire science that flows from this revision of scientific understanding is tantamount to calling astronomy since Copernicus "geocentrism" or quantum physics "the physics of certainty." To put it in simpler terms, this new understanding revealed that a hidden order lurks in these complex, apparently chaotic systems. Even James Gleick betrays the extent to which the word is a misnomer in his book popularizing the term and the science for the public imagination:

either deterministic mathematics produced steady behavior, or random external noise produced random behavior. That was the choice.

In the context of that debate, chaos brought an astonishing message: simple deterministic models could produce what looked like random behavior. The behavior actually had an exquisite fine structure, yet any piece of it seemed indistinguishable from noise.[4] (78–79)

Although Prigogine's term "dissipative structures" hasn't sold as well, it is in most senses much more accurate. First, it focuses on the dynamic system which undergoes the sudden transformation from *apparently chaotic* to *increasingly ordered* on the other side of the bifurcation point. Second, it implies the structure in Prigogine's mathematical model which specifies when such orderliness is not only possible but even likely to arise.

In the scientific community, that is the primary significance of Prigogine's work, for which he was awarded the Nobel Prize in 1977. A *New York Times* article about the award to Prigogine announced, "A Chemist Told How Life Could Defy Physics Laws."[5] Here at last was a scientific reconciliation between the purer, more rational, microscopic world of particles and forces and entropy with the seemingly unruly macroscopic world that we all experience and inhabit, a world teeming with variety and a confusing array of complex, interwoven, hungry structures that grow, and grow more complicated, willy nilly, feeding in an open exchange with the world around them. The biosphere is imperialistic and dynamically unstable. It fills niches, abhors a vacuum. In the realm opened by our unmediated human senses (i.e., without cloud chambers or radio telescopes), small changes in the beginning conditions of things (I forgot my car keys . . .) can have enormous consequences at the end (. . . and so I missed the interview, lost the job, became poor, went crazy, assassinated a president, changed the course of history). The macroscopic world is always in process, just as our hearts beat toward that tick when they will stop. Large things become subject to entropy's effects. Gear rubs against gear, muscle against bone. And as we well know, time moves in one direction only. We can't grow young and plants don't turn back into seeds. We experience a world of time-bound dissipative structures, not a world of elegantly predictable mechanical collisions and reversible, symmetric reactions. Any study of this world, Prigogine asserts, requires a science of *becoming*.

The self-organizing system of biology probably started in a complex chemical soup that engendered microbes. Prigogine's model explains how such an initial leap to life might have occurred. In fact, certain self-organizing chemical reactions have already been elucidated which imitate the process.[6] These primordial microbes in turn engendered more complex organic systems which in turn engendered nervous systems, which in turn engendered intelligences, which in turn became, in human form, imperialistic and greedy, using much more than their share of energy, information, and material resources (and, consequently, disturbing the natural balance through the dissipation that turns up as heat, noise, and chemical pollution in our environment).

One of intelligence's projects, it is now clear in this technological age, is to extend itself artificially, to remake nature in its own image or, failing that, to create a non-nature of artificial things that rivals nature in power and complexity. A bleaker way to view this project is as a form of parasitism: intelligence's craving for superforms of organization establishes an imbalanced exchange of resources with nature. But another of intelligence's projects is to describe and communicate, to discourse about reality. This brings us to the heart of postmodernism's concerns and to the concerns of this essay: *because Prigogine's model challenges classical science's assumptions about the locale of reality, it also indicts the insufficiency of classical science's discourse about reality.* As such, it is part of postmodernism's three-pronged attack on classical scientific discourse.

1. Quantum mechanics preaches the essentially stochastic or probabilistic nature of nature at the subatomic level; there is no way, claim Heisenberg, Bohr, and the Copenhagen interpretation their theories inspired, to know reality at the subatomic level. An electron, a photon, a positron, a quark lack certain definite qualities until we observe them: they are definitely fuzzy. On the contrary, our discourses of observation determine their qualities.

2. Analyses of the sociology, philosophy, history, and discourse of science by such writers as Kuhn, Feyerabend, and Gadamer show that normal science itself is vulnerable to irrational forces and is trapped by the assumptions and language of its own discourse.

3. Prigogine's theory transfers the locus of reality from the microscopic world, where the goal is still simplicity (the sort of simplicity that would be indicated in a unified field theory—the simplicity of *being*) to the macroscopic, where the emphasis is on complexity and time's arrow (the science of *becoming*). The distinction between *complicated* and *complex* is important: quantum theory gives a picture of a logically complicated subatomic reality. But the world we experience is not only complicated, it cannot be described by simple mathematics. Rather, it

requires a complicated (nonlinear partial differential) mathematics. Even when the mathematics of *becoming* (complexity, dissipative structures) is brought to bear on microscopic processes, it brings the microscopic world into alignment with the reality—the way time and complexity work—at the level of our macroscopic experience, not vice versa. So the fundamental question is not the now rather irrelevent one of whether reality resides at the macroscopic or microscopic level, for the two views collapse into one larger version of the complexity of reality at any level. Rather, the fundamental question becomes a typically postmodern one: *which level of description, which sort of discourse is more universally applicable, more epistemologically potent?* Prigogine's work provides very strong evidence in favor of the discourse of becoming.

Prigogine himself, in collaboration with philosopher of science Isabelle Stengers, claims that his theory holds consequences for "a new synthesis" or, as they titled their book in its original French edition, *La Nouvelle Alliance* (the English translation is entitled *Order Out of Chaos: Man's New Dialogue with Nature*). Others have also commented on the new paradigm, the new form of discourse between humans and nature, suggested by Prigogine's work.[7] In my analysis, there is a further important consequence of Prigogine's thinking: *a collapse of the distinction between natural and artificial.* These syntheses are part of a larger postmodern reshaping of our understanding of the universe, and of the role of our description and observation in it, that was begun in quantum physics and emerges in other postmodern discourses, including the scientific discourses of cybernetics (which directly confronts the role of the human observer in terms of information) and genetics (which drives mechanically towards the decoding of the distinction between artificial and natural life). In this view, those scientific discourses are in partnership with the nonscientific discourses of literary postmodernism.

In brief, one of the most important consequences of this postmodern shift is a reevaluation of the relative veracity, or epistemological potency, of literary discourse versus scientific discourse. To put it another way, in my view there is another synthesis consequent upon Prigogine's theory: a collapse of the epistemological value—though not stylistic or generic distinctions—of scientific over fictive discourse.

William Marshall's *Roadshow*: Novel as Traffic Jam

Prigogine's model has been applied to a number of familiar, even mundane phenomena, from social and biological evolution, to genetic morphology, to tidal action on geological formations, to social divisions in ant colonies, to tribal ritual transactions, to population dynamics, to the invention of cities, to the progress of international science policy, to consumer choices in an open market, to the action of cortical neurons, and most recently, to the spontaneous formation of paired subatomic particles out of a quantum vacuum by electrons. But for the purposes of this essay, what is most interesting is Prigogine's view of the apparently spontaneous formation of traffic jams.[8] Colleagues working with Prigogine's model of the dissipative structure have offered a predictive model of traffic flow. They note that traffic can be treated as a fluid. Certain turbulent fluids under the right conditions give rise spontaneously to coherent, self-organizing systems, vortices that grow more organized. Perhaps the most dramatic of these coherent self-organizing systems in fluid is Jupiter's red spot, which was revealed by the recent Voyager flyby to be a centuries-old storm system. Traffic, like a tornado, represents a leap to a new order of organization, despite the havoc it wreaks in human terms.

In what follows I try to show how one particular fiction about traffic jams behaves like a dissipative structure. The analysis is helped by the coincidence of the subject of the novel with Prigogine's interests in traffic. My goal is to show that not only does this one fiction behave as a dissipative structure, but that fictional discourse in general may best express the various syntheses implied by Prigogine's theories. From this point of view, any text can now be seen as a "biosocial phenomenon," subject to the same laws that govern certain naturally occurring phenomena (dissipative structures).

In this argument, fiction begins with two advantages over other "biosocial phenomena." First, novels describe the world of our senses, the reality of the macroscopic world on which Prigogine's description of nature places new emphasis. Second, Prigogine challenges the veracity of classical science's descriptions of a rational nature. As a consequence, fiction in general—and postmodern fiction in particular—emerges as a powerful alternative mode of epistemological discourse that captures a reality forbidden to science by virtue of its own restrictive style.[9]

Postmodernist fiction is recognizable by two important features. As many other high literary forms have done, it uses the style or formal structure of the work to mirror its plot, theme, imagery and/or subject.[10] That is, it self-consciously manipulates its own artifice, drawing the reader or audience into a collaboration of pretenses. However, postmodernist fiction is also afflicted by a typically civilized malaise: a distrust of its own artifice and a deep doubt as to whether the author can communicate anything at all to the reader through systematic discourse. In some instances this distrust expresses itself as a willful undermining of the linguistic/expressive system (Beckett's *The Lost Ones* and such Burroughs' novels as *Nova Express* and *The Soft Machine* are good examples). In others, it emerges as a purposeful multiplication of competing and exclusive interpretations that never achieve resolution (Thomas Pynchon has mastered this technique in *The Crying of Lot 49* and *Gravity's Rainbow*). In still others, it can be seen as an ironic or paradoxical creation of complex, hyperevolved, but self-dismantling or self-undermining systems: John Barth, Robert Coover, and Joseph McElroy all return to this technique throughout their fiction.

While we can find examples of both features—self-reflexiveness and antisystem—expressed in some texts written before the twentieth century, I call these "pre-postmodernism." Shakespeare's *King Lear*, and to a lesser extent the middle scenes of *The Taming of the Shrew* and *A Midsummer Night's Dream*, for example, flirt with the nihilistic idea that language cannot really communicate. Indeed, the whole project of the Royal Society in the seventeenth century (Sprat and Wilkins, following Bacon's lead) devoted to finding an Adamic, pure language of notation for Nature[11] proceeded from a distrust of natural language that postmodernists appreciate, although it sought codified solutions that postmodernists could not endorse. The Talmudic system of layering exegesis upon exegesis—Midrash—derived from competing systems that finally swamp information from the original text. Indeed the very layout of the traditional Talmudic page, with its apotheosis of marginalia, signals a welter of competing interpretations that would be familiar to readers of postmodern fiction. And of course, the eighteenth-century novel self-consciously fiddled with its own narrative form in a growing experiment that culminates in *Tristram Shandy*. So postmodernism is perhaps best understood as a characteristic disposition of a discourse to its own authority, an ultimate commitment to eschewing ultimate commitments, a rejection and ironic demolition of all systems including its own, a distrust of codes, an indictment of organization (particularly, official organization), all of which it views as conduits to dehumanizing and decontextualizing meaning. But postmodern fiction achieves particular poignancy and urgency as a counterstatement to that scientific system of systems, cybernetics, which explicitly claims to be able to design a deterministic description for human communication and thought.

From this point of view, we can see why the traffic jam would lend itself to postmodern treatment.

Along with Don DeLillo's *White Noise*, there is one other recent novel which captures the strange, Prigoginesque sense of system in the traffic jam: William Marshall's *Roadshow* (1985). This is a funny, quirky book not easy to categorize. *Roadshow* is "A Yellowthread Street Mystery," one of a series of novels by Marshall about a police precinct in the fictitious Hong Bay district of Hong Kong. Yet it develops, along with several other novels in the series (there are nine so far), a characteristic machinery of narration that I think of as "pop postmodernism." It takes for granted—without any of the intellectual accoutrements or tacit philosophizings common in less popular and less accessible postmodern texts—the machinery or style of discontinuous narration, quick cuts to parallel strands of the story, lacunae, self-conscious manipulation of language and punctuation, layers of conflicting interpretations, and more importantly, a distrust of making sense out of the world through the application of system. Like many other postmodern texts, *Roadshow* creates a palimpsest of multiple and sometimes contradictory views of a centrally elusive truth, but it does so without explicitly worrying about the nature of truth or the relationship of discourse to reality. At the same time, this worry is so deeply written into postmodern life that it emerges as a consequence of the story, willy-nilly.

One way to read *Roadshow* is as an epistemological thriller. A number of individual detectives each follows his own systematic science, creating multiple and competing views of a single set of phenomena. By accident, the chaos and turbulence of events brings them to the same conclusion at the same time, a form of synchronicity. We have seen this postmodern methodology in other novels more clearly and consciously concerned with epistemological questions, like Pynchon's *V.* and Barth's *Giles Goat-Boy*. In many of Marshall's novels, suspense derives from the clever intertwining of stories and narrative fragments that accelerate and become briefer as they more closely intertwine toward the climax. In *Roadshow* however, this technique achieves a certain elegance and resonance because it mirrors so consciously the story itself, which essentially looks like this.

A man and a woman copulate illicitly, early in the morning, in a car on a back street of Hong Bay. The Embarrassment Man, who makes his living by catching couples *in flagrante delicto* and wordlessly extorting money from them in exchange for his silence, appears at the side of their car. The embarrassed man furiously pays off the Embarrassment Man. A bomb goes off under the street, leaving the couple luckily unharmed (but embarrassed) and the entire road a wreck. The explosion was not an accident. The policemen of the Yellowthread Police precinct and the Bomb Squad begin investigating. But "there was no rhyme or reason to it" (15). In a lovely irony, the police interrogate the Embarrassment Man and learn that he is kept at the "top of his profession" by his cataracts: he is legally blind.

In General Gordon Street, in the heart of the business district, Inspector Lo tries to keep the traffic moving as all sorts of vehicles maneuver to be near parking meters at exactly 8:00 A.M. so they can park strategically for the day. As he leans against one of the meters, chastising a motorist, all the meters explode in a daisy-chain, killing Lo and several others.

Meanwhile, Detectives Spencer and Auden (!) investigate bullet holes in the roof of the Dharma Datu Zen Buddhist monastery.

At the same time, Sergeant O'Yee mans the desk back at the precinct and adopts a stray mongrel dog.

Matthews, the bomb expert, observes the damage and notes, "There might have been sermons in stone but in explosives there was poetry. In the Isandula Street there had been an entire Epithalamion" (29). Despite the fact that the explosions cause chaos in human terms, and the motives for them remain mysterious until the very last pages, they are obviously the work of a very nifty, very neat, very orderly, and in Matthew's terms, "artistic" mind: "'They wanted it, they planned, it, they laid it, and they did it,'" says Matthews. There is an ingenious plot here, on Yellowthread Street, and here, in the novel. In speculating about the intelligence behind the explosions, one character thinks later that "if you had that kind of mind a city was one enormous playground" (42).

Kyle-Foxby, a young, English-university-trained urban planner, has come to Hong Bay to study the traffic because it is reputedly the worst in the world. He is trying to devise a system to make it flow. He notes with dismay as he watches his friend and subordinate, Lo, die, that the traffic in General Gordon Street "was [now] flowing in a planned, orderly, controlled manner" (39–40), an observation made curious by the fact that, before the explosion, General Gordon Street was notorious for its perpetual traffic jam, despite Kyle-Foxby's planning and Lo's physical intervention in directing traffic. The explosion has paradoxically sped things up.

We get a glimpse of the anonymous mastermind. He is having his ears cleaned by the Ear Wax Man, because he's suffered a series of explosions lately that have pressed the wax in his ear against the eardrum. The author calls him "the man of Limited Time."

Roadshow is pop postmodernism in part because the number of stylistic and thematic flourishes that are strongly identified with postmodernism seem to be delivered as a matter of course: there is an aura of self-reflexiveness that is unself-conscious, a disingenuous self-consciousness, if you will. For example, Feiffer, more or less the central character, looks at a map of the Hong Bay area and notes that "it represented nothing" (48). This line becomes a sign of Feiffer's misreading of the mystery. He has the right map but the wrong interpretive method; he's looking for a map of clues, when in fact, as we learn later, he ought to be looking at a map of streets, taking things much more literally. Yet the notion of a meaningless or unreadable text is familiar to aficionados of postmodern fiction. Similarly, Feiffer's nihilism at this point in the mystery is offhanded, second nature: "'For all I know, they're not trying to do anything,'" he tells another detective (48). "It was nothing. It was all he had" (49). His despair, and the style the author uses to signify it, are aspects of a trickle-down postmodern viewpoint that suffuses our culture.

Later, Feiffer pursues his poststructuralist discourse about the meaning (or meaninglessness) of names of streets on maps with Dr. Albert Nonte, an American, formerly attached to the Chinese embassy, who now by his own admission is, in a way, hiding. "'I'm a toponomist. From the Greek *Topo:* a place and *-nomist*—I'm a namer of goddamn roads and streets, that's where I've ended up'" (62). Feiffer visits him out of desperation, thinking (wrongly) that the meaning of the names of the streets will be the connection in the bombings, and that the mastermind is sending some sort of message to the authorities by his choice of targets. Nonte rejects the notion as absurd:

"No one—with the possible exception of me—would even have enough information about the meaning of half the street names to start getting symbolic about them, let alone start killing people in them. . . . If you think I'm going to tell you that, yes, it's some sort of dark deep secret symbolic assault on the thoroughfares of life, you can forget it." (60–61)

As an alternative, he suggests, "'Maybe all they want to do is stop the traffic.'" (61). Aficionados of detective stories from the time of Sophocles' *Oedipus Rex* are familiar with the figure of the self-indicting criminal.

Another explosion goes off. Kyle-Foxby's dismay grows.

If he had a month of computer time he could not have arranged it better; the closed streets made no difference at all—for all the difference they made they could have been totally non-existent.

It was crazy. It was impossible to close a street in Hong Kong and not have the traffic all over the island grind to an instant, solid halt. It was impossible. All the studies proved it. (75)

Spencer and Auden pursue clues in the monastery to an unrelated mystery. They function as comic relief. O'Yee struggles between his affection for and desire to get rid of the dog. He phones the CIA to pursue his instinct that somehow Nonte is involved. Feiffer still insists that the signs on the map mean "nothing." Nonte laments that his life had amounted to nothing. The author informs us that Hong Kong had come from nothing and was headed towards nothing, an aberrant product of a nineteenth century colonial empire that had never progressed. The word "nothing" tolls through the book as it does in *King Lear*.

Then Kyle-Foxby, the traffic cyberneticist, thoroughly mystified by the phenomenon of traffic speeding up rather than slowing down, with each disruption, roams the streets at night. He sees someone bending over a hole in the middle of the street. He's just passed an intersection where the stoplights are red in all four directions. Something is seriously amiss. Rather than make an illegal u-turn, he makes three right-hand turns, following the rule of the one-way traffic system he himself devised. By the time he gets back to the intersection, the trouble he could have prevented has started. A man fires a gun and bombs go off, severing the body of one of the conspirators, a man from the Department of Main Roads.

This is the turning point in the plot. What postmodern philosophy of discourse Matthews had courted up to now is hereafter subordinated to the compulsory needs of the detective story to wrap things up, solve the mystery, ensure that the systems of deduction and sense work to right a wrong. At the same time, this is where the novel becomes most notably reflexive: the traffic metaphor and subject dictate a mode of discourse that coincide.

Feiffer picks up clues through old-style procedural footwork that lead him to deduce that a relatively secluded intersection of streets, Moore's Pocket, is the object of the pattern of bombings, though in precisely what way is unclear. Simultaneously, Kyle-Foxby, by applying systems analysis, comes to the same conclusion. At the same time, Matthews of the Bomb Squad narrows down the suspects to a small group who learned bombmaking from his own hands (a twisted wire acts as an idiosyncratic signature on the bombs, the signature being his own). Meanwhile, O'Yee nails his suspect with a clue from his CIA connection in Hong Kong. He calls to Spencer and Auden, the comic acrobat/athletes, in anticipation of a scene at Moore's Pocket.

The narrative speeds up at this point, in imitation of the traffic it describes and the cascading revelations it brings. In the terms of statistical thermodynamics employed by Prigogine, we would call this a *bifurcation point*, the origin of the dissipative structure—a system-shattering moment when the previous, simpler organization can no longer support the intensity or frequency of its own fluctuations, and either disintegrates or jumps to a new level of order and integration. At this new level of order, the particulars of change are nondeterministic. One cannot tell where the system will be at any future moment, given certain prior conditions, a feature of nonlinear partial differential equations (see note 13). In essence, dissipative structuration illustrates, is exemplary of, the antideterministic values of postmodern literature. In *Roadshow* scenes play back and forth among each other in quick-cut style. All the players converge on Moore's Pocket, knowing the next bomb is going to explode soon. Marshall's prose turns subtly reflexive here:

> Somehow at the far extremity of the system, synchronized lights on the other streets were holding it, organizing it, and it was flowing. It was still working. Between each car and truck and bus and taxi there was a measured space: three feet, maybe less, never more—the system was still balanced, organized—it was speeding up, getting faster and faster, still holding the gap.

A bomb goes off and, instantaneously, "It was gridlock. It was the final, ultimate traffic nightmare: the total, complete shutdown when no vehicle, nothing—nothing at all—could move a single inch in any direction" (171). "Everything was ruined, going crazy, concertinaing." The Yellowthread Street police all converge on Moore's Pocket, which turns out to be hiding a fabulous stash of South African gold and diamonds carefully guarded and kept secret. Nonte and his lover, a former member of the bomb squad (as Matthews deduced) had planned the perfect crime, along with the now-dead fellow from Main Roads. They were going to stop all the traffic so they could rob the South African cache and make away on foot before any help could arrive. However, the separate but equally effective systems of readings employed by Matthews, O'Yee, Kyle-Foxby, and Feiffer have all solved the mystery, simultaneously, in a convergence, a growth of revelation, a sudden organization of meaning out of a chaotic welter of data ("it was nothing"). From an array of bits and pieces of data comes a sudden new order of understanding, and the system of the characters' understandings, as well as the system of the narrative itself, leap into a higher plane, another form of spontaneous self-organization. The moment of insight acts like a dissipative structure. The explosively meaningful reorganization and addition of information in the reader's mind, the revelation, is entirely consonant with the activity of a dissipative structure.

Roadshow as Dissipative Structure

There are a few instances in recent science fiction where Prigogine's theories have already been incorporated into the subject, though not the form, of the novel. Bruce Sterling, one of the newer New Wave science fiction writers identified with the cyberpunk subgenre, gives us the best examples. In *Schismatrix* he shows us a far future that assumes entire technologies and social organizations based on (what are by then) well-proven theories of Prigogine. The manipulation of social and biological forms takes place through the application of technologies based on Prigoginian principles: "None of [the] material advances matched the social impact of the progress of the sciences. Breakthroughs in statistical physics proved the objective existence of the four Prigoginic levels of complexity" (237; also, see epigraph to this chapter).

In another projection of Prigogine's influence, A. A. Attanasio's *Radix* describes Rubeus, a hyperintelligent and malevolent computer mind whose creative power comes from a madness Attanasio describes as an analog of the dissipation in a dissipative structure. "He was indeed mad [Dissipative]," writes Attanasio,

> and that joyed him so thoroughly an oblique smile creased his cheeks. [Madness is the supreme strategy.] To free himself from the Delph's programming—to be free—Rubeus had to break out of the mind. His mental fluctuations generated a Prigogine effect: They increased the number of interactions among psychic systems and brought them into contact with each other in new and sometimes creative ways. Given enough time, Rubeus thought his mind would create higher-order equilibrium: a new Mind, bigger and more aware—capable of out-thinking Creation. [Life is a pattern.] He thought that. (372; brackets in the original)

In even more direct ways, this entirely imaginative description provides a model for the subject of this chapter: how the fictional text breaks out of its programming, the grammars of generic and linguistic constraints, to create "a new Mind, bigger and more aware—capable of out-thinking Creation." Yet, Attanasio remains content to describe such a possibility rather than to actuate it in his own discourse.

Roadshow is not the most sophisticated of novels; nor is there any evidence that Marshall, in constructing it, was aware of or applying Prigogine's theories. Nevertheless, the novel demonstrates the incorporation of Prigoginian principles in both the theme and the form of its discourse. At the same time, though its plot is rather clever and it is a good exemplar of the mystery novel, it is does not invite speculation about the nature of discourse or engage the reader in any special philosophizing. It is in some senses a very typical procedural mystery. Yet, its very typicalness broadens the applicability of an analysis of *Roadshow* as a dissipative structure to the novel in general. Let me reiterate similarities between this novel and the characteristics of dissipative structures before widening the orbit of these assertions.

Dissipative structures arise in open systems . . . : Hong Bay, the fictional district of Hong Kong, is clearly in commerce with the rest of Hong Kong, just as the novel *Roadshow* clearly is in commerce with the reader's deductive and intuitive capacities to keep its tension. Indeed, Hong Bay's traffic system in particular relies upon the flow of vehicles elsewhere in Hong Kong, as Kyle-Foxby notes at several points. That is, *Roadshow* depicts and is an example of the elaborations (or "reiterative states") of an open system. It helps to make this point by contrasting its "open-system mystery" with closed-door mysteries popular in the early and middle part of this century. In several Agatha Christie novels, or in the prototype of the subgenre, Poe's "The Purloined Letter," for instance, all the information needed for the solution of the problem is present in the room; these texts create closed systems.[12] In fact, the mystery and the traffic jam both rely on input from outside to maintain their nonlinear fluctuations far from equilibrium (see below) and gain momentum (perhaps that is the wrong phrase to describe what happens to traffic when more cars enter the system, though not for what happens to interpretations near the bifurcation point in a novel).

. . . that exist far from equilibrium and fluctuate nonlinearly:[13] All of Hong Kong is due, as Marshall notes in a prefatory remark, to revert to ownership by mainland (i.e. communist) China in 1997. In the interim, it remains suspended in a limbo, a capitalist aberration with a destabilized economy tolerated by a communist government. In fact, such a view of economics as a thermodynamic system is not merely metaphorical, and some researchers have already found fruitful results when applying the model of dissipative structures to economic systems.[14] Prigogine himself remarks that social phenomena are best described by nonlinear equations, especially when they are kept far from equilibrium or can be described as such. Hong Bay is a highly unstable place: "the tourist brochures advise you not to go there after dark" (9). Its traffic is particu-

larly chaotic: "Lo, the single sentinel at the edge of chaos in the gutter, waved his hand to keep the traffic moving" (16). And the narrative leaps about with unexplained hiatuses and discontinuities in discrete packets of uneven length (fluctuates nonlinearly) following no apparent order, except that time moves only forward: no flashbacks or recursions are given (see *irreversibility*, below).

Self-organization and irreversibility: Roadshow is narrated through a series of disconnected fragments which gradually grow more intertwined, related, and organized. At the same time, the story focuses on a series of bombs, placed at strategic points, which disrupts the traffic flow of the fictional Yellowthread district of Hong Kong. After a certain point in the telling, all three (manner of telling, impulse of the characters, movement of traffic towards gridlock) achieve a momentum of their own to the extent that events seem to plunge on by themselves. The detectives seem compelled to pursue the trail of their criminals willy nilly, just as the traffic jam and the resolution of the mystery seem inevitable, three intertwined self-organizing events so large that no human intervention could possibly prevent them. The forward movement of time is crucial in this regard. First, the antagonist is kept anonymous, except insofar as the author refers to him as "The Man of Limited Time." Marshall also uses the cheap thriller device of telling us the time of day at intermittent points: ("8:50 a.m. In Sepoy Street, ..."). Finally, the timing of the bombs is crucial in the history of the traffic system.

These intertwined systems display sensitive dependence upon initial conditions (The Butterfly Effect): Small, local inputs of information and coincidences (both of which are entropic processes) at the front produce global consequences for the entire system at the end. Detectives stumble upon very small clues (a twisted wire, a hint in a phone conversation, a coincidence in the records of a murdered woman) that lead to very large revelations. Small explosions of little apparent consequence in side streets in Hong Bay eventually produce massive consequences for the entire traffic system of Hong Kong. Foreshadowing hints early in the novel (Nonte's remark, "'Perhaps they're trying to stop the traffic'") loom large in the reader's deciphering of the mystery and understanding of Nonte's motivations as it draws to a close.

Bifurcation points: Like most good procedural mysteries, *Roadshow* moves from instability to instability. At each instability, in keeping with the craft of a good mystery, it is extremely difficult, if not impossible, to determine the next state of the system based on previous information about the system. Either things look deterministic but are stochastic (detectives are following the wrong trail), or inversely, things look chaotic and noisy but have a deeper structure and meaning (insignificant, casual, or random events hold the key to unlocking the mystery). As information is added, these events grow even less comprehensible, and the system of motives behind them no clearer, until the narration reaches a very striking point of epiphany or breakthrough. Then the entire nature of the work changes: the narrative coils in on a single point and the characters all simultaneously solve the puzzle. The moment of greatest instability in an open system far from equilibrium is what statistical thermodynamicists call a "bifurcation point"—from which the system generally leaps to one of two or more possible states of higher organization.

Attractors: An *attractor* is a node or event that looks (on a graph of its mathematics) like the hole in the bottom of a basin. A dissipative structure tends towards an attractor, which acts like a precipitant or irritant around which crystals begin to form in a supersaturated solution. Once the crystallization process begins, it will continue as long as the system remains open to more input. Indeed, a form of crystallizing process known as the Bénard instability inspired Prigogine's first intuitive leap to his theory of dissipative structures. A Bénard instability occurs when a specially prepared liquid is heated from below. As the heat is turned up, the solution begins to organize itself spontaneously in a very striking manner: ovals of colors arrange themselves kaleidoscopically and under the proper conditions will continue to grow more organized. This phenomenon intrigued Prigogine because this pattern of growth resembles that found in living cells; i.e., such spontaneous self-organization of matter looked like what might have happened in that primordial jump from inanimate to animate.[15]

The crystallization of traffic around the attractor formed by the bombing in Moore's Pocket is clearly this sort of event. As Prigogine noted in an interview,

> data on highway driving show there actually is a transition to a different phase when the critical concentration is reached. Driving is a very good example of bifurcation. In effect, there's a phase change from a group of individuals driving to a coherent structure: the highway as a whole. (Weintraub 312)

In other words, the traffic jam actually attracts other cars by involving them in a larger system than that created by their individual intentions or patterns (trips). Cars that would have been occupying other, more randomly distributed spaces in the road system at any given moment now find themselves *occupying a place in the larger organization ordained by the traffic jam* (in this case, someone else's organization!).

Holographic phenomena: As these last comments imply, recognizing phenomena as dissipative structures entails recognizing a new, nonclassical relationship between part and whole. By Prigogine's own estimation, one of the most important aspects of his theory is that it implies a necessary connection between part and whole in the system. In the view of classical dynamics, the system could be analyzed into its working parts and described deterministically by equations for motion, force, etc. In a dissipative structure, however, we cannot view the part sensibly out of the context of the whole. Stengers points to a holographic or ecological process at work by which individual actions, especially unexpected ones, can be explained (mathematically determined) but not anticipated (mathematically predicted) by "social interactions under non-equilibrium conditions." This describes quite neatly the eventual motivation of the criminal at work (Nonte and his homosexual Chinese lover seek their fortune at any expense) and largely the actions of the detectives.

Order out of disorder: All in all, the pleasing qualities of this novel, as with many other novels, its intrigue and complexity and dark humor, arise from a vision of a larger social system that exists through a fragile and harsh conspiracy between the forces of order and the forces of disorder. Even in microstructures of the novel's language and metaphors, this conspiracy or collaboration is implied. The dissipative structure as Prigogine defines it creates a crafty metaphor for how order (the traffic jam, the consummation of Nonte's plot, the convergence of detectives on the scene) arises from disorder (the bombs that speed up traffic, the offhand denials and despair by Nonte, the individual flounderings by the detectives) in social systems. In this novel, order arises out of disorder on three levels: a world-circling traffic jam organizes itself spontaneously; the befuddled detectives solve the mystery of who is behind the plot; and the narrative itself becomes more systematically ordered according to the events it portrays. Whether or not this qualifies *Roadshow* as a dissipative structure in more than metaphorical ways remains to be seen.

More than a Metaphor: Literature as a Dissipative Structure

It would be pretty to think that *Roadshow*'s structure and concern with traffic so remarkably coincide with Prigogine's work because William

Marshall was aware of Prigogine's work. But in the absence of any evidence to suggest such direct knowledge, we must abandon notions of intentionality or influence to explain the neatness with which *Roadshow* illustrates Prigogine's principles, and turn elsewhere to explain the correspondence.

In *Roadshow*, the relationship between the traffic described in the novel and the form of that description is a sort of semantic or stylistic metaphor. The structure of the narrative, the arrangement of apparent time controlled by the author, the decisions to switch more rapidly among the various parallel points of view have an *artificial* relationship to this story about a well-orchestrated traffic jam: while there is something artfully "natural" about such a choice, there is nothing *necessary* in it. We can well imagine the same story told in several other fashions. However, when Prigogine and his collaborators describe the traffic jam itself as a dissipative structure, they mean it literally. And I also mean it literally when I describe the novel as fitting the model of dissipative structure. Traffic and fiction are true analogues of each other.

Part of the fascination of the traffic jam is its ambiguity. Is it an artificial event, only conceivable as the product of a hyper-evolved hypercomplex urban system, which itself presupposes systems of biology, intelligence, urbanization, technology? Is it man-made? Or is it in some crucial sense a naturally occurring event? What role does human choice play in traffic jams? Is there any room in Prigogine's system for the role of human intention, or is it averaged out according to some statistical method? These questions are interesting precisely because of the analogue between the literary text and the traffic jam.

The fact is, Prigogine does take into account the role of human decisions in contributing to traffic jams. He and his collaborators have found that individual decisions behave as linear equations, while "decisions taken under the influence of the environment (friends, media, etc.)"—or in the case of traffic, competition—behave as nonlinear equations ("Order through Fluctuation" 122).[16] But even traffic considered as the product of intelligence "arises spontaneously from a far-from-equilibrium system fluctuating non-linearly." So the model with which we choose to study it makes its artificiality moot.

In the same way, when we choose to view certain fictions as belonging to a class of phenomena like traffic jams—making the same leap Marshall does in *Roadshow*'s style—then we have swept away the question of their artificiality in favor of questions about the nature of their organization. To take another tack, compare a fictional text to a chemical reaction. The Bénard instability, for instance, is artificial, yet it isolates some essential aspect of nature. Even the city itself as a complex, evolving system that distributes population and structures (and vehicles) in a certain fashion seems to be described neatly by the mathematics of dissipative structures.[17] But is it natural or artificial? Fractal geometry, to which the science of chaos is intimately related, also is highly artificial, yet it does something hitherto inconceivable, generating natural geometries through combinations of deterministic and randomizing equations, to the point that Hollywood moviemakers use fractals to generate naturalistic landscapes as backdrops.[18] So, too, the novel can easily be viewed as this sort of human product: a highly complex, designed system, in this case one of artfully arranged information produced by a constant exchange between the mind and hand of the author.

At first glance, such a definition is horrifying to a traditional humanist. It portrays the novel as a system of information; it also sounds rather clinical, with its reference to "the mind of the author." What seems most threatening is the idea that the novel could be modeled in mathematics, because mathematics raises the specter of loss of free will. But not this mathematics. The beauty and import of Prigogine's model is that it frees us from the determinism of total predictability when we try to describe the behavior of complex macroscopic systems, whether social, natural, or chemical. At the same time, it accepts a way of talking about necessity, about those constraints on a system that can be calculated, familiar not only to scientists who desire to define aspects of a phenomenon that *are* empirically measurable, but also to literary critics who want to specify the code of the text. Prigogine's model gives us a good way to describe the peculiar balance between chance (the stochastic) and necessity (the predictable), or say it how you will: free invention within a structure of constraints, or mutation from within a genetically coded population, or the choice of an individual utterance from within a grammar of constrained possibilities that literature, and indeed most human behavior, exhibits. Prigogine writes, "The social fact finds its expression in constraints imposed upon an individual" ("Order through Fluctuation" 123). At one point in *Order Out of Chaos*, Prigogine and Stengers even seem to consider that the course of scientific progress behaves as a succession of dissipative structures, giving rise to each other in the "natural" course of events (307ff.). They view scientific discourse as one of these social facts.

So the path is paved for viewing the act of literature as a dissipative structure, like other products of the human mind and behavior, neither artificial nor natural. If anything, the literary text is best viewed as the result of the intersection of the author's mind with a very peculiar technology (a sort of antimechanistic technology) designed in its most advanced forms to capture the evanescent movements and fluctuations of the mind itself. Thus, the literary text is the trace or result of one dissipative structure giving rise to another dissipative structure. In this way, literary texts are like neurons, which some researchers also have shown fit the model of a dissipative structure.[19] If we accept as a general principle that novels give clear and unparalleled access to human cognition in the act of organizing information, then *Roadshow* shows intelligence acting as a dissipative structure. Though *Roadshow* is unique in the remarkable coincidence of its subject with Prigogine's concerns (traffic jams), it illustrates how literature generally may act as a self-organizing system, growing willy nilly through bifurcation points towards higher orders of systematic organization. Far from taking the fun out of reading, this growth towards structure is thrilling and mysterious, since the reader is a participant, urging the novel onward in its headlong rush towards revelation.

There are two further possible supports for this temporary scaffolding between Prigogine's model and the nature of the literary text. One possibility is that the postmodern novel is at a point in the evolution of literary technique that it actually is able to reflect and capture how phenomena like the mind or social organization or human decision-making processes or traffic work. So the new scientific models suggest that the world operates more like novels tell us it does than like classical physics tells us it does. The second possibility is that though there may be no direct influence between Prigogine and Marshall, both are responding, partly unconsciously, to a larger "postmodern condition" or paradigm which influences the way we see natural and human systems ordering themselves.

All this discussion of the literary text operating according to some scientific model would still seem somewhat clinical were it not for the fact that understanding the movement of the mind has (again) become the most crucial project for science, since it seems that limitations in our understanding of the universe, or perhaps even the shape of that universe itself, is bound up with the nature of our minds.[20]

A New View of Literature's Epistemological Potency

Whether or not we understand literature in general or certain postmodern fictions in particular as dissipative structures, it is still obvious that Prigogine's theories empower literature in a new way.

From a global-historical perspective, literary discourse represents an alternative to science's epistemological discourse. As science becomes better and better at what it does, literary discourse finds new and more hypertrophic avenues for its expression. In some aspects or in some artistic hands, this is a retreat. Literature, as John Barth, Philip Roth, and others suggested in the early 1960s, becomes exhausted by its competition with reality, and is pushed into narrower and more sterile byways of linguistic, stylistic, and formalistic experimentation until it gives up any claims to representation of reality altogether in favor of a position as demon or imp of the perverse or, as Donald Barthelme put it, "the leading edge of the trash phenomenon."

In other hands, this complementarity of scientific and literary expression presses literature to adopt the guise of the competition, to use scientific language and postures to undermine science's claim to superior veracity with antiformalist formalisms and ironic self-deconstructions. This is what occurs in a class of fiction I've called elsewhere "cybernetic fiction" (*The Soft Machine*). In other words, literature generally, but postmodern fiction in particular, struggles to establish itself as that alternative which Prigogine alludes to as "a new mode of description in which time and freedom, rather than determinism, would play a fundamental role." In postmodern fiction and in many passages which might exemplify "pre-postmodernism," we find an emphatic divergence from the naturalism and mechanical plots found in the nineteenth-century novel (Austen, Dickens, Stendhal, Eliot, et al.), and a definite break from the mock determinism we find in modernist novels by Joyce, Raymond Roussel, and Kafka. Instead, the postmodern novel and its precursors emphasize the contingent, the random, the systematic, the irrational, the unmechanistic, and the subjectivistic experience of time opposed to enslavement to clockwork regularity and chronology.

Prigogine's new paradigm empowers fiction as that "new mode of description" which eludes Whitehead and Kant. If he is correct, then science now is forced to grant that reality exists at a level of human experience that literary tools are best, and historically most practiced, at describing. As a result, even by science's own terms, literary discourse must be understood as a superior form of describing what we know. Ironically, while many literary theorists beat a retreat from literature, science itself reinvests the literary work with a new value.

There is another recent attempt to define the literary text as a dissipative structure. William Paulson's remarkable work, *The Noise of Culture*, follows Michel Serres's, Henri Atlan's, and Prigogine and Stengers's interrelated programs for the creation of a thoroughly interdisciplinary view of discourse to focus more closely on the relationships among the concepts (and associated disciplines) of information theory, deconstruction, and self-organization. Paulson particularly shows the fruitfulness of viewing the literary text as Prigogine suggests we view other phenomena: as an open system, never completely decipherable or interpretable, one kept in fluctuation by an exchange of noise with its own unstable system of codes, the reader's interpretation, and the culture at large.

Paulson's argument confronts, but never gets past, the problem of the distinction between natural and artificial as regards the literary text. For that reason his argument struggles with the role of the mind of the author. For it is when the literary text makes that magical leap from the mind to erupt onto the page that we have crossed the threshold from natural to artificial in a sharper and more distinct way than in any other kind of organized behavior save speech, perhaps. And if one cannot resolve the dichotomy, as I think Prigogine's theories do directly, then one cannot get past deconstruction's stance.

Paulson's reading of this stance is incisive: deconstruction claims that the meaning of the text is impossible to recover unambiguously; consequently, the "very impossibility of doing so becomes the only 'meaning' literally recoverable." The notion of an authoritative meaning sent by the author which literary scholars position themselves to receive is thus, if not demolished, then dismantled beyond repair (93ff.). Similarly, while "it is not hard to show that in deconstructive readings, for example the production of meaning—considered as an autonomous self-organizing process—does not come to a halt" (95), such a reading dynamic clearly values the production of ambiguities and complexities through what might best be called *anti-interpretation*.

Paulson adopts the deconstructive framework for understanding the power of literature in terms given by information theory: the literary text, (holding it in suspense, (understanding it as incomprehensible (as "irreducible to any schema other than itself" (184), (holding positive meaning in abeyance in favor of complexity (or in favor of the positive value of rendering no positive interpretations)))) is a powerfully short route to the creation of new information, new meaning, out of noise. Literature becomes empowered because it continually revives the culture. But it does so from the margins, as a subversive, an irritant, a native alien:

> Literature is not and will not ever again be at the center of culture, if indeed it ever was. There is no use in either claiming or debunking its central position. Literature is the noise of culture, the rich and indeterminate margin into which

messages are sent off, never to return the same, in which signals received are not quite like anything emitted. (180)

From the viewpoint of a discourse mandated by the science of dissipative structures, such a stance no longer seems either necessary or even tenable. There is a difference between being in the center of culture, as measured by some banal statistic about how many people actually read a particular book, versus being in the center of the most important fight a culture can join: the fight over how to view and discourse about reality. For instance, very few people read (and even fewer understand) quantum theory, that sister project that has helped create postmodern consciousness. But quantum theory has been in the center of one of this century's most important epistemological controversies for the last fifty years. Though the pop culture may find it easy to ignore in some explicit way, it already has been influenced by quantum theory in a subtle and pervasive fashion, not to mention the technological fallout from quantum mechanics (emerging in computer design, for instance). Postmodern literature has joined the battle and has had a similar effect. In fact, it has continued to receive succor, (I believe mistakenly) from quantum theory itself[21] and now, in a more explicit and justifiable fashion, from Prigogine's chaos theory.

As a result of this I am not willing, after all, to assist literature in its abdication of the center, nor to espouse its epistemological impotence, nor to arrange its "graceful exit from the era of *Literaturwissenschaft*" as Paulson himself describes his project (185) or as Leslie Fiedler partly does in *What Was Literature?*

My reasons for resisting this fashionable turn are several. While I agree that literature will rarely move a culture as swiftly and immediately and profoundly as it once might have, and cannot hope to compete with other narrative forms like film and television in popularity, it still breaks new ground in documenting our culture's understanding of itself, its relationship to individual imaginations, and the relationship between those imaginations and the universe that feeds it. Rather than resort to a wry archaeological metaphor—literature resides atop "the rubbish heap" of culture, inviting readers to "scrounge" in it for its treasures—I am convinced that literature represents the potential of a culture to reorganize that trash heap as it goes along, making the next stratum possible, thinkable, conceivable. Not all literature does that all the time, of course. In fact, most literature is incapable of doing anything more than add to the rubbish most of the time. But if we gauge the effectiveness of literature by its ultimate influence, then postmodern texts by such writers as

Pynchon, Beckett, Burroughs, Barth, Vonnegut, Calvino, and DeLillo finally do alter our culture's relationship to reality. Certainly, they are no slower in doing so than quantum physics, which has been at it since 1927 or so.

Judging by the effectiveness of these advanced forms of literature, the path is open to a somewhat different and more positive conception of literature's place in the culture. The consequences of Prigogine's relocation of scientific realism into the discourse of complexity as opposed to the discourse of simplicity is to place literature beyond deconstruction and beyond the embrace of powerlessness which, in Paulson's description, is all that is left to empower it. Instead, the literary project no longer need feel itself to be in competition with science, nor need it put itself beyond science, and adopt a posture of indifference to its own relationship to reality. Rather, literary discourse becomes a model for a kind of knowing that best communicates the "opacity"—or at least the complex unpredictability—of the universe. This is Prigogine and Stengers' term for a kind of knowing that literary discourse gets at best and scientific discourse does not capture at all. In the very last pages of *Order Out of Chaos*, they quote philosopher of science Hermann Weyl to this effect: "Scientists would be wrong to ignore the fact that theoretical construction is not the only approach to the phenomena of life; another way, that of understanding from within (interpretation) is open to us."[22]

In our postmodern era, where the furthest-flung sciences are showing us this essential limitation on that sweet, complex, turbulent slip between mind and universe, literary discourses are well-placed to serve as avatars for a new science, a new way of knowing, unlike any which Western culture has embraced during the last three hundred years.

Notes

1. The rest of this description is worth reading for its complete and brilliant exploitation of this exegesis of traffic as apocalypse. It proceeds painstakingly and garishly for nine pages (119–28).

2. See Prigogine and Herman, *Kinetic Theory* and "Two-Fluid Approach."

3. George Eliot's well-worn parable of the pier glass in *Middlemarch* is a good example of early literary uses of the microscopic to formulate interesting metaphors.

4. In this context, it is hard to understand why Gleick's book contains virtually no mention of Prigogine. The omission seems willful. The scientists (like Mitchell Feigenbaum) whom Gleick credits with discovering this new science of chaos are investigating the same phenomena as Prigogine: far-from-equilibrium systems that leap to new orders of complexity and organization on the other side

of bifurcation points. The *New York Times* reported extensively on Prigogine during and after his reception of the Nobel Prize in 1977, and Gleick has been an editor and reporter there since 1978. (See *NYT*, Wednesday, October 12, 1977, D19.) Both authors appeal to Mandelbrot's mathematics of fractal geometry to explain certain concepts in self-organization and complexity. Both authors are aware of the paradigm-changing impact of this branch of statistical dynamics and its influence on other disciplines, including population ecology, fluid dynamics, etc. In fact, both appeal strongly to Kuhn, with some important differences: Gleick obviously wrote his book to confirm the model of paradigm change that Kuhn explicates in *The Structure of Scientific Revolutions*. Prigogine entertains the notion that scientific change is a sort of dissipative process and takes Kuhn to task, a bit, for not recognizing the extent to which scientific change can come from the normal processes of science, and that no "revelation" is necessary. He and Stengers cite the communal process by which a theory of dissipative structures came into being: "On the contrary, this development clearly reflects the internal logic of science and the cultural and social context of our time" (*Order Out of Chaos* 309).

What is even stranger is that Gleick recently wrote a long article for the *New York Times Sunday Magazine* devoted to systems analysis of traffic flow in several United States cities (see March 1988).

5. *New York Times*, Wednesday, October 12, 1977, D19.

6. Several chemical reactions that organize themselves into higher levels of complexity have been explored from the point of view of Prigoginian dynamics. In fact, Prigogine predicted in the 1940s that such reactions were mathematically possible. It wasn't until they were discovered independently in the 1970s that he was awarded the Nobel Prize, since his powerful theory had an enormous intuitive appeal (to nonclassical scientists) but no substantive proof. A well-studied chemical reaction involving citric acid, potassium bromate, and the ceric-cerous ion couple, the Belousov-Zhabotinskii reaction, follows the predictive model of dissipative structures by producing an oscillating reaction of three compounds that continue to produce each other in cycling reactions. The mechanism of the reaction has been elucidated by Richard Noyes, who called it an *Oregonator*, a kind of chemical clock. See R. Noyes and R. J. Field, *Annual Review of Physics* (1974) 25: 95. For a briefer discussion of this mechanism, see Prigogine, *Becoming* 120–21.

7. See Jeremy Rifkin's *Algeny*; Michael Talbot, *Beyond the Quantum* (New York: Bantam, 1988), 133–37; Alvin Toffler in his introduction to Prigogine and Stengers.

8. For a good collection of essays running the gamut of Prigoginan applications, see Schieve and Allen.

9. For a longer commentary on this restrictive style of scientific discourse, see Prigogine and Stengers's postface to Serres's *Hermes: Literature, Science, Philosophy*, ed. Josué V. Harari and David F. Bell (Baltimore: Johns Hopkins University Press, 1982). See also my "Literary Text as a Dissipative Structure: Ilya Prigogine's Theory and the Postmodern Synthesis," in *Literature and Technology*,

ed. Mark Greenberg and Lance Schachterle (Lehigh: Lehigh University Press), in press.

10. Such self-conscious manipulation of formalisms to reflect content (such as it is) is coextensive with literature itself. The changing rhythms and formulas of heroic poetry, the dance of the chorus in Greek drama, the manipulation of type in seventeenth-century poetry and in Laurence Sterne, and self-referential discourses on genre by Fielding in *Joseph Andrews* are but a few examples of this turn.

11. See Thomas Sprat, *The History of the Royal Society of London for the Improving of Natural Knowledge* (London, 1667) and John Wilkins's *Essay towards a Real Character and a Philosophical Language* (London, 1668).

12. Perhaps the consummate creator of closed-system texts was Raymond Roussel, who uses Poe-like devices, which he learned from Jules Verne, in hypertrophic form to create incredible verbal devices, filled with puns, word-play, verbal conundrums, palimpsests and rebuses, usually enshrouding some central mystery. For an elaborate discussion of this technique, see Porush, chap. 2.

13. Nonlinear equations are particularly difficult equations to solve. They typically are used to describe a whole class of phenomena where, given all the information one needs to know about a system (that is, the system is deterministic), it is virtually impossible to predict where that system will end, at a specified number of iterations or times later. For a long discussion of how a new science grew up around these phenomena described by nonlinear partial differential equations, see Gleick.

On the matter of nonlinear equations representing a sort of discourse about reality that classical physicists were trained to ignore or discount, Gleick quotes physicist and chaotician spokesman, Doyne Farmer:

> *Nonlinear* was a word that you only encountered in the back of the book. A physics student would take a math course and the last chapter would be on nonlinear equations. . . . We had no concept of the real difference that nonlinearity makes in a model. The idea that an equation could bounce around in an apparently random way—that was pretty exciting. . . . It seemed like something [randomness coming out of determinism] for nothing, or something out of nothing. (250–51)

14. For the application of Prigogine's model of dissipative structures to economic systems, see R. Stephen Berry and Bjarne Andresen, "Thermodynamic Constraints in Economic Analysis," and Russel Davidson, "Economic Dynamics," in Schieve and Allen.

15. See "Wizard of Time: Ilya Prigogine," in Weintraub 335–36.

16. See also Prigogine and Herman. The authors remark,

> The purpose of the present work is to examine the features of a system in which each individual member, while trying to achieve his own optimum solution, has characteristics that depend on his being imbedded in a group. Of course, the characterization of the group then depends on the interaction between its members. The competition between individuals must, in the end, lead to either anarchy or some type of collective strategy. The latter is beautifully exemplified by the magnificently coordinated flight of a large flock of birds or the remarkable darting collective motion of a school of fish. In the same way, . . . no

theory of vehicular traffic can ignore the problem of driver strategy. For this reason, traffic theory is a discipline that combines in a unique fashion concepts of mathematical statistics, physical theory and human behavior. (xi)

17. See P. M. Allen, "Self-Organization in the Urban System," in Schieve and Allen, 132–58. See also P. M. Allen, "Order by Fluctuation and the Urban System," in *New Quantitative Techniques for Economic Analysis*, ed. Giorgio P. Szego (New York: Academic Press, 1982), 115–52.

18. As a consequence, filmmakers such as George Lucas quickly recognized the value of fractals in generating alien but natural-seeming landscapes in animation. See Gleick 114.

19. G. Schoner and J. A. S. Kelso, "Dynamic Pattern Generation in Behavioral and Neural Systems," *Science* 239 (March 25, 1988): 1513.

20. Into this gap has rushed an extraordinary number of books, ranging from silly-mystical to essential, many combining physics, philosophy, and metaphysics into one engaging stew. Most do a very good job of recounting the basic history of the New Physics which has brought us to this pass: Heisenberg's uncertainty principle, Einstein's challenge to Bohr, the Einstein-Podolsky-Rosen gedanken, Bell's Theorem, and most recently, the Aspect experiments. See: Fritjof Capra, *The Tao of Physics* (Berkeley: Shambhala, 1975); Denis Postle, *Fabric of the Universe* (New York: Crown, 1976); Gary Zukav, *The Dancing Wu Li Masters* (New York: Morrow, 1979); Richard Morris, *Dismantling the Universe* (New York: Simon & Schuster, 1983); Paul Davies, *Other Worlds* (New York: Simon & Schuster, 1980); Paul Davies, *God and the New Physics* (J. M. Dent & Sons, 1983); Werner Heisenberg, *Physics and Beyond* (New York: Harper & Row, 1971); Nick Herbert, *Quantum Reality* (New York: Doubleday, 1985); Fred Alan Wolf, *Taking the Quantum Leap* (New York: Harper & Row, 1981); Heinz R. Pagel, *The Cosmic Code* (New York: Simon & Schuster, 1982); Bernard D'Espagnat, *The Sense of Reality* (Berlin: Springer-Verlag, 1983); John Gribbin, *In Search of Schrodinger's Cat* (New York: Bantam, 1984); Michael Talbot, *Beyond the Quantum* (New York: Bantam, 1988); David Bohm, *Wholeness and the Implicate Order* (London: Routledge & Kegan Paul, 1981).

21. I say "mistakenly" because many literary critics leaped upon Heisenberg's philosophical assertions, made under the influence of his friend Heidegger, that the observer and the observed were intimately intertwined in some fundamental epistemological way. The trouble with this view, which Heisenberg expounded in *Physics and Philosophy* and *Physics and Beyond*, is that it commits a fundamental error of scale. The quantum effect which is so confounding to classical physicists is confined to the subatomic level. At the level of our daily experience, at the level of objects of macroscopic size, quantum effects become vanishingly small (infinitesimal) and therefore insignificant. So for any practical or even measurable purpose the observer has no effect on the observed. By contrast, Prigogine's theory attacks conventional scientific views at the macroscopic level itself.

22. Hermann Weyl, *Philosophy of Mathematics and Natural Science* (Princeton: Princeton University Press, 1949), quoted in Prigogine and Stengers 311.

Works Cited

Attanasio, A. A. *Radix*. New York: William Morrow & Co, 1981.

Calvino, Italo. "The Night Driver." *t-zero*. Trans. William Weaver. New York: Harcourt Brace Jovanovich, 1969. 128–36.

DeLillo, Don. *White Noise*. New York: Penguin, 1986.

Girard, René. *La violence et le sacré*. Paris: Editions Bernard Grasset, 1972.

Gleick, James. *Chaos: Making a New Science*. New York: Viking, 1987.

Marshall, William. *Roadshow*. New York: Henry Holt & Co, 1985.

Paulson, William. *The Noise of Culture: Literary Texts in a World of Information*. Ithaca, NY: Cornell University Press, 1988.

Porush, David. *The Soft Machine: Cybernetic Fiction*. New York & London: Methuen, 1985.

Prigogine, Ilya. *From Being to Becoming: Time and Complexity in the Physical Sciences*. San Francisco, CA: W. H. Freeman, 1980.

———. "Order through Fluctuation: Self-Organization and Social System." In *Evolution and Consciousness*. Ed. E. Jantsch and C. H. Waddington. Ontario: Addison-Wellesley, 1976. 93–126.

Prigogine, Ilya, and Robert Herman. *Kinetic Theory of Vehicular Traffic*. New York & Amsterdam: Elsevier, 1971.

———. "A Two-Fluid Approach to Town Traffic." *Science* 204 (1979): 148–51.

Prigogine, Ilya, and Isabelle Stengers. *Order Out of Chaos: Man's New Dialogue with Nature*. New York: Bantam, 1984. Trans. of *La Nouvelle alliance: Métamorphose de la science*. Paris: Gallimard, 1974.

Schieve, William C., and Peter M. Allen, eds. *Self-Organisation and Dissipative Structures*. Austin: University of Texas Press, 1982.

Sterling, Bruce. *Schismatrix*. New York: Ace Science Fiction, 1986.

Weintraub, Pamela, ed. *The Omni Interviews*. New York: Ticknor & Fields, 1984.

4

The Chaos of Metafiction

Peter Stoicheff

It seems, at first, curious that metafictional texts possess characteristics of chaotic systems in the phenomenal world, but they do. I want to present four main characteristics shared by them: nonlinearity, self-reflexivity, irreversibility, and self-organization. As much as possible, I want to avoid using chaotic structures merely as a convenient metaphor or allegory for the structures contained within metafiction. We could easily lend the narratives we employ for the phenomenal world, or psychology, or politics, or metaphysics, to the investigation of texts. In fact, our instincts as critics have often led us in the past to do just that, to speak of a text as an organism which gives birth to meanings, which has a life of its own, which is fathered by so and so, which contains an Oedipal antagonism toward its progenitors, and so on. (Even to say that a text "contains" something is to give it a spatial dimensionality it only metaphorically has.) The crucial purpose in exposing the chaos and complexity of metafiction is not to provide another vocabulary through which to speak of a text; nor is it to suggest that the dynamics of metafiction are *like* those of chaos or of complex systems. Instead, it is to show that metafiction displays the properties located in what science calls chaos, and that a metafiction text *is* a complex system. It is also to show that metafiction and scientific chaos are embraced by a larger revolution in contemporary thought that examines the similar roles of narrative, and of investigative procedure, in our "reading" or knowledge of the world.

A fiction text contains many strategies for metamorphosing the apparent chaos or randomness of phenomenal reality into an order comprehensible to its reader. Usually, a text employs these strategies covertly, and thereby sustains the illusion that it does not mediate between reader and world, but opens a neutral window onto that world for the reader. As a consequence, the strategies recede beneath the surface of the text's significant intentions, to counsel calmly and imperceptibly the reader's impression of the text's neutrality as the reading process continues, maintaining what Roland Barthes skeptically terms "the totalitarian ideology of the referent" ("To Write" 138). The mimetic text must maintain the reader's happy ignorance of the illusion in which he is enmeshed, and not disrupt his intuitive belief that it is permitting a linear transmission of reality to him. "Really, universally, relations stop nowhere," writes Henry James, "and the exquisite problem of the artist is eternally to draw, by a geometry of his own, the circle in which they shall happily *appear* to do so" (vii; James's emphasis).

This linear transmission occurs across the margin between two supposedly separate entities, text and world, that share a boundary of language whose dynamics exhibit strange and frequently unpredictable behavior.[1] Post-Saussurian investigation has revealed how that margin is not an unproblematic prism of decoding but a highly chaotic site where the indeterminacy of language proliferates. Previously, the writer's and reader's reliance on the linearity of the process of semantic production tended to produce texts whose ideologies reflected that linearity. A traditional nineteenth-century novel of verisimilitude, we liked to think, translated this reliance into a historical determinism, in which the language of fiction, its sequence, causality, and closure, fully articulated the past and revealed its consequences for the present. Language, like mimetic novels' fabula, had a determinable and stable effect, capable of accurate manipulation by the writer, and obedient interpretation by the reader; and this intuitive belief in the stability of the medium resulted in fiction whose social, political, sexual, and metaphysical ideologies reflected it. The measurement system of language, in other words, dictated its own results just as, in physics, the reliance on the linearity of systems created results, and paradigms, which reflected that linearity.

If language is recognized not as a neutral occasion for the direct transferral of meaning, but as a chaotic generator of significance whose interpretations are multiple, then the determinacy of the text it occasions becomes somewhat troubled. In effect, the text ceases to transmit the exterior world, and interrogates its own medium of transmission instead. This self-reflexive moment is a characteristic indicator of a metafictional text: it is a moment at which the subject under investigation is neither the ontological nor the phenomenal worlds external to it, but the complex margin of language strategy that separates the text from them.[2] To investigate its indeterminacy is akin to exploring the chaotic mannerisms of deterministic disorder; one witnesses both a bewildering randomness and an elusive order.

To varying degrees many fiction texts, not solely postmodern ones, are actually metafictional.[3] The novel, a product of the suspicion of genre itself, was pressured into existence as an escape from the accrustation of conventional form. It is no coincidence that one of the first, *Tristram Shandy*, remains one of the clearest examples of a metafictional text, for it contains the anxiety of an absence of predecessors, and a consequently obsessive self-inspection that is one of its subjects, and delights. Metafiction, in other words, is a synchronic response to the chaotic nature of its medium, and the structures inherent in it are a consequence of the dynamics of language, not a reflection of the chaotic phenomenal or ontological realities that all art, in some way, engages. We could say, then, that metafiction is an investigation of the chaos of meaning's production.

The random patterning within a simple phenomenal system is creative, because it generates "richly organized patterns, sometimes stable and sometimes unstable" (Gleick 43). The system of the metafictional text is creative as well, producing what Barthes terms the "jouissance" of an inexhaustible possibility of interpretations.[4] A simple example of this is in John Cheever's *Bullet Park*. Through a series of circumstances, both diabolical and innocent, two characters previously unknown to each other become adversaries. One is named Hammer, the other Nailles. This "mysterious power of nomenclature," as Cheever terms it, causes the reader to question the text's relationship to the actual world he knows. If the text does not present him with too many of these coincidences, he is likely to be forgiving, to say they lie (so to speak) within the realm of possibility, and to persist in reading it as a mimetic novel of realism. If the text presents the reader with several linguistic coincidences like this one (and *Bullet Park* does) the reader is forced to recognize that it is deliberately disrupting his assumptions concerning the linear relationship between text and familiar world. Cheever's naming of names reveals the conventionality of that process, and also reverses expectations of causality, for here a name is contingent upon its character's function in the text, not upon a familiar corpus of names in reality. This disconnects the text from any stable external referent, and calls attention to its artifice.[5] It is a maneuver that demands the reader's creative interplay with the text, for in the absence of any realistic model for nomenclature, the reader must generate possible motives himself. If we were to spend more time with *Bullet Park*, we would see how this creative process is both enlisted and frustrated, for the text refuses to confer significances upon the characters that correspond to their names (as in *David Copperfield*, for instance, where Agnes is the element of purity in the hero's life) or that are simple ironic inversions of that correspondence (as in *The Great Gatsby*, where Daisy is the opposite of innocence).

One result of this is a disruption of our familiar system of cause and effect (in the case of a text, the "cause" is the author's intention, and/or the donnée of the phenomenal world, and the "effect" is the text's obedient transmission of it); replacing this system in the metafictional text is the very provisional status of cause, origin, presence, and absolute truth. This amounts to a refusal of the hierarchical structure of meaning dependent upon an authorial creation and a passive readerly consumption, a refusal of "God and his hypostases—reason, science, law" (Barthes, "Death" 147), and undoes the one-way transmission of the author-reader connection. Metafiction exploits the understanding that a text cannot be an author's wind-up watch confidently demarcating the universe, submissively consulted by the reader, and that instead it is a chaotic system created by the text's limitless potential for interpretation, and the author's relinquished power. This diminished status of the author precludes a magnetic north of truth in the text, and frees signification to disseminate in ever-burgeoning patterns, as we shall see.

This is one way of saying that within the finite space of any text are an infinite number of possible meanings, whose hierarchy metafiction refuses to arbitrate. It can do this by disregarding the codes that we rely on to guide our interpretation of a text, or by exposing them as artificial, the product of conventions which themselves have only a very tenuous (and usually arbitrary or nonlinear) relationship with the signified.[6] With this refusal to arbitrate between an infinite number of possible meanings comes a metafictional text that is ostensibly meaningless for the reader of mimetic fiction, or at least ostensibly so random in its structure that it allows no communication—a chaotic text that contains no differentiation and hence no significance, anarchistically makes and shatters framing structures, and dissolves the primacy of the referent. Yet, as in the phenomenal world, hidden structures exist within this apparent chaos, which emerge when instability is given its due.[7]

The labyrinth is one of the metaphors that metafiction summons most frequently in the face of structure's absence. In its crossings and recrossings that rupture straight lines (by which we traditionally image "plot") it represents metafiction's disruption of linearity, and hence of fiction's previous hierarchical systems. (It also obscures the intuitive Western concept of linear genealogy; one example of this is Cheever's Hammer and Nailles, but Nabokov's parody of genealogical filiation in Humbert Humbert's history, Pynchon's opaque etymology of Oedipa Maas, Eco's William of Baskerville are others). The activity of the labyrinth involves an infinite repetition or retracing that obstructs memory, and hence origin, and in this way is metonymic of fiction's violated descent from reality that metafiction makes explicit. When extended to the dimensions of time, or voice, or causality, or symbolism, or closure, its relevance multiplies exponentially. In the mythic template of the labyrinth story, its relevance is evident too. In one happy version of the Dionysus-Ariadne romance he says to her, "I am your labyrinth"; a connection exists here between the labyrinthine patterning of fiction that finds no mimetic release, and Barthes' concept of "jouissance." It is Thesean logic which allows escape from the labyrinth, but as J. Hillis Miller points out, even Theseus "cannot free himself from his bondage with Ariadne herself, however much he tries to forget her" (62). Daedalus, the labyrinth maker, accidentally identifies himself to his enemy Minos while solving a labyrinth problem of how to negotiate a thread through the chambers of a seashell, thus imprisoning himself once again. As Miller writes, "[t]hread and labyrinth, thread intricately crinkled to and fro as the retracing of the labyrinth which defeats the labyrinth but makes another intricate web at the same time—pattern is here superimposed on pattern" (61).[8] Both Borges's *Labyrinths* and Barth's *Lost in the Funhouse* play with these various associations.

The Lorenz "strange attractor," in its display of an infinite number of paths in a finite space,[9] is analogous to Barth's description of the labyrinth as "a place in which, ideally, all the possibilities of choice (and direction, in this case) are embodied, and . . . must be exhausted before one reaches the heart" (Barth, "Literature of Exhaustion" 34).[10] Ambrose, the enmeshed protagonist of Barth's short story "Lost in the Funhouse," recognizes some of these possibilities at the end of the fiction in which he is enmeshed, lamenting that "the plot doesn't rise by meaningful steps but winds upon itself, digresses, retreats, hesitates, sighs, collapses, expires" (96), and witnessing "how readily he deceived himself into supposing he was a person. He even foresaw, wincing at his dreadful self-knowledge, that he would repeat the deception, at ever-rarer intervals all his wretched life" (93). And Puig's *Kiss of the Spider Woman* places the tantalus-like figure at the center of the intricate, and self-replicating, web of fiction and history from which there is no escape.

Barth's labyrinth is in the guise of a house of mirrors, which allows him to point out that if fiction's mimetic lineage from reality is necessarily severed, one of its ensuing patterns involves self-reflexivity. Mandelbrot's studies of chaotic systems revealed just such patterns, beautifully embodied in fractals, which "above all . . . meant self-similar" (Gleick 103). Self-reflexivity in metafiction is the product of its desire to expose the covert structures that allow fiction to masquerade as reality; it is always involved in the simultaneous processes of manufacturing illusion and revealing its artifice. It thus becomes an eternal system of creating and deconstructing,[11] whose self-interpreting pattern is realized in the *mise-en-abyme* that eternally defers the revelation of truth or knowledge, again raising the specter of Lorenz's strange attractor, which "displays infinite regress, like an unending sequence of Russian dolls one inside the other" (Gleick 150). In "The Circular Ruins" Borges makes this order, latent in the chaos of text production, apparent in a dialectic of dream and shattered image that emerges from the chaos of fire. In the story a man desires to dream another into fictional existence, and is assisted by "Fire . . . [who] animate[s] the phantom dreamt by the wizard in such wise that all creatures . . . would believe the phantom to be a man of flesh and blood" (72). Before long, the dreamer fears that his creation will "meditate on his abnormal privilege and somehow discover his condition of mere simulacrum. Not to be a man, to be a projection of another man's dream—what incomparable humiliation, what vertigo!" At the point of death (which takes place at the center of the circular ruins, in fire), he realizes that he too is dreamed; the creator is himself a fiction, imprisoned in a world whose author is, by extension, no less fabricated. The paradox is not limited to the musings of a metafictionist. In *Order Out Of Chaos* Prigogine and Stengers ponder whether we are "mere fictions produced by our imperfect senses? Is the distinction between life and death an illusion?" (252).

The vertiginous *mise-en-abyme* of metafiction is sustained by its perpetual dialectic of interpretation and deconstruction. It creates a pattern that stretches, not toward revelation, but around it. To reach the center of Borges's "circular ruins" would be to abolish the mediation of the imagination and the forms such as language that it assumes, and to apprehend directly, to be consumed by the referent as by fire (to reach the center of the labyrinth would be similarly fatal, and paradoxical, for it holds the minotaur, "dual and ambiguous," as Foucault terms both it and language [14]). But this cannot be done, except in solipsism (which the blind Jorge, Eco's reverent parody of Borges in *The Name of the Rose*, approaches, though he too is ultimately consumed by fire). Instead, man exists within, and is defined by his own burgeoning circles of interpretation, within the circular ruins of metaphor. The more information he gathers, the greater the number of intervening circles of language to carry it, the larger the indeterminacy, the more complex the interpretation, and the wider the abyss whose circumference he travels. As Nabokov's Humbert laments to his Lolita (but at the same time to his reader, both objects of his desire), "I have only words to play with!"[12] The Lorenz-like infinite tracings of interpretation suggest the Lolita-like elusiveness of truth. They also postulate a universe of appearance, interpreted into being by the observer-reader, whose private "labyrinth of lines traces the image" of his own face (Borges, *Dreamtigers* 93).

Many metafictional texts, styled with deceptive naiveté, attempt (with a premonition of their own defeat) to discover simplicity, to abolish redundancy and reduce communication to determinate elements of expression that would filter metaphor out and release a text from its linguistic imprisonment. The "simpler" a text becomes, however, the more it is aware of its own medium, because the discrepancy between itself and the multifaceted world is larger and calls attention to the artifice of the text. A text is inevitably involved in some kind of reduction process. From the chaos of the world outside itself, it processes the smallest amount of information that is necessary to communicate it. This is, as Foucault points out, because of the "simple, fundamental fact of language, that there are fewer terms of designation than there are things to designate" (14). Yet its result is always in the form of language, and hence always a metaphor, and metaphor always generates the labyrinthine structure of the text.[13] The indeterminate process of meaning-production in metafiction is a complex system, for it organizes itself into a pattern which is responsive to the metaphoric character of language. At the center of the pattern, around which text and reader carry on their inexorable dance, would be the abolition of metaphor, the unmediated truth whose guise is logic. Theseus's logic allowed him escape from the labyrinth, but just as he was doomed to replay the chaos of his desire, so is logic fated to break against the paradox of metaphor, which simultaneously asserts and denies identity by stating the impossible lie "this is that."

Frequently, metafictional texts are constructed so as to frustrate and reorient the logical reader they posit. In Nabokov's story "Signs and Symbols," a central (and anonymous) character is inflicted with a condition called "referential mania," an inability to refrain from recognizing metaphoric relationships between the phenomenal world and himself. He

> imagines that everything happening around him is a veiled reference to his personality and existence. He excludes real people from the conspiracy—because he considers himself to be so much more intelligent than other men. Phenomenal nature shadows him wherever he goes. Clouds in the staring sky transmit to one another, by means of slow signs, incredibly detailed information regarding him. His inmost thoughts are discussed at nightfall, in manual alphabet, by darkly gesticulating trees. Pebbles or stains or sun flecks form patterns representing in some awful way messages which he must intercept. (69)

Because he excludes real people from the conspiracy, we gather that he is a reader, and the world is his text. This text of the world is permeated by chaotic matter (clouds, wind-tossed trees, stains, strewn pebbles, and flecks of sunlight) which contains messages he cannot but devote "every minute and module of his life to the decoding of . . . . " He is a reader caught in the unending enigma of a text's interpretation and deconstruction, and thus his labyrinthine world of the text, like a fractal, is infinitely self-reflexive and not bound by scale: "If only the interest he provokes were limited to his immediate surroundings—but alas it is not! With distance the torrents of wild scandal increase in volume and volubility" (70). His readerly solution is to try to break through the fabric of the chaotic text that surrounds him, "to tear a hole in his world and escape" (69).

The reader of the story attempts, among other tasks, to decide whether or not the boy is successful in doing this. Although the reader may not suspect it, he is interested in the problem because, as a reader himself, he shares the boy's fate. Hence the reader is much like the man in Borges's "The Circular Ruins" who recognizes his own identity at his peril. (In fact, just as Borges's man is immolated in the fire of chaos at the point when he realizes his own fictitious status, so does the sun permeate "Signs and Symbols" in various verbal disguises, whose decipherment would reveal to the reader his identity with the referential maniac.) The reader is thus compelled to follow various paths of possible disclosure in the text, none of which are conclusive.[14] Because we read early on in Nabokov's story of the (provisional) connection between the boy and a bird, a following description of a dying bird "helplessly twitching in a puddle" seems foreboding (68). At the end of the story, the parents' phone rings ominously three times at night. Is this the hospital calling to say their son has died? These possibilities, two of many imbedded in the text,[15] reverse the story's tendency to achieve closure. Out of the chaos of indeterminate meaning the reader arranges possibilities that sustain his disseminating journey around the unknowable truth of the boy that lies at its center. Operating as a complex system, "Signs and Symbols" defeats closure by eternally adapting to new interpretations, which are essential to its generative process.[16] Disorder, randomness, and nonlinearity, all characteristics of a metafictional text's understanding of language, can create their own order. This will in turn give way to another phase of deconstruction, for that order is dependent upon a reading of it, which will inevitably be interrogated. As Prigogine and Stengers write, "nonlinearities may produce an order out of the chaos of elementary processes and still, under different circumstances, be responsible for the destruction of this same order, eventually producing a new coherence beyond another bifurcation" (206).

Hence, the traditional supremacy of order over disorder is negated in chaotic systems and in metafiction, and the hierarchy is neutralized to a dialogue between the two which generates the patterns of the natural world and of the text. In the case of metafiction, this neutralization prevents New Criticism's dream of organic unity, in which components (such as symbol) can be isolated and then extrapolated to serve as the text's message. It also disturbs the pre-Saussurean dream of unified meaning over multiple interpretations. The "irreversibility" Prigogine and Stengers postulate is manifested in a metafictional text's resistance to the reader's urge to trace back from a text's local detail to its comprehensive meaning. This urge in the reading process, which creates a hierarchy among components in the text to the point where disclosure of truth is achieved, defers in metafiction to the continual bifurcations of construction and deconstruction, any of which are comprehensible on the local level but incapable of extrapolation to a unified meaning. Order is therefore manifested in pattern, and hence the metafictional text's absorption with style, and suspicion of "the whole and the one . . . of the transparent and the communicable experience" (Lyotard 81). Any literary text, as Barthes claims, "is not a line of words releasing a single 'theological meaning' (the 'message' of the Author-God) but a multidimensional space in which a variety of writings, none of them original, blend and clash" ("Death" 146). Metafiction, in making this explicit, sustains a complexity which is initially experienced by the reader as disorder or noise, incapable of reduction to authorial intention or ordered message. To "understand" a metafictional text, one must reject seeing it as a vertical organization of a text's components into a closed order that is interpreted as meaning. Rather, one replaces this view with the recognition of lateral patterns in which disorder becomes order, mystery becomes illumination and then fragments into a new disorder. This pattern generates the reader's continual interpretation of the metafictional text and also the text's self-generation, for the pattern is a manifestation of the text's reading of itself.

The metafictional text alerts its reader to the possibilities of self-generative readings that are latent in any text; in fact it produces a multiply-interpretative and highly self-conscious reader. Marked by an absence of theological "meaning" and a celebration of superficial pattern over significant depth, metafiction sensitizes the reader to transcoding rather than to certainty. Its "order" is contingent upon the disorder whence it emerged and is always susceptible to fragmentation, whence a new order must arise. Inevitably, this consequence of "reading" a metafictional text reorients subsequent reading of nonmetafictional texts for that reader. Metafiction's reciprocity between order and disorder, whose source is the language which is shared by all texts, stimulates the rereading of a supposedly realistic text. It unravels the stability of the ordered happy ending, for example, and sees in its place the "unresolved tensions of dialogic oppositions" (Garret 221) that masquerade as closure. Ultimately, it wakens readers from the Romantic dream of the linear relationship between a text's microscopic and macroscopic aspects and alerts them to the "recurrent tensions between individual and general perspectives in [for example] Victorian multiplot novels" (Garrett 223). Ironically, the literary legacy of metafiction is not so much to fiction as yet unwritten as it is to previous texts; it retrospectively causes readers to see, in the limits to readings that a text erects, the possibility for new readings that examine what is excluded, and why.

Another of its legacies must concern our reading of our world. This takes us into the territory of narratology, which I only want to touch on, because it is discussed more fully in this volume by Paulson, Knoespel, and Hughes and Lund. If we were now to take into consideration the possibility that the phenomenal world itself is chaotic, infinitely self-replicating and fractally ordered, the antimimetic properties in metafiction bespeak a nice irony. Metafiction begins with a distaste for purportedly referential or realistic literature and yet, in its reaction against such, nevertheless manifests the structures of the chaotic phenomenal world. The difference is that metafiction generates those structures in itself, and so, although not mimetic of a chaotic world, nevertheless shares its aspects. Just as "the approximation of a coastline to a fractal is better than its approximation to a smooth curve" (Davies 60), so are reality's contours closer to metafiction's than to mimetic fiction's. All of which raises the problem posed by O. E. Rossler in "How Chaotic is the Universe?": "It could turn out . . . that a universe that is chaotic itself *ceases* to be chaotic as soon as it is observed by an observer who is chaotic himself" (317). A mimetic fiction text masquerades as a copy of the world but is not. Metafiction reveals, on the other hand, how the emergent chaos of a text and of the world are one. This is made explicit in Borges's "The Library of Babel," where the world is postulated as a library of texts, where the margin between text and world, in other words, does not exist.[17] Like phenomenal reality, the library is infinite, self-reflexive, and chaotic: "If an eternal traveler were to cross it in any direction, after centuries he would see that the same volumes were repeated in the same disorder . . ." (58). And reminiscent of Rossler's postulate that chaos is interpreted as order by the chaotic observer, Borges writes that to this traveler who is within the "disorder" of the universe (or in our case the text) and hence chaotic himself, his surroundings would become "an order, the Order" (58). For Borges, Pynchon, Nabokov, and most writers of metafiction, the world is a text that is read, and our interpretation of our world is a function of our reading of texts.

One aspect of this relationship is our impulse to enlist narrative structures for our understanding of life; to see our lives, or our social and political histories, or the dynamics of our phenomenal world,[18] reassuringly as narrative emplotments that we are practiced at reading on the page or hearing told. It is inconceivable, then, that metafiction's distortion of sequential narrative, of the relationship between author and reader, of causality and of closure, will not have consequences for our narrativistic reading of our world, our history, and our lives.

Metafiction's reluctance to allow narrative to exist independent of its medium might also illuminate the anthropomorphic profile of our worldly narratives. The noise that any narrative contains is contained by these larger narratives as well, though we try to disregard it in our desire for sequential and closed structures. It is precisely this desire that also makes the initial reading of metafiction difficult. Our worldly narratives, through which we construct what we think of as reality, are themselves a tissue of previous narrative texts with which they blend and clash, and which we choose to interpret in various ways. Whatever we call reality is revealed to us through the narratives we compose or, as Prigogine and Stengers word it, "through the active construction in which we participate" (55). A Euclidean narrative produces a Euclidean understanding of a Euclidean world. The metafictional "narrative of chaos" produces a metafictional understanding of a metafictional or chaotic world. The difference is that in the latter the process of self-interrogation is built into the narrative, freeing it from the tautological determinism that inhabits earlier narratives. This is why the introduction of a metafictional narrative into our world is a juncture that, like a complex system's development, is irreversible. With it, we operate in a cultural *mise-en-abyme*, where we can neither escape the self-conscious reading of our own imposed narratives nor fail to see, as Borges might have put it, the recognition of our own face in our narrative tracings.

Notes

1. David Porush, in the preceding Chapter 3, discusses a different boundary, that between the writer's "mind" and "the page," which Porush regards as "a magical leap" that "crosse[s] the threshold from natural to artificial in a sharper and more distinct way than in any other kind of organized behaviour save speech, per-

haps." It is partly because he has already discussed that threshold that I restrict myself to the next two in the sequence of a text's production—the boundary between the text and the world it communicates and, later in the chapter, the boundary between the text and the reader—but more importantly because metafiction self-consciously engages with these boundaries more relentlessly.

2. Waugh suggests three different types of texts occasioned by what I would term "self-reflexive moments": the "self-begetting novel," the "new realism" novel, and "those fictions which . . . posit the world as a fabrication of competing semiotic systems which never correspond to material conditions" (18–19).

However, it becomes difficult to make these distinctions clearly; a metafictional text is usually a hybrid of all of them.

3. Eco even undercuts the historical category of postmodernism: it is "not a trend to be chronologically defined, but, rather, an ideal category—or better still, a *Kunstwollen*, a way of operating. We could say that every period has its own postmodernism, just as every period would have its own mannerism (and, in fact, I wonder if postmodernism is not the modern name for mannerism as a metahistorical category" (*Postscript* 66). See, also, Fredric Jameson's foreword (to Lyotard) where he writes that Lyotard characterizes "postmodernism . . . not as that which follows modernism and its particular legitimation crisis, but rather as a cyclical moment that returns before the emergence of ever *new* modernisms" (in Lyotard xvi; Jameson's emphasis).

4. Eco refers to texts that do this as "open texts" which "have been planned to invite their Model Readers to reproduce their own processes of deconstruction by a plurality of free interpretive choices" (*Role Reader* 40). Or, from a different angle, see William Carlos Williams's *Paterson*: "A dissonance / in the valence of Uranium / led to the discovery // Dissonance / (if you're interested) / leads to discovery" (Book IV, 176).

5. This frame-break is doubled, in fact, when Nailles recognizes the absurdity of their names, as if he is becoming suspicious of his own existence as fictional artifice: "Lying in bed that night Nailles thought: Hammer and Nailles, spaghetti and meatballs, salt and pepper, oil and vinegar, Romeo and Juliet, block and tackle, thunder and lightning, bacon and eggs, corned beef and cabbage, ham and cheese, curb and snaffle, shoes and socks, line and sinker" (Cheever 54).

6. A famous example of this is the opening paragraph of John Barth's "Lost in the Funhouse":

> For whom is the funhouse fun? Perhaps for lovers. For Ambrose it is *a place of fear and confusion*. He has come to the seashore with his family for the holiday, *the occasion of their visit is Independence Day, the most important secular holiday of the United States of America*. A single straight underline is the manuscript mark for italic type, *which in turn* is the printed equivalent to oral emphasis of words and phrases as well as the customary type for titles of complete works, not to mention. (72)

7. I take the phrase "given its due" from Gleick, who writes that "almost no one in the classical era suspected the chaos that could lurk in dynamical systems if nonlinearity was given its due" (42).

8. The reader might also be interested in Miller's "Narrative Middles: A Preliminary Outline" (*Genre* 11, no. 3 [1978]: 375–87) and *Fiction and Repetition: Seven English Novels* (Cambridge: Harvard University Press, 1982) for later discussions and expansions of these concepts.

9. See Gleick: "[The attractor's] loops and spirals were infinitely deep, never quite joining, never intersecting. Yet they stayed inside a finite space, confined by a box. How could that be? How could infinitely many paths lie in a finite space?" (140).

10. Barth is playing on the word "heart" here. It is not only the center of the labyrinth, but of desire. The Dionysus-Ariadne myth, of course, illuminates the connection in another way. In a recently published interview, Barth again conflates the two, when he says that "The formal energy of the [*Funhouse*] stories comes from [their self-reflexive] concerns . . . which are really of a fairly technical nature. But the juice [jouissance?] of the fiction, the heart of the fiction, comes from such other concerns. . . . Those *are* love stories, no question about it" (in Lampkin 488).

11. Or, as Waugh writes, metafiction "breaks down the distinctions between 'creation' and 'criticism' and merges them into the concepts of 'interpretation' and 'deconstruction' " (6).

12. Borges playfully points to the infinite imperfection of language and its interpretations in "The Library of Babel": "for every sensible line of straightforward statement, there are leagues of senseless cacophonies, verbal jumbles and incoherences" (53). As Borges also frequently shows in his texts, language's degeneration of its own preexisting forms is an occasion for the generation of significance. As Porush argues in this regard: "The result is never-ending loops of interpretation and ever-widening circles of meaningfulness" (134).

13. The search is doomed not only because language itself is inescapably metaphorical, but because the mind, conscious of itself being conscious, operates inside its own *mise-en-abyme*. Metafiction, that is, reveals as much about consciousness as it does about fiction and language. As Gabriel Josipovici points out in a discussion of fictional framing: "we are condemned to see through frames. And recognition of this brings with it a kind of freedom, for it stops us from falling into the trap of thinking that meaning inheres in words, objects or events" (296). And Paul Ricoeur paraphrases Sartre: "to imagine is to address oneself to what is not" (152).

14. See Barthes, *S/Z* 210. Curiously, in the hermeneutic code the apparently opposite terms "closure" and "disclosure" beget one another and become interchangable.

15. In fact, the reader's urge to create determinate meaning out of chaos in "Signs and Symbols" inevitably leads him to read it symbolically. One impoverished, indeed parodic, result of this is the tantalizing mythological origin of the entrapped son and the father who seeks to design his son's escape from the sanatorium. The implicit early connection of both of them with a fallen bird, coupled with the pervasive use of the word "sol" (sun) in almost all surnames in

the story, recalls the proud decoder to the Daedalus and Icarus myth. Yet that too merely raises the spectre of the labyrinth.

16. Eco defines an "open text" as "a paramount instance of a syntactic-semantico-pragmatic device whose foreseen interpretation is part of its generative process" (*Role Reader* 3).

17. See Hayles (151–52) for a full discussion of this passage in "The Library of Babel." She argues that the story is one example of Borges's questioning of "the assumption that there is a 'reality' to reflect."

18. ". . . a scientist is before anything else a person who 'tells stories.' The only difference is that he is duty bound to verify them" (Lyotard 60).

Works Cited

Barth, John. "Lost in the Funhouse." *Lost in the Funhouse: Fiction For Print, Tape, Live Voice.* New York: Doubleday, 1988. 72–97.

———. "The Literature of Exhaustion." *Atlantic Monthly* 220, no. 2 (1967): 29–34.

Barthes, Roland. *S/Z.* Trans. Richard Miller. New York: Hill and Wang, 1974.

———. "The Death of the Author." *Image, Music, Text.* Ed. and trans. Stephen Heath. New York: Hill and Wang, 1977. 142–48.

———. "To Write: An Intransitive Verb?" *The Structuralist Controversy: The Languages of Criticism and the Science of Man.* Ed. R. Macksey and E. Donato. Baltimore, MD: Johns Hopkins University Press, 1972. 134–56.

Borges, Jorge Luis. *Dreamtigers.* Austin, TX: University of Texas Press, 1964.

———. *Labyrinths: Selected Stories and Other Writings.* Ed. Donald A. Yates and James E. Irby. New York: New Directions, 1964.

———. "The Circular Ruins." *Labyrinths* 45–50.

———. "The Library of Babel." *Labyrinths* 51–58.

Cheever, John. *Bullet Park: A Novel.* New York: Knopf, 1969.

Davies, Paul. *The Cosmic Blueprint.* London: Heinemann, 1987.

Eco, Umberto. *Postscript to the Name of the Rose.* New York: Harcourt Brace Jovanovich, 1984.

———. *The Name of the Rose.* Trans. William Weaver. San Diego: Harcourt Brace Jovanovich, 1983.

———. *The Role of the Reader: Explorations in the Semiotics of Texts.* Bloomington: Indiana University Press, 1979.

Foucault, Michel. *Death and the Labyrinth: The World of Raymond Roussel.* Trans. Charles Ruas. Berkeley: University California Press, 1986.

Garrett, Peter K. *The Multiplot Novel: Studies in Dialogic Form.* New Haven: Yale University Press, 1980.

Gleick, James. *Chaos: Making a New Science.* New York: Viking, 1987.

Hayles, N. Katherine. *The Cosmic Web: Scientific Field Models and Literary Strategies in the Twentieth Century.* Ithaca: Cornell University Press, 1984.

James, Henry. Preface. *Roderick Hudson*. By James. New York: Scribner's Sons, 1907.

Josipovici, Gabriel. *The World and the Book: A Study of Modern Fiction*. Stanford: Stanford University Press, 1971.

Lampkin, Loretta M. "An Interview with John Barth." *Contemporary Literature* 29, no.4 (1988): 485–97.

Lyotard, Jean-François. *The Postmodern Condition: A Report on Knowledge*. Trans. Geoff Bennington and Brian Massumi. Foreword by Frederic Jameson. Minneapolis: University of Minnesota Press, 1984.

Miller, J. Hillis. "Ariadne's Thread: Repetition and the Narrative Line." *Critical Inquiry* 3, no.1 (1976): 58–77.

Nabokov, Vladimir. *Lolita*. New York: Putnam's Sons, 1955.

———. "Signs and Symbols." *Nabokov's Dozen: A Collection of Thirteen Stories*. New York: Doubleday, 1958. 67–74.

Porush, David. *The Soft Machine: Cybernetic Fiction*. New York: Methuen, 1985.

Prigogine, Ilya, and Isabelle Stengers. *Order Out of Chaos: Man's New Dialogue With Nature*. Toronto: Bantam, 1984.

Puig, Manuel. *The Kiss of the Spider Woman*. New York: Knopf, 1979.

Ricoeur, Paul. "The Metaphorical Process." In *On Metaphor*. Ed. Sheldon Sacks. Chicago: Chicago University Press, 1981. 141–57.

Rossler, O. E. "How Chaotic is the Universe?" *Chaos*. Ed. Arun Holden. Princeton: Princeton University Press, 1986. 315–20.

Waugh, Patricia. *Metafiction: The Theory and Practice of Self-Conscious Fiction*. New York: Methuen, 1984.

Williams, William Carlos. *Paterson*. New York: New Directions, 1963.

5

The Emplotment of Chaos: Instability and Narrative Order

Kenneth J. Knoespel

A recent cartoon portrays postmodernism as a carnival arcade barely contained by the conventional frame of the cartoon image.[1] As caricatures of Foucault, Kristeva, and Bloom hawk their wares, workers with helmets labeled "deconstructors" break through the inky margins with jackhammers. With slight adjustment we can imagine a similar scene in which the mathematical practitioners of chaos theory open margins not with the linguistic gear of the deconstructionist but with iterative equations plotted on color monitors. The conflation of such images should not surprise us, for each has come to appear as a theoretical seismometer registering inconsistencies in systems whose logical integrity previously went unquestioned. Today even the casual observer of academic frontiers registers a shift from an unquestioned faith in the consistency of metaphysical systems and mathematical logic to a hypercritical expectation that perturbations may be detected in all systems of thought. Where critical inquiry previously assumed stability, it now explores instability and confronts complexity previously ignored or simply unseen. Whether we turn to deconstruction in philosophy or chaos theory in mathematics, we discover strategies that challenge the profession of grand logical systems to explore procedures which we may think of as experimental. The comparison of such widely divergent disciplines as literary theory and applied mathematics—amply indicated by essays in this book—also follows from our renewed interest in the methodological commensurability of disciplines that have long appeared separated. As Jerome Bruner has noticed, we live in a world where we no longer need to fix on the objects that distinguish different fields of inquiry but where we may identify the *processes* of understanding that bring us together (44).

The study of narrative is one way for approaching the epistemological framework behind different disciplines. Narrative theory has challenged literary critics to recognize not only the various strategies used to configure particular texts within the literary canon, but to realize how forms of discourse in the natural and human sciences are themselves ordered as narratives. In effect narrative theory invites us to think of all discourse as taking the form of a story. From such a vantage point, we virtually negotiate our way through life by telling stories that explain who we are and what we are doing, and by having these stories grafted onto the stories told by others. For theoreticians, the examination of such narrative networks offers a means for detecting how individuals and disciplines account for themselves, as well as how entire periods have become framed through metanarratives. From Northrop Frye's archetypes to Fredric Jameson's political unconscious, from Hayden White's historical tropes to Michel Foucault's epochal episteme, theoreticians have become accustomed to the formation and application of overarching narratives that function within critical discourse as framing devices.

The reception of Thomas Kuhn's *Structure of Scientific Revolutions* shows how attractive such metanarratives are for describing the sciences as well.[2] But here we must be careful. While a narrative matrix certainly draws the sciences and the humanities together, it does not necessarily function on the level of the metanarrative. The cognitively fragmented world in which we live may provoke longings for metanarratives that, nevertheless, turn out to give only illusions of mending our fragmentation. Today theorists subvert metatheories one moment and erect new ones the next. While Jean-François Lyotard defines postmodernism through the dissolution of overarching metanarratives, Jameson saves them through internalization.[3] What was once systematic metaphysical order has shifted to a psychological order that may be drawn out through the study of ideology. And even the theory of a discourse of fragmentation becomes itself an overarching narrative. The neo-Kantian formulation of spatial-temporal *transcendence*, once used to affirm the complementarity of the natural order and human science, has been replaced by a new form of transcendence in the synthetic visions offered by metanarrative.[4]

While such universalizing strategies promote far-reaching inquiry, they also simplify the ways that meaning is negotiated within different disciplines. Katherine Hayles deserves much credit for reorienting the approach to interdisciplinary studies by challenging "impact" studies that would register the influence of science on literature.[5] But by replacing the horizontal "impact" model with another that assumes the presence of an underlying cultural model, she anticipates a totalizing vision that obscures the importance of her local examples. We may use the matrix of information theory to compare Mitchell Feigenbaum and Jacques Derrida without transforming information theory into a metanarrative that reveals "underlying forces at work within culture." Hayles does not need to authorize her valuable inquiry by lending transcendent validation to information theory.

Caution is necessary in another quarter as well. Chaos theory and deconstruction have such radically different institutional implications that it can be misleading to emphasize their complementarity. While deconstruction subverts efforts to make itself into a universal system, chaos theory—as the term itself suggests—is expectantly regarded as a basis for a new foundational synthesis. Deconstruction challenges the impulse to build an overarching system found in the very inclination to view chaos theory as the entrance to a new scientific orientation. The difference should be stressed, for it calls into question efforts to exaggerate the affinity of the two approaches. Rather than stressing an all-inclusive metatheoretical position, I want to consider how local narratives render stable the destabilizing methods made available by deconstruction and chaos theory. At a time when much inquiry draws attention to the big picture, I want to look at the little pictures. The emplotment of chaos arises less from universal expectations than from the demonstration of its applicability to a range of local problems.

In this essay I limit the narrative grounds for comparing chaos theory and deconstruction. Rather than pursuing the impulse to metanarrative, I redirect attention to local narratives that are the subjects of the investigations of chaos theory and deconstruction and bring stability to their destabilizing inquiry. My discussion has three parts. In the first I notice how the very methods used to explore instability in chaos theory and deconstruction formulate a rhetoric of stability by appealing to other notational systems. Each critical enterprise, whether situated in ordinary language or in mathematical notation, engages a form of metaphorical emplotment that works to legitimate the enterprise. Each notational system maps itself with reference to the other. While chaos theory authorizes itself through logocentric arguments, deconstruction delineates itself through an appeal to a "geocentric" or "matheocentric" vocabulary. In the second part, I open a discussion of local emplotment by noticing how *enthymemes* or examples map discourse within each discipline. In the third I look at the play of examples in chaos theory and deconstruction and argue that they generate both stability and instability. No matter how universal the profession of instability becomes, it remains accompanied by stabilizing narratives which appear whenever the decentering modes are applied. In the simplest sense, logical instability remains stabilized through local narrative.

I. From Plot to Emplotment

In the conclusion to his book *Science in Action*, Bruno Latour compares scientific discourse to a process of continuous mapping. Whether the scientist maps the earth's surface, microorganisms within its geological structure, or the gaseous clouds of a supernova, the process involves the ongoing inscription of information. Latour's comparison presents scientific discourse as a process generating a vast collection of narrative overlays in which each inscription has a bearing for what comes before and after it. The comparison works so well because it draws upon the kind of conceptualizing metaphors that Lakoff and Johnson have shown are seated deeply in our linguistic experience. We may puruse the idea of metaphoric mapping even further by following the ways deconstruction and chaos theory map themselves. A single word opens a configuration that shows how mathematical and linguistic systems authorize themselves in reference to each other. The passport that takes us into such a conceptual network is the Anglo-Saxon word "plot."

The English word "plot" places an array of disciplinary interests before us, including Aristotle's *Poetics*, the navigator finding a location in the middle of water or stars, a mathematical practitioner mapping an equation with Cartesian coordinates, not to forget the space in which Farmer McGregor plants his cabbages. Of course, plot also bears an idea of intrigue often associated with figures like Guy Fawkes or Lee Harvey Oswald. Etymologically, "plot" derives from the Latin *plattus* and the Greek *platus*, both designating a flat or plane surface.[6] The earliest use of plot and its related form *plat* designates a small patch set off from surrounding ground or terrain. Its use as a noun for a particular place or ground explains its subsequent use as a verb for marking or measuring a piece of land and accounts for its more specialized extension into map-making and assimilation into geometry.[7] In its more technical sense, "plot" signifies the location of a point by means of coordinates on a map or within a geometric configuration. "Plot" appears for the first time as a plan or sketch for writing in the sixteenth century.[8] "Plot" comes to engage an even more complex tradition of discourse when it is used to translate the Latin word *fabula*, itself used to render the Greek words *logos* and *muthos*.[9] I have emphasized the metaphoric connections of plot because it offers grounds for looking at how chaos theory and deconstruction have mapped their own inquiries.

A simple way to describe chaos theory is in regard to plotting points within a system of coordinates. Where previous studies of the curves generated by nonlinear equations suspended their work when patterns seemed to disappear and randomness took over, chaos theory discovered highly complex fluctuations. Mitchell Feigenbaum has suggested that we may even think of chaos theory as part of an evolving analysis of trajectories that began with Niccolò Tartaglia and Galileo.[10] The mapping of the flight of cannon balls is related to the study of aperiodic trajectories generated by nonlinear equations for the simple reason that both appear as inscriptions within a system of coordinates. The difference between plotting the trajectory of cannon balls and the bifurcating trajectories in chaos theory is a radical difference in complexity. Using coordinate systems of greater complexity than the classical Cartesian grid, chaos theory makes visible as patterns numerical data previously regarded as random or inconsequential "noise." It is not, however, numerical plotting that I would emphasize but the kind of linguistic plotting or emplotment provoked by the mathematical activity.

In addition to engendering a new way of plotting trajectories, chaos theory generates metanarratives to describe its implications. In their recent book *Order Out of Chaos*, Ilya Prigogine and Isabelle Stengers emphasize that the study of disordered systems has already allowed such articulate explanation of natural phenomena that it has been compared to the formulation of Heisenberg's uncertainty principle (9, 218–32).[11] While their declarations draw attention to the new field—as does the Foreword by Alvin Toffler—they also misrepresent it. Just as quantum mechanics does not offer scientific justification for the elaboration of a transdisciplinary view of relativism, so chaos theory does not authorize a promiscuous expectation of chaos arising from a transcendent idea of disorder. Contrary to its popularized versions, chaos theory has not constituted a metamathematical shift. What chaos theory and the uncertainty principle share more than anything else is the sophisticated extension of mathematics within carefully controlled parameters. While popular presentations of chaos theory allude to its foundational aura, the work undertaken within the field does not pertain to metaphysical speculation on primordial origins and cannot be equated with a prophecy of universal disorder. When applied to cosmology, chaos theory is experimental and hardly validates metaphysical theory (Gunzig et al.). Aware of the term's ambiguity, some mathematicians have added the seemingly contradictory adjective "deterministic" to chaos theory to indicate its limited function.[12] In fact, Prigogine and Mandelbrot avoid using "chaos" in their professional publications. At present it is best to think of chaos theory as an ongoing articulation of mathematical theory. As one scientist says, we should

> consider chaos as a kind of order without periodicity. Within generally chaotic regimes one can discover patterns of ordered motion interspersed with chaos at smaller scales, provided sufficiently high resolving power is reached in numerical or laboratory experiments. Instead of the usual spatial or temporal periodicity, there appears some kind of scale invariance which opens the possibility for renormalizing group considerations in studying chaotic transitions. (Bai-Lin 5)

The experimental orientation of chaos theory reminds us that such inquiry is separated from the analytical or theorem-proof orientation of traditional mathematics. Even when applied to cosmology, chaos theory may be thought of justifying itself through application.

The very use of "chaos theory" to denote mathematical research concerned with the analysis of fluctuations in reiterating non-linear equations marks an intriguing example of the mythification of scientific work. By labeling such work as chaos theory, investigation is placed against the mythic background of early Greek philosophy, where chaos functions as a foundational concept in cosmology.[13] In effect, such labeling places chaos theory within the context of metaphysics and hints that it will uncover as yet hidden origins. Never mind that the term was playfully introduced into mathematics in 1968 (and earlier in the nineteenth century by Ludwig Boltzmann in the context of thermodynamics) and that most scientists would deny such a metaphysical association.[14] The term remains and continues to be perpetuated by the very books and articles written to clarify the research. James Gleick's recent bestseller is a case in point. At the same time that Gleick provides an intelligent survey of the new science, he validates it by situating it in the received traditions of the culture. Moving from one chapter to the next, the reader encounters a gallery of quotations from literature (John Updike, Stephen Spender, Wallace Stevens, Marlowe, Conrad Aiken, Herman Melville) that legitimate the wonders of the new science through canonical expressions of Anglo-American culture. Such decorous comparisons are not restricted to popular discussions. The introduction to a recent scientific publication on chaos includes an excerpt from Ovid's description of the cosmos (probably indebted to Lucretius) in the *Metamorphoses*: "Before the ocean was, or earth, or heaven, / Nature was all alike, a shapelessness, / Chaos, so-called, all rude and lumpy matter, / Nothing but bulk, inert, in whose confusion / Discordant atoms warred"

[I.5–9] (Schuster 1).[15] A collection of seminal papers on chaos theory prepared in Beijing is prefaced by the simple declaration: "The Emperor of Center was called Hundun (Chaos)" (Bai-Lin, title page). At my home institution I have listened to the physicist Joe Ford initiate a faculty lecture on chaos theory with a reading from Genesis.

At a time when chaos theory and deconstruction are approached as sister projects at national conferences, there is more than a bit of irony in underscoring how the Derridean enterprise would respond to the metaphysical setting provided for chaos theory. Having challenged literary theorists and philosophers to recognize the ways their own discourse is bounded by metaphysical assumptions inherent in language, deconstruction is also able to detect the logocentric assumptions present in science. For Derrida, the ceremonial narratives used to frame chaos theory would exemplify the way mathematics has been used "to complete and confirm a logocentric theology" (Derrida, *Positions* 35). Such acts are not innocent but remind us how frequently mathematics has situated itself within logocentric narratives in order to validate its accomplishments and proclaim its universal potential. Deconstruction cannot be restricted to literary theory or philosophy; it also challenges scientific discourse to acknowledge the ways that it remythologizes itself by appealing to foundational myths present within Western tradition or to myths generated by the history of science.

Although Derrida challenges us to notice how mathematics may be emplotted by metaphysical narratives, mathematics also allows us to recognize how deconstruction itself is emplotted. Rather than legitimating its inquiry through metaphysics, deconstruction explains its own formulation of a science of writing or grammatology through mathematics. In Derrida's effort to criticize how phonetic writing generates appeals for transcendental validation, mathematics offers a strategic means for standing outside ordinary phonetic writing. In effect, the notational independence of mathematics offers Derrida a locus for thinking beyond metaphysical systems that invariably seek to find closure. From his commentary on Husserl's *Origins of Geometry* to more recent work such as *The Truth in Painting*, Derrida describes his work in reference to mathematics. The reasons for such a strategy are not arcane. In contrast to ordinary phonetic language which bears a myriad of transcendental operations, mathematics offers a neutral system of notation that makes no appeal for metaphysical closure. In contrast to chaos theory, which emplots itself within logocentric narratives, deconstruction emplots itself within what we might call geocentric or matheocentric narratives.

Consider the following examples. For Derrida, a problem avoided by Husserl—the difference between phonetic and nonphonetic writing—becomes central. The problem signaled in *The Origin of Geometry* and *Speech and Phenomena* becomes foregrounded in *Grammatology*, where the formulation of a science of writing is made possible by the expanding complexity of nonphonetic notational systems. "The science of writing should therefore look for its object at the roots of scientificity" (Derrida 27). Fittingly, the eighteenth-century schemas for universal languages reviewed in *Grammatology* register not only a frustration with the inadequacy of enclosing metaphysical systems, but acknowledge the possibilities offered by mathematics. An important *prise de conscience* regarding mathematics appears in an exchange between Derrida and Julia Kristeva in *Positions*:

> A grammatology that would break with this system of presuppositions, then, must in effect liberate the mathematization of language, and must also declare that the practice of science in fact has never ceased to protest the imperialism of the *logos*, for example by calling upon, from all time, and more and more, nonphonetic writing. Everything that has always linked *logos* to *phone* has been limited by mathematics, whose progress is in absolute solidarity with the practice of a nonphonetic inscription. (34)

The linkage between deconstruction's defiance of metaphysical closure and the development of mathematics is made even more explicit as the discussion continues.

> The effective progress of mathematical notation thus goes along with the deconstruction of metaphysics, with the profound renewal of mathematics itself, and the concept of science for which mathematics has always been the model. (35)

The passages are important, for they indicate how completely Derrida privileges mathematics. At the same time that mathematics offers a model for the radical destabilization of phoneticism, it provides grammatology with a stabilizing force because it denotes grammatology's goal. The reference to the "imperialism of the *logos*" set against the "effective progress of mathematical notation" dramatizes a clash between notational systems. While Derrida challenges his readers to deconstruct logocentric narratives, he uses mathematics—and particular the radically different notational system of mathematics—to emplot his own critical inquiry.

The place of mathematics in Derrida's work extends well beyond overt references. The references to mathematics, especially as a sign system that escapes metaphysical enclosure, permit us to see how Derrida has generated a sign system that may also subvert metaphysical enclosure. Such a system emerges not from mathematical notation but in the metaphorical play that abounds within Derrida's work. In effect, Derrida's neologisms, which cause such problems for some readers, mark a step toward a new calculus that would interrogate our propensity to build and enforce systems. In contrast to metaphysical discourse that would enclose and defend, Derrida provokes continual openings. Critical figures—*différance, supplement, aporia, trace*—register such boundary work. Deconstruction's abundant deployment of terms is a graphic mechanism for provoking the reader to stand out and question how a text achieves meaning. While chaos theory inscribes its work on an evolving body of mathematical practice (particularly within phase space or Poincaré sections), deconstruction has no such established heuristic form. Indeed, in the absence of such a form, deconstruction plays with the possibility that the inscriptions of ordinary language work as a kind of phase space being stretched and folded on the rectangle of the written or printed page. Derrida's terms are not promiscuous agents but are like the graphic markers in mathematical equations. Each pertains not to a single universalized function but to a particular setting.

The foregoing discussion could certainly be extended. For now I can only make a general observation. Derrida's appeal to mathematics as a force to counter the imperial inscription of logocentrism seems to rely on a idealized conception of "pure" mathematics. Given the applied context of mathematics present throughout science and engineering, Derrida seems to exaggerate its subversive power as a nonphonetic system of inscription. All mathematics may be viewed as intersecting with phonetic language. In fact, the logocentric emplotment of chaos theory noted above challenges Derrida's supposition that "[t]he effective progress of mathematical notation . . . goes along with the deconstruction of metaphysics" (35). Although Derrida's own description of mathematics makes it appear decontextualized from its phonetic setting, he seems to use to it forecast grammatology's own unfolding story. Rather than pursue such questions any further, I will now turn my attention to the narrative functions that are at work in chaos theory and deconstruction on the local level.

II. Examples as Forms of Local Narrative

So far I have noticed how deconstruction and chaos theory emplot their inquiry. Whereas chaos theory becomes emplotted through logocentric narratives by practitioners who would validate their enterprise, Derrida emplots deconstruction by literally asking his reader to think of writing as a form of mathematical notation. The local narratives present in both writing and mathematics offer a means to develop the consideration of emplotment even further. By local narrative, I mean the shorter narrative operations present in disciplines as well as in our daily activity. These narratives include everything from instructions or directions (medicine bottles, telephone books, computer manuals) to cartoons on postmodernism in *The Village Voice*. Such mundane forms occupy our attention much of the time and have become a focal point in disciplines like psychology, linguistics, and artificial intelligence because they provide access to assumptions present in more complex discourse. While postmodernism has become criticized for being an ideological justification for a bric-a-brac culture, its inquiry also emerges from the recognition that the most fragmented events provide access to cognitive acts at all levels of cultural discourse. The spectacle of talking and dancing food in the refrigerator of Pee Wee's Playhouse reminds us of the unacknowledged "stories" present in all objects that surround us. An awareness of the myriad of narratives that surround us at every moment caused Mikhail Bahktin to observe that he "hears voices everywhere" (quoted in Todorov 21). In what follows I will concentrate on a function central to all local narratives: exemplification.

Our use of language relies on examples. We learn language through examples and continually use examples to orient ourselves within discourse. If we wish, we may think of examples as providing surveying tools that help orient our discourse with others. Practically, examples comprise appeals that would establish a common terrain for listeners or readers. Seen in this way, communication functions through an elaborate network of exemplification. Examples help us make points and enforce stability, but they also open discourse by challenging an audience to revise the maps they have used to plot experience. Finally, examples remind us that all understanding is temporally mediated.

We may formulate the major narrative functions associated with exemplification as follows:

1. *Examples promote closure*: Closure does not occur in a simple manner but has teleological implications on several levels. Closure may appear as a specific answer, affirm the coherence of an abstract system, and even bring credibility to the person responsible for the solution. At the same time, closure does not mark the termination of a rational process but invites the problem solver to test the result by working the answer back through the preceding narrative.

2. *Examples provoke openings*: Comprehension of an example pro-

motes its extension to other phenomena. Once an example is understood, it may be adapted to multiple settings with the result that each application further demonstrates the authority of the abstraction.

3. *Examples may subvert the system they are intended to affirm*: If solutions other than those dictated by the system are discovered, the entire system may be tested, resulting either in its elaboration or refutation.

Within the tradition of Western rhetoric, *enthymemes* are regarded as instruments that may contribute to a larger argument. In effect, the student of Aristotle or Cicero is taught to think of the example as a disciplined foot-soldier to be used in the execution of a larger strategy. The enthymemes that contribute so vitally to the discussion in Aristotle's *Poetics* and *Politics* become themselves the object of inquiry in the *Rhetoric*. Here discussion centers not on the overarching argument but on the evidence a speaker uses to forward a position. By selecting an example that is appropriate for an audience, the speaker may control the point of the example. Conversely, an ill-chosen example may blunt the speaker's argument by diffusing or scattering the audience's attention. For Aristotle, an enthymene comprises a miniature argument and should be thought of as a syllogism. Aristotle distinguishes two primary enthymemes.

> There are two primary species of enthymemes, namely: 1) Demonstrative Enthymemes, which prove that a thing is, or is not, so and so; and 2) Refutative Enthymemes, [which controvert the Demonstrative.] The difference between the two kinds is the same as that between syllogistic proof and disproof in dialectic. By the demonstrative enthymeme we draw a conclusion from consistent propositions; by the refutative we draw a conclusion from inconsistent propositions. (*Rhetoric* 2:22)

In later rhetorical manuals such as the *Rhetorica ad Herennium*, the demonstrative enthymeme becomes ground not only for closure but for testing premises. "*Exempla* are not distinguished for their ability to give proof or witness to particular causes, but for their ability to expound these causes."[16] While an exemplum is graphically abbreviated, its rational expansion takes place in silence in the mind of the listener or reader. In essence, the figure works as a muted logical statement that supports the more developed argument of a larger narrative through abbreviation.

The definition of exemplum notices an assumption about its use that is frequently ignored. Use of an exemplum should direct attention not to the result, but to the cause. Often we mistakenly assume that an exemplum would simplify discourse, when it actually works towards its complication. Rather than working as a simple syllogism, the figure works as a vehicle for elaboration. Salvatore Battaglia notices that the exemplum provided the Middle Ages with a means to investigate daily life and for this reason may even be regarded as a counterforce to the Bible (467). The remarkable expansion of scientific discourse in the seventeenth century came about through a proliferation of examples. The examples of natural phenomena that the Middle Ages and Renaissance controlled with compendia provoked renewed investigation by the seventeenth century. In fact, the exploration of physical phenomena may be compared to the proliferation of commentaries on *fabulae*. Phenomena included in Aristotle's *Parva naturalia* or Ovid's *Metamorphoses* are no longer moments of definitive closure, but sites for expansion (Knoespel). In the study of physics and mythography alike, the reader must not only resolve the meaning of an example but simultaneously locate it within the experience of the evolving narrative. Response to the example requires that attention given to the causes figured within the short narrative also extend to exploration of their presence within the evolving narrative as well as the already experienced narrative. Examples are narrative ligatures essential in making sense of our progression within texts. The three levels of narratival representation discussed by Paul Ricoeur—prefiguration, configuration, refiguration—offer a helpful way for thinking about the expectancy that accompanies examples. Anticipating how an example will fit a particular context marks a moment of prefiguration which is subsequently followed by stages in which its prefigured significance melds into the evolving configuration of the text. Finally the example itself may bring about refiguration of the entire text(52–87). The example is an advantageous figure with which to consider reading because it not only disrupts our viewpoint by making us refigure the fable but also offers a potential frame for ordering our assimilation of the story. The example functions something like a scenic overview that we may turn into while driving through the countryside. Like the overview which invites the driver to temporarily suspend his progress to consider the terrain he has crossed, the example invites the reader to consider his progress within the text.

The importance of *exempla* in the Middle Ages and Renaissance identifies an impetus for the renewed examination of exemplification within discourse. Their central position within classical rhetorical tradition explains their prevalent position in all forms of discourse before the rising preoccupation with building and maintaining physical and metaphysical systems in the eighteenth century. Today we witness a renewed interest in the local event simply because we have learned of the limited efficiency of universal claims. Work by André Jolles and Hans Robert Jauss on medieval discourse may be extended to contemporary discourse as well. Lyotard's work suggests that we should recognize far more than we do the place of the shorter narrative forms within scientific discourse (22). Similar arguments appear in the discussions that use biological metaphors to describe networks of communication. When Gilles Deleuze, Félix Guattari, and Michel Serres use biological metaphors such as the "rhizome" or "parasite," they insinuate the presence of narrative units as rhizomes that affect the nature of discourse as a whole (*Thousand Plateaus, Parasite*). Local narratives have not disappeared but are becoming more crucial to our critical understanding of all forms of discourse.

Enthymemes are hardly insignificant within discourse. They are narrative forms central to the ways we test and assimilate experience in the human as well as the natural sciences. Any experience presents itself as a potential example. An example may consist of a single word in a novel used to demonstrate a point or may comprise the entire novel itself. While one day we may work to build a complex argument with an exemplary experience, the next day we may collapse the grandest argument into an example. A major feature of discussion within the human and natural sciences is a reliance on the exemplification of highly complicated theoretical matter. While such abbreviation is a sign of scientific discourse, it may also lead to a suspension of communication if the audience does not understand the abbreviations used. The consideration of examples reminds us that our communication virtually consists of relaying examples that we would use to support and test abstractions. It is not an exaggeration to think of examples as forming the very currency of our exchanges.

The sciences are made up of examples. If we think of all experiments as examples that would test or demonstrate the validity of natural principles, we find that, rather than repudiating examples, the sciences privilege them. In the introduction, I noticed the proclivity of contemporary discourse to exaggerate metanarratives. Overarching narratives are hardly removed from examples but bear assumptions about their significance. Consider Kuhnian paradigms. Although the paradigm would order an array of examples (normal science), it is also through the anomalous example that the paradigm is transformed. Although scientific discourse may be simplified by describing it as paradigm dependent, its complex activity becomes more apparent if we approach it through the examples on which it depends. Bruno Latour makes this point repeatedly in *Science in Action*. In contrast to Kuhn who identifies the paradigm shifts in the history of science and the ways they order the myriad of local narratives, Latour concentrates on the exemplary inscriptions that actually make up the practice of scientific discourse. By looking at science in action rather than the history of science—laboratory science rather than museum science—Latour is able to show scientific activity as a contest involving the generation and coupling of examples.

Word problems in mathematics are a particularly advantageous place to explore scientific exemplification. Viewed historically, story or word problems form an ever-present body of short narratives. They are a fundamental vehicle for teaching mathematics and remind us that science as we know it cannot proceed without them. Pedagogically, story problems may be compared with the examples used in a sermon or the anecdotes used in a classroom because they furnish us with a means for dealing with abstractions. Their importance extends beyond formal education; we generate word problems in applying mathematical principles to the world. Such narratives make solutions possible, for they define the phenomena under consideration, formulate relationships between different phenomena, and generally prepare for the translation of linguistic configurations into mathematical notation. The pervasive presence of such forms within education and application indicates the extent to which mathematics may be regarded as a discipline that is committed to the world. While "plot" bears an idea of measurement when used to describe narrative structure, it also challenges us to recognize how the action of "plotting" within a coordinate system denotes an expectancy that has already been narratively encoded. Examples, and more particularly story problems, constitute a crucial narrative mode for negotiating understanding in all disciplines. Just how crucial they are becomes apparent when we look at their place in deconstruction and chaos theory.

III. Playing with Examples

In the first part of my discussion I noticed that efforts to draw together chaos theory and deconstruction are challenged by deconstruction's impulse to call into question the logocentric narratives used to validate the mathematical theory. In the second part I suggested that local narratives offer a hermeneutically rich setting for looking at the way audiences within the human or natural sciences negotiate understanding. In the final section I will suggest that the relation between deconstruction and chaos theory opened in the first section can most profitably be pursued by considering the way these theories deal with local narratives.

One of the obvious yet overlooked aspects of chaos theory is its practical application. Although chaos theory certainly bears implications for mathematical theory, it has attracted attention within scientific disciplines because it provides a means for describing an array of natural phenomena. Research on deterministic chaos has already had great impact on several fields, including chemistry, biology, and physics. It has a bearing on fluids near the onset of turbulence, lasers, chemical reactions, particle accelerators, biological models for population dynamics, and cardiac care. Benoit Mandelbrot's computer-enhanced images of iterative functions have extended its application. These applications of chaos theory makes it into a virtual generator of examples. My colleague Ron Fox has observed that

> the general picture of nonlinearity got a lot of people's attention—slowly at first, but increasingly. . . . Everybody that looked at it, it bore fruit for. You now look at any problem you looked at before, no matter what science you're in. There was a place where you quit looking at it because it became nonlinear. Now you know how to look at it and you go back. (Quoted in Gleick 305–6)

Chaos theory does not only direct attention to a specialized area of metamathematics but redirects attention to specific local phenomena. One important implication of such work is the way it has challenged mathematics to become less of an idealized and more of an experimental science.

The invitation for pedagogical play marks more than a renewal of applied mathematics. It challenges the top-down model of teaching mathematics and promotes the convergence of mathematical intuition with everyday experience. Chaos theory, as Steven Wolfram has suggested, challenges mathematics from the ground up.[17] The "evangelical" conclusion to Robert May's seminal article on chaos from 1976 makes a similar point.[18] Benoit Mandelbrot's work—and indeed his career—is an example of the reorientation of applied mathematics. He has called *Fractal Geometry of Nature* "a manifesto and a casebook" (104). The book provokes expectation that phenomena hitherto ignored in chemistry, biology, and physics can be approached through the study of complexity and suggests that the history of science itself may contain evidence of earlier efforts to comprehend such complexity. After meeting Mandelbrot at a Cornell seminar in the fall of 1986, I shared with him a project exploring the ways the Renaissance mathematicians approached Euclidean geometry through the practical geometry of Archimedes. By the next day, Mandelbrot was asking detailed questions about Archimedes and looking through medieval and Renaissance treatises himself.

The pedagogical play provoked by chaos theory also appears in deconstruction's analysis of texts. Indeed, such play negotiated through the local narratives provoked by each theory may be a more useful means of comparing such forms of mathematical and linguistic analysis than excessive emphasis on instability. Deconstruction, like chaos theory, is a form of analysis that begins with a local problem. Each engages in logical analysis for the study of problematic phenomena and identifies habits of investigation that have come to privilege themselves. By analyzing marginal or boundary situations, these modes bring into range phenomena that have been avoided. The broad range of Derrida's own investigations offer an indication; they all are *examples*. Moreover, they are examples in a significant manner. Rather than demonstrating closure as in a syllogism, they open inquiry. Consider Derrida's observation at the very beginning of *Grammatology*. Referring to the second part of the book, concerned with Rousseau's *Essay on the Origin of Language*, he writes:

> This is the moment, as it were, of the example, although strictly speaking, that notion is not acceptable within my argument. I have tried to defend, patiently and at length, the choice of these examples (as I have called them for the sake of convenience) and the necessity for their presentation. (lxxxix)

Derrida's qualified use of "example" comes from an awareness that examples are frequently used to demonstrate a closed system. Whether one considers collections of biblical parables, mythographic handbooks from antiquity through the Renaissance, histories of philosophy or of science, or even the Platonic theory of forms itself one discovers *exempla*—indeed *paradigma*—placed in containers and ready to use as prescribed. In complete contrast to "canned" examples applied with definite prescriptions, Derrida initiates a move away from closure and towards opening and extension.

As we compare chaos theory and deconstruction, we notice in each the crucial importance not of axiomatic or systematic statements but of examples. For each, examples provide a stable means for exploring instability. Their stability, however, is not conventional in the sense that it would enforce allegiance. Rather it is the stability that accompanies anything used as a heuristic devise. Examples become phenomenologically rich sources, a means not only of simple affirmation but also of extending inquiry. We may follow such extension by comparing it to the play of surplus meaning or information that follows an interpretive act.[19]

Surplus meaning refers to the complex psycho-linguistic phenomena which are generated by interpretive acts but which remain unacknowledged in the formulation of a response. Surplus information pertains to data which may be quantified but not necessarily comprehended through a single formulation. In the case of nonlinear equations, the erratic behavior of a dynamic system in one scale may urge the scientist to alter the parameters used to deploy and interpret the data. In the case of deconstruction, figurative language hitherto read into a traditionally enforced meaning of a text is allowed to play within a larger spectrum of meaning. Far from explicating the text for itself, deconstruction asks how the language of a particular text becomes the register or reservoir for astonishingly absent ideas such as being, the mind, or the subconscious. In chaos theory and deconstruction alike, the expectation of surplus meaning works to open the system to reinterpretation. Each project holds open the possibility of extending the strategies available for analysis. It does so not through a single interpretive act but through a process of iteration. Deconstruction, rather than reading a single text a single time, promotes the reading of many texts many times for an ongoing confessional comprehension of how meaning is generated.

In their efforts to articulate surplus meaning and surplus information, deconstruction and chaos theory account for ongoing phenomena. Chaos theory describes a new kind of order but finally also relies on previous mathematical models as heuristic tools. Deconstruction also relies on models, but models of reading and interpretation that have become habitual and even unconscious. For each, an understanding of how inquiry validates itself through limits promotes an awareness that symbolic systems also outrun preexistent meaning. Merleau-Ponty referred to critical acts as "that paradoxical operation through which by using words of a given sense, and already available meanings, we try to follow up an *intention* which necessarily outstrips, modifies and in the last analysis stabilizes the meanings of the words which translate it" (quoted in Norris 52). Deconstruction's work has been directed toward the amplification of such surplus. Rather than repressing such surplus or assuming that it fits into a preconceived system, Derrida would use it as a vehicle for asking how a text works. An idea of surplus works within disordered systems studied by chaos theory as well; such surplus becomes apparent when we recognize the watchfulness with which apparently chaotic systems are regarded. Where disorder was seen in the past, new forms of order now manifest themselves. Changing forms of representation—the replacement of classical Cartesian coordinates with the time-series diagrams of phase space—contribute to an expectancy for new patterns. Probably the most graphic images of "surplus" within chaos theory come from the paisley-like patterns of Mandelbrot's fractals which swirl into ever-new openings. From such a position we may understand the important affinity between deconstruction and mathematics. Both presume a continuous proliferation. By articulating disorder, by decentering what appears as a privileged text, chaos theory and deconstruction intervene to make us aware of other forms of order. Just as chaos theory seeks to define order which has hitherto remained undecipherable, deconstruction exposes experience which has been "ignored in order to preserve the illusion of truth as perfectly self-contained and self-sufficient presence" (Kearney 106).

IV. Openings

Deconstruction and chaos theory complement as well as challenge each other. Deconstruction cannot simply embrace chaos theory as a destabilizing ally but requires practitioners of chaos theory to recognize how the emplotment of their discipline contributes to its stabilization and mythification. Chaos theory in turn challenges practitioners of deconstruction to recognize how thoroughly Derrida uses mathematics to stabilize his grammatology. By following the ways each discipline emplots and authorizes itself through reference to the other, we discern a crucial way that disorder becomes ordered. But while we learn much by identifying how the disciplines map themselves into each other, we learn even more by looking at the way each is occupied with exemplification.

Rather than using examples to bring about closure, each maps its enterprise through the proliferation of local narratives that attest to the instability within discourse. Examples, however, also provide heuristic stability by marking moments in an evolving discourse. The instability discerned and analyzed by deconstruction and chaos theory should not be fetishized through overarching metanarratives. Inquiry with the natural and human sciences is not simply a matter of theoretical paradigms but involves the engagement with and generation of a myriad of local narratives. Science, as well as cultural discourse, cannot afford to orient itself through theoretical master plans alone but needs to consider the smaller narratives that bring order out chaos.

At a time when it is tempting to emphasize the complementarity of deconstruction and chaos theory, we must also acknowledge their differences. While deconstruction subverts efforts to make itself into a universal system, chaos theory is expectantly regarded as a frontier for a new foundational synthesis. The fact that deconstruction challenges the impulse to build overarching systems separates it from the inclination to view chaos theory as the entry way to a new scientific orientation. The difference should be stressed, for it calls into question efforts to exaggerate their affinity on logical grounds. Rather than being caught in a longing for an all-inclusive metatheoretical position, we should work to understand how local narratives render stable the destabilizing methods made available by deconstruction and chaos theory. Ultimately, the emplotment of chaos arises less from universal expectations than through a multitude of local narratives.

In a well-known paragraph in his *Philosophical Investigations*, Wittgenstein notices that because all languages are incomplete, they may be compared to a still evolving city where the suburb's "regular streets and uniform houses" represent the scientific languages that have evolved from ordinary language.

> Our language can be seen as an ancient city: a maze of little streets and squares, of old and new houses, and of houses with additions from various periods; and this surrounded by a multitude of new boroughs with straight regular streets and uniform houses. (Paragraph 18)

Where in Wittgenstein's evolving notational metropolis should we map chaos theory and deconstruction? However we answer, their presence has a bearing on how we negotiate our ways in the the city, for each engages us in a process of remapping. Each reminds us as well that maps are not made all at once but emerge from continually evolving social acts. The landscape becomes refigured not through a single universal act but one block at a time, so that each house or street becomes a marker for an ongoing enterprise. In effect, each local narrative becomes the act of a surveyor which brings forth order out of a chaos. By providing plottings for an evolving configuration, the essays in this book are participants in this process. Finally, they too are examples that lead not to closure but to new openings.

Notes

I want to thank Katherine Hayles for her support and criticism of this essay at every stage. For critical commentary and insightful discussion, I also want to thank my colleague Richard Grusin.

1. *The Village Voice Literary Supplement*, October 1988.

2. For a recent account of Kuhn's reception, see Larry Laudan, *Science and Values: The Aims of Science and Their Role in Scientific Debate* (Berkeley: University of California Press, 1984); for a review of various "stories" about scientific change, see I. Bernard Cohen, *Revolution in Science* (Cambridge: Harvard University Press, 1985).

3. The narrative function is losing its functors, its great hero, its great dangers, its great voyages, its great goal. It is being dispersed in clouds of narrative language elements—narrative, but also denotative, prescriptive, descriptive, and so on. Conveyed within each cloud are pragmatic valencies specific to its kind. Each of us lives at the intersection of many of these. However, we do not necessarily establish stable language combinations, and the properties of the ones we do establish are not necessarily communicable. (Lyotard 28).

Also: "Do we not ourselves, at this moment, feel obliged to mount a narrative of scientific knowledge in the West in order to clarify its status?" (Lyotard 28).

In his introduction to Lyotard's book, Jameson makes the following challenge:

> Lyotard does indeed characterize one recent innovation in the analysis of science as a view of scientific experiments as so many smaller narratives or stories to be worked out. . . . Lyotard seems unwilling [to take a further step] in the present text, namely to posit, not the disappearance of the great master-narratives, but their passage underground as it were, their continuing but now *unconscious* effectivity as a way of "thinking about" and acting in our current situation. (xi–xii)

4. While we may no longer accept the theoretical synthesis offered by neo-Kantian philosophy or phenomenology, we may still learn from their rigorous grounding in science. Ernst Cassirer's work on relativity and quantum mechanics (*Substance and Function* and *Einstein's Theory of Relativity* [bound together] (New York: Dover Publications, 1953 [1923]) remains a model of interdisciplinary literacy. An unfortunate consequence of phenomenology's assimilation into Anglo-American literary theory is the presumption with which it would speak about the assimilation of knowledge at the same time that it is virtually silent before science. In its American setting, phenomenology has turned from inquiry grounded in mathematics and science and emphasized psycho-linguistic epistemology. The arrogance present in certain literary theory that would absorb virtually everything in a handy epistemological schema exemplifies the filtered reception of such philosophical work. We live in a period that has come close to proclaiming universal theories of knowledge based on the analysis of a single literary or philosophical text. Under these circumstances it is not the sciences that threaten to "explain" the humanities but theorizing humanists who make claims to account for the sciences. I want to acknowledge such problems at the outset because they challenge us to formulate even more rigorous grounds for interdisciplinary work between the natural and human sciences. See also John Michael Krois, *Cassirer: Symbolic Forms and History* (New Haven: Yale University Press, 1987).

5. See also *The Cosmic Web: Scientific Field Models and Literary Strategies in the 20th Century* (Ithaca: Cornell University Press, 1984); for Hayles's most recent account of chaos theory and deconstruction see "Chaos as Orderly Disorder: Shifting Ground in Contemporary Literature and Science," *New Literary History* 20:2 (Winter 1989), 305–22.

6. The noun "plot" referring to a small area or small piece of ground comes from late Old English probably before 1100. The noun "plot," referring to a secret plan or conspiracy, is traced to the Old French word *complot.* See *OED* and *The Barnhart Dictionary of Etymology* ed. Robert K. Barnhart (New York: H. W. Wilson, 1988); for the relation of "plot" as a small piece of ground or patch and "plane" and subsequent links to Greek and Latin, see Eric Partridge, *Origins: A Short Etymological Dictionary of Modern English* (London: Routledge & Kegan Paul, 1958).

7. The earliest reference to "plot" within the context of mapping appears in Robert Recorde's *Pathyway to Knowledge* (London, 1551). Referring to the geometric quadrant, Recorde writes that you may use the instrument "not onely (to) measure the distance at ones of all places that you can see togyther, howe muche

eche one is from you, and every one from other, but also therby to drawe the plotte of any countreie that you shall come in" (fols. a3 verso to a4 recto).

8. Sir Philip Sidney refers to narration "as an imaginative ground plot of profitable invention" (*An Apology for Poetry* [1592 (1583)], ed. Forrest G. Robinson [Indianapolis: Bobbs-Merrill, 1970], 58). Edmund Spenser uses "plot" in *A Viewe of the Presente State of Irelande* (1596) to refer to political plans but not in the sense of conspiracy: "I woulde rather thinke the Cause of this evell which hangeth uppon the Countrie, to proceed rather of the unsoundnes of the Counsells and Plottes, which youe saie have bynne often tymes laied for her reformacions or of faintnes in followinge and effectinge the same" (*Works of Edmund Spenser* vol. 9, *The Prose Works*, ed. Edwin Greenslaw et al. [Baltimore: Johns Hopkins University Press, 1949], 44; see also 43). See also Francis Bacon in *Advancement of Learning* (1605): "Wherefore I will now attempt to make a general and faithful perambulation of learning with an inquiry what parts thereof lie fresh and waste and now improved and converted by the industry of man, to the end that such a plot made and recorded to memory may both minister light to any public designation, and also serve to excite voluntary endeavours" (*Essays, Advancement of Learning, New Atlantis, and Other Pieces*, ed. Richard J. Jones [New York: Odyssey Press, 1937], 234). Bacon's use of "plot" for the material encompassed in his work also indicates a strong association with surveying.

9. For reference to *fabula* for *muthos*, see Francesco Robortello, *In Librum Aristotelis De Arte Poetica Explicationes* (Florence, 1548) and Pietro Vettori, *Commentarii in Primum Librum Aristotelis De Arte Poetarum* (Florence, 1560). See also Reuben A. Brower, "The Heresy of Plot," *Aristotle's Poetics and English Literature* ed. Elder Olson (Chicago: University of Chicago Press, 1965), 157–74; 161.

10. Comments made by M. J. Feigenbaum in a paper delivered at Cornell University at a symposium, "The Inchoate: Complex Interrelations in the Humanities and Sciences," April 16–18, 1987, prompted me to look at the evolving analysis of trajectories. For trajectories, see also Ilya Prigogine and Isabelle Stengers, "Postface: Dynamics from Leibniz to Lucretius," in *Hermes: Literature, Science, Philosophy*, by Michel Serres (Baltimore: Johns Hopkins University Press, 1982), 135–55; for trajectories in phase space, see James P. Crutchfield et al., "Chaos," *Scientific American* 255:5 (1986), 46–57.

11. See also Ilya Prigogine, *From Being to Becoming: Time and Complexity in the Physical Sciences* (New York: W. H. Freeman, 1980).

12. Schuster defines deterministic chaos as follows: "[D]eterministic chaos denotes the irregular or chaotic motion which is generated by nonlinear systems whose dynamical laws uniquely determine the time evolution of a state of the system from a knowledge of its previous history" (1).

13. See Chalcidius, *Platonis timaeus interprete chalcidio cum eiusdem commentario*, ed. J. Wrobel (Leipzig: B. G. Teubner, 1876).

14. Tien-Yien Li and James A. Yorke appear to be the first to use the term "chaos" in relation to nonlinear dynamics in "Period Three Implies Chaos," *American Mathematical Monthly* 82:10 (December 1975), 985–92.

15. Prigogine cites Goethe's *Faust* at the beginning of *From Being to Becoming*, p. ix.

16. My translation from the Latin cited in Battaglia: "Primum omnium exempla ponuntur nec confirmandi neque testificandi causa, sed demonstrandi" (455).

17. Steven Wolfram emphasized this point in a conversation with the author at Cornell in November 1986.

18. The most important applications, however, may be pedagogical. The elegant body of mathematical theory pertaining to linear systems (Fourier analysis, orthogonal functions, and so on) and its successful application to many fundamentally linear problems in the physical sciences, tends to dominate even moderately advanced University courses in mathematics and theoretical physics. The mathematical intuition so developed ill equips the student to confront the bizarre behaviour exhibited by the simplest of discrete nonlinear systems. . . . Yet such nonlinear systems are surely the rule, not the exception, outside the physical sciences. Not only in research, but also in the everyday world of politics and economics, we would all be better off if more people realized that simple nonlinear systems do not necessarily possess simple dynamical properties.

May refers to this conclusion as "evangelical" on page 459.

19. See Hayles for a detailed comparison of deconstruction and information theory.

Works Cited

Aristotle. *The Rhetoric of Aristotle*. Trans. Lane Cooper. Englewood Cliffs: Prentice-Hall, 1960.

Bai-Lin, Hao. *Chaos*. Singapore: World Scientific Publishing, 1984.

Battaglia, Salvatore. "L'Esempio Medievale." *La Coscienza Letteraria del Medievo*. N.p.: Editore Liguori, 1965. 447–85.

Bruner, Jerome. *Actual Minds, Possible Worlds*. Cambridge: Harvard University Press, 1986.

Deleuze, Gilles, and Félix Guattari. *A Thousand Plateaus: Capitalism and Schizophrenia*. Trans. Brian Massumi. Minneapolis: University of Minnesota Press, 1987.

Derrida, Jacques. *Edmund Husserl's Origin of Geometry*: An *Introduction*. Stony Brook: Nicholas Hays Ltd, 1978.

———. *Of Grammatology*. Baltimore: Johns Hopkins University Press, 1976.

———. *Positions*. Chicago: University of Chicago Press, 1981.

———. *Speech and Phenomena*. Evanston: Northwestern University Press, 1973.

———. *The Truth in Painting*. Chicago: University of Chicago Press, 1987.

Gleick, James. *Chaos: Making a New Science*. New York: Viking, 1987.

Gunzig, Edgard, Jules Geheniau, Ilya Prigogine. "Entropy and Cosmology." *Nature* 330 (1987): 621–24.

Hayles, N. Katherine. "Information or Noise? Economy of Explanation in Barthes's *S/Z* and Shannon's Information Theory." In *One Culture: Essays in Science and Literature*. Ed. George Levine. Madison: University of Wisconsin Press, 1987. 119–42.

Jauss, Hans Robert. *Toward an Aesthetics of Reception*. Minneapolis: University of Minnesota Press, 1982.

Jolles, André. *Einfache Formen*. Tübingen: Max Niemeyer, 1982.

Kearney, Richard. Prefatory Note to discussion with Jacques Derrida. *Dialogues with Contemporary Continental Thinkers: The Phenomenological Inheritance*. Manchester: Manchester University Press, 1984. 105–6.

Knoespel, Kenneth J. "Medieval Ovidian Commentary." *Narcissus and the Invention of Personal History*. New York: Garland, 1985. 23–58.

Lakoff, George. *Women, Fire, and Dangerous Things: What Categories Reveal About the Mind*. Chicago: University of Chicago Press, 1987.

Lakoff, George, and Mark Johnson. *Metaphors We Live By*. Chicago: University Chicago Press, 1980.

Latour, Bruno. *Science in Action: How to Follow Scientists and Engineers through Society*. Cambridge: Harvard University Press, 1987.

Lyotard, Jean-François. *The Postmodern Condition: A Report on Knowledge*. Trans. Geoff Bennington and Brian Massumi. Minneapolis: University of Minnesota Press, 1984.

Mandelbrot, Benoit B. *The Fractal Geometry of Nature*. Rev. ed. New York: W. H. Freeman, 1983.

May, Robert M. "Simple Mathematical Models With Very Complicated Dynamics." *Nature* 261 (1976): 459–67.

Norris, Christopher. *Deconstruction: Theory and Practice*. London: Methuen, 1982.

Prigogine, Ilya and Isabelle Stengers. *Order Out of Chaos: Man's New Dialogue with Nature*. New York: Bantam, 1984.

Ricoeur, Paul. *Time and Narrative*. Vol. 1. Chicago: University of Chicago Press, 1984.

Schuster, Heinz Georg. *Deterministic Chaos*. Weinheim.: Physik-Verlag, 1984.

Serres, Michel. *Le Parasite*. Paris: Editions Grasset, 1980.

Todorov, Tzvetan. *Mikhail Bakhtin: The Dialogical Principle*. Minneapolis: University of Minnesota Press, 1984.

Wittgenstein, Ludwig. *Philosphical Investigations*. 3d ed. New York: Macmillan, 1958.

II

ORDER: REVISIONING FORM

6

Representing Order: Natural Philosophy, Mathematics, and Theology in the Newtonian Revolution

Robert Markley

Philosophy is written in that vast book which stands forever open before our eyes, I mean the universe; but it cannot be read until we have learned the language and become familiar with the characters in which it was written. It is written in mathematical language, and the letters are triangles, circles and other geometrical figures, without which means it is humanly impossible to comprehend a single word.

Galileo, *The Assayer*

Galileo's vision of classical geometry as the "mathematical language" of a divinely authored and ordered universe underwent almost constant revision and recalibration during the course of the seventeenth century. By the early eighteenth century, the time of Newton's third edition of the *Principia*, the mathematical language of the universe had become more complex than the "triangles" and "circles" he envisioned; so, too, had accounts of the relationship between mathematics and the physical world (see Whiteside 116). Although Galileo's image of a mathematically ordered universe continued to be invoked and updated, its straightforward equation of a representable and coherent creation with divine intention belied the ideational complexity of Newtonian mathematics, what Morris Kline (following Husserl) calls the "mathematization" of nature—the use of "pure," abstract mathematical reasoning to describe theoretically the structures and operations of physical phenomena (*Loss* 51–60).[1] If the growing sophistication of both experimental science and mathematics during the seventeenth century led to increasingly flexible and complex conceptions of physical and metaphysical order, it also required an increasingly complex series of discourses to explicate the function and significance of new discoveries in mathematics and natural philosophy. Paradoxically, the very complexity of the technical languages of Newtonian science led to efforts to popularize its "system of the world" by simplifying its ambiguities, by generally treating it as merely a grander version of the assurance that Galileo's rhetoric (if not his mechanics) had displayed a century earlier.

My purpose in this essay is to study some of the ways in which discursive elaborations of "order" emerge within and among the various languages—the various semiotic systems—of natural philosophy between roughly 1660 and 1750. In this respect, my aim is not to offer a linear history of "order" as a concept, scientific or otherwise; in fact, my approach calls into question some of the basic assumptions that historians and philosophers of science have traditionally made about the relationships between mathematics and "order." To investigate the discursive contexts of mathematics is to call into question some of the bedrock, essentialist philosophical assumptions that are still used to celebrate what we might call the ideology of certainty, the set of beliefs that valorize mathematical descriptions as elegant, beautiful, and self-sufficient on the one hand, and useful, pragmatic, and accurate on the other. I shall concentrate on the problem of mathematics as a form of representation in Newtonian science, on the questions it raises about the relationship of practice and theory, of the physical universe and mathematical "principles," that are used to define it. In contrast to those historians of science who claim that Newton and his contemporaries reflexively identified order with mathematics (for example, see Westfall, "Newton and Order"), I shall argue that the discourses of "order" before and during the Newtonian revolution are the sites of complex attempts both to describe accurately and to idealize the natural world. The notions of scientific "order" advanced by Boyle, Newton, and their successors are at least in part the unstable products of the cultural, political, and theological discourses that impinge upon and shape the two primary projects of seventeenth- and eighteenth-century natural philosophy: to observe, record, and classify phenomena and to celebrate this process as an approximation of a metaphysical revelation that justifies the works of God to man. The crucial problem for seventeenth and eighteenth-century natural philosophers is to theorize a basis for "order"—to discover the kinds of principles that can mediate between an imperfect physical reality (the legacy of Calvinist perceptions of the world that underlie the efforts of Protestant experimentalists to redeem a fallen nature)[2] and the perfection and dominion of the "Author of Nature" (Boyle, *Style* 50). This Sisyphean undertaking, as Boyle and Newton, in different ways, are both well aware, must always remain imperfect and incomplete precisely because God himself is beyond the reach of scientific reason. Therefore, to describe accurately, say, Newton's conception of order, one would have to add as an appendix all of his published and unpublished works. "Order" is ultimately the ideal that lies beyond and motivates the total of Newton's various researches, the "supplemental" knowledge that his investigations can only imperfectly reflect.

In seventeenth- and eighteenth-century natural philosophy, then, "order" does not describe a linearly evolving scientific concept but a discontinuous narrative that must be elaborated, explained, justified, and defended—a would-be meta-discourse that is both generated by and transcends a variety of competing historical utterances.[3] Notions of scientific "order" during the period are implicated in the discourses of politics, history, and theology, in various attempts to stabilize not only scientific rhetoric but a host of other semiotic schemes—among them alchemy, chronology, history, and theology—that seek to represent as unambiguously as possible the natural world. For Boyle, Newton, and their contemporaries, the attempt to order the universe becomes an epistemological quest to invent or discover an authoritative—and often self-consciously performative—semiotics. In this respect, the discourses of natural philosophy between 1660 and 1750, including mathematics, are not disinterested efforts to render the inchoate orderly but attempts to enact schemes of hierarchical structuring that trope the discrete phenomena of the natural world as parts of a coherent design. Yet the metalanguage that natural philosophers seek can never, in a post-lapsarian world, be written. For the Christian virtuosos of the late seventeenth and eighteenth centuries, "order" is ultimately deferred to a millenarian future of divine revelation, to an ideal semiotics that can be only incompletely and imperfectly realized in the experimental philosophies and mathematical investigations in which they are engaged.[4]

1

Seventeenth-century natural philosophy, as most historians recognize, was based on the constitutive metaphor of the two books: the book of nature and the Bible.[5] For Boyle, this analogy structured both the method and the ends of scientific investigation: "the Book of Grace," he assures his readers, "doth resemble the Book of Nature; wherein the Stars . . . are not more Nicely nor Methodically plac'd than the Passages of Scripture" (*Style* 53). By equating the radical epistemological practices of experimental science with what he considers the authoritative word of the Bible, Boyle, like Bacon and other seventeenth-century philosophers, turns the natural world into a vast text that can be studied and deciphered to reveal its divinely ordained coherence. The metaphor of the two books, in this regard, lends credibility to experimental philosophy as it seeks to minimize the epistemological problems of scientific investigation by justifying natural philosophy as a means to a theological end. The "Endlesse Progress" of science (a phrase that Boyle borrowed from Bacon's *Advancement of Learning*) offers natural philosophers numerous opportunities to celebrate divine wisdom and, as Thomas Sprat, John Wilkins, John Ray, and Joseph Glanvill assert, to buttress the authoritarian, patrilineal political ideology of Restoration England (Boyle, *Excellency* 63).[6]

What Boyle never explicitly acknowledges in his defenses of natural philsophy, however, is that the Bible itself had, by the mid-seventeenth century, become the site of intense political, ideological, and theological contention.[7] The Bible's authority—its claim to be the revealed word of God—had been challenged by political radicals in the 1640s and 1650s, including Gerrard Winstanley, Samuel Fisher, and Clement Writer, and by political philosophers, including Hobbes and Spinoza, who continued to write after 1660.[8] Moreover, the thousands of Protestant and Catholic explications of the Bible written during the seventeenth century created a number of competing and often contradictory interpretations that effectively destabilized the notion of an authoritative scriptural language capable of uniting Christians in a single set of beliefs. Far from being an unassailable means to validate experimental science, the Bible had become in the seventeenth century the focus of seemingly endless debates, the occasion for whole series of polemical interpretations that underscored the "corrupted," politicized nature of theological discourse.[9] In this respect (the propagandistic rhetoric of the Royal Society to the contrary), no single, straightforward "method" of interpretation could be gleaned from the study of the Bible; if anything, invoking particular interpretations of the Bible, as Samuel Parker did in defending the Royal Society and Henry Stubbe did in attacking it, politicized the rhetoric of natural philosophy. Boyle's "latitudinarianism," in this light, may be seen as an attempt to find the broadest common ground, to suppress theological differences in invocations of Christian, scientific unity.[10]

The destabilizing of the Bible as a metaphysical guarantee of authoritative meaning helps to explain why seventeenth-century writers from Comenius, to John Wilkins, to Isaac Newton were fascinated by the possibility of "real characters" and "philosophical languages" that would replace the fallen—that is, dialogical—language of seventeenth-century Europe with an idealized semiotics that ordered natural phenomena by the simple act of naming them. In his treatise on a natural language, George Dalgarno noted the common interests of scientists who "uncover the nature of things by careful scrutiny of similarities and differences and by collection and arrangement according to the method and order among them," and those who, in their efforts to name "things naturally, introduce into that chaos, the form, beauty, and order of an ideal world existing in the mind, by a logical creation" (37; my trans. differs slightly from Slaughter's, *Taxonomy* 143–44). As Dalgarno suggests, the essential problem confronting practitioners of seventeenth-century "natural philosophy" was not simply asserting that order existed in the universe—this was, after all, a theological and political commonplace—but enacting or embodying that order in stable, "authoritative" systems of representation.[11] For Wilkins, the semiotics of natural philosophy—its systems of taxonomy, recordings of raw data, encyclopedic surveys, and wide-ranging "theories"—were intended to be performative rather than simply descriptive: "the proper end and design of the several branches of Philosophy," he notes, "[are] to reduce all Things and notions unto such a frame, as may express their natural order, dependence, and relations" and thereby to overcome "the Curse of *Babel*, namely, the multitude and variety of *Languages*" (1, 13).

Yet at the same time that Wilkins, Boyle, Sprat, Glanvill, and others were proselytizing for the Royal Society's version of natural philosophy, they were acutely aware of the limitations of their efforts. Having spent nearly three hundred pages laying out his hierarchically arranged taxonomies, Wilkins is forced to acknowledge those things which his real character cannot express, "all such as are appropriated to particular *Places* or *Times*" (including titles of rank, office, and profession) and things which "are continually altering, according to several ages and times" (clothing, games, food, tools, and so on)—in short, everything that is not reducible to a general "essence" (295–96). The tables and taxonomies that fill the pages of both Wilkins' *Essay* and the Royal Society's *Philosophical Transactions* in the 1660s and 1670s were intended to be heuristic rather than comprehensive. They did not elaborate theories so much as offer instances of provisional schemes of organization which could and would be revised to accommodate the results of future experiments. In his "Proemial Essay" to *Certain Physiological Essays* (1661), Boyle argues that scientific theories are simply "superstructures" dependent on experimental evidence and should be "look'd upon only as temporary . . . not entirely to be acquiesced in, as absolutely perfect, or uncapable of improving Alterations" (9). Theories and taxonomies, for Boyle and his fellows in the Royal Society, are seen as heuristic rather than systemic.

In natural philosophy, then, taxonomers like Ray and Wilkins were faced with what we might call the crisis of Baconianism, the breakdown of elaborate hierarchical structures (based on broadly Aristotelian assumptions) into incomplete or contingent systems whose demonstrable "order" did not lead to deducing universal laws but to polemical defenses of theological and political doctrines. Their ultimate justification for their endeavors was not that they had constructed perfect representations of the natural world but that they had imperfectly and incompletely represented divine perfection. Ray, in describing the primitive chaos, locates order not in the physical structure of matter but in God's intention: "these [particles were] variously and confusedly commixed, as though they had been carelessly shaken and shuffled together; yet not so, but that there was order observed by the most Wise Creator in the disposition of them" (6). Thus we have paradoxical claims by Bacon, Ray, and Boyle that they are making "endlesse progress" in their scientific endeavors and that these inquiries can never achieve a complete understanding of God's creation. The ideology of progress in the 1660s and 1670s paradoxically depends on the assertion that natural philosophy is *inherently* imperfect and incomplete; it is completed epistemologically and ontologically *only* by God. The strongest statement of this view is Boyle's *Excellency of Theology*, published in 1674 but written in 1665 as a mortification of scientific pride. In this crucial articulation of the theoretical assumptions of seventeenth-century science, Boyle contrasts "the doubtfulness and incompleatness of Natural Philosophy" (143) to the certain truths of theology and revelation. "If knowledge be, as some Philosophers have styl'd it, the Aliment of the Rational Soul," he says, "I fear I may too truly say, that the Naturalist is usually fain to live upon Sallads and Sauces, which though they yield some nourishment, excite more appetite than they satisfie, and give us indeed the pleasure of eating with a good stomach, but then reduce us to an unwelcome necessity of always rising hungry from the Table" (120–21). Boyle's arguments for theology from the design of nature sanction scientific investigation and the kind of "endless progress" that Bacon had envisioned in *The Advancement of Learning*, but they also make clear that natural philosophy has no internal mechanism of its own to propel "progress" or to create order. Order can be apprehended *only* by revelation. In this respect, the metaphor of the two books encourages a "close reading" of the natural world, a study of it analogous, even parallel, to biblical exegesis, but philosophical investigations ultimately depend on teleological assumptions—what Boyle calls "the Universal Hypothesis" (*Excellency* 52)—that the universe is perfectly and exquisitely ordered.

Consequently, the natural philosophy of Ray and Boyle lacks an internal means to theorize its schemes of organization; Boyle, among the most astute and candid of the early natural philosophers, repeatedly emphasized the provisional nature of his findings and attacked those philosophers who formulated grandiose systems on the basis of limited data:

> when men by having diligently study'd Chymistry, Anatomy, Botanicks, or some other particular part of Physiology . . . have thought themselves thereby qualify'd to publish compleat Systems of Natural Philosophy, they have found themselves by nature of their undertaking, and the Laws of Method, engag'd to write of several other things than those wherein they had made themselves Proficients, and thereby have been reduc'd, either idly to repeat what has been already, though perhaps but impertinently enough, written by others on the same Subject, or else to say any thing on them rather than nothing. (*Physiological Essays* 3–4)

Significantly, Boyle attacks the genre of system-writing, the universalist projects of those who would order everything in the natural world on the basis of a handful of observations. His emphasis is instructive; even as he questions the systematizing tendencies of Descartes and his followers, Boyle implicitly acknowledges a nostalgic vision for an Aristotelian certainty that philosophy can, by the 1660s, no longer provide. Without a theory—a narrative or a semiotic structure—to provide a framework for its experimental results, the natural philosophy practiced by Boyle and Ray threatens to become a gimcrack collection of miscellaneous data that have no observable order independent of theological justifications propped up by the metaphor of the two books and its corollary, Boyle's argument from design. The attack by Henry Stubbe on the Royal Society for its irreligion, in this sense, is not merely an opportunistic means to discredit the conservative ideology of the Restoration intellectual establishment but a historically important response to what natural philosophy by 1670 had accomplished: by undermining Aristotelian notions of taxonomy—of the idea of system itself—Boyle's experimental science had effectively destroyed the internal theoretical mechanisms by which latter-day Aristotelians in the Renaissance had ordered their research programs. In brief, natural philosophy in the 1670s might be described as an epistemology in search of a theoretical structure that did not depend on overtly metaphysical or narrowly utilitarian justifications.

2

In an important sense, Boyle's experimental natural philosophy lacked what the Continental mathematical tradition—represented most frequently in seventeenth-century English scientific writing by Galileo and Descartes—could offer: a means to theorize rigorously the governing principles of an ordered universe. Descartes' mathematicizing of nature demonstrated the potential of mathematics as an ideal semiotics, as both a means to demonstrate and an unproblematic reflection of an ordered and perfect universe (Kline, *Loss* 42–45). In this respect, Cartesian mathematics offered itself as a complete, self-sufficient, and self-enclosed system. Descartes displaces the authority of the ancients and, claimed his English detractors after 1687, the authority of experimental observation onto the proofs offered by geometry, onto mathematics itself. As Otto Mayr notes, Cartesian descriptions of the mathematicized universe are based on mechanistic metaphors; their methodologies are founded on the assumption that the "vast book" of nature can be read as though it were a series of mutually reinforcing blueprints rather than as an infinitely problematic and ultimately undecipherable text (Mayr 62–70). Descartes becomes the villain for later eighteenth-century writers like John Keill and Benjamin Martin precisely because his materialism and rigorously mathematical descriptions of nature seemed to exclude both the kind of experimentation fostered by the Royal Society and the theological appeals made by Boyle, Newton, Ray, and others to the active role of God in maintaining his creation. In one respect, the challenge facing Newton and his followers was to reconcile the epistemological tradition of experimental science with the rigor and predictive capabilities of mathematics, as Newton implies in his preface to the first edition of the *Principia* and as Cotes argues in his preface to the second (Newton, *Principia* xvii-xviii, xx-xxi, xxvi-xxvii). In seeking to reconcile these traditions, Newton and his followers tested and redefined the limits of mathematics' ability to represent increasingly complex conceptions of the order of creation.

In the two decades between Newton's first concentrated work in mathematics and the publication of the *Principia* in 1687, his pursuits in alchemy, optics, chronology, and the interpretation of biblical prophecy suggest that he effected a crucial displacement of both the vehicle and tenor of the metaphor of the two books.[12] Boyle's argument from design was ultimately based on his conception of the Bible's "Complication [of] Rhetorick and Mystery" (*Style* 50) as a reflection of God's inscrutable purposes; even if many of its intricacies lay beyond human understanding, the Bible—and hence the natural world—was perfectly ordered. But, for Newton, the bulk of the Bible—all of the Old and New Testaments save the prophetic books and Revelation—was a flawed, imperfect text that had been corrupted by ignorance, problems of translation and linguistic transmission, and the willful deceit of trinitarians like Jerome and Athanasius (Markley, "Theological Writings"). Again and again in his unpublished writings, Newton seeks to detail the processes of the Bible's "corruption" by Christian polemicists. In his attacks Newton calls into question not merely the validity of specific passages but at least implicitly the asssumptions on which the metaphor of the two books rests. In effect, Newton rejects the idea of the Bible as divine logos, thereby implicitly subverting logocentric justifications for scientific research that depend on the belief that investigating the natural world serves the same ends as studying the word of God in the Bible.

In this regard, Newton's implicit rejection of the constitutive metaphor of the two books necessitates a far-ranging theoretical quest to find, derive, or demonstrate a new grounding for his investigations, a new authorizing language to redeem the logocentric tradition of a transcendent teleology. Frequently, scholars of Newton's works, like Manuel (103), have called attention to his compelling drive to establish order in his undertakings; but they usually make this assertion to demonstrate that the "same" principles are at work in both his "scientific" and "nonscientific" work. By seeking to demonstrate the method in Newton's apparent "madness," these historians use their notion of his quest for order as a means to validate modern conceptions of scientific inquiry. In contrast, I would emphasize that this "quest" entails Newton's redefinition of order as a complex, overlapping series of semiotic endeavors, none of which in and of itself is adequate to explain a physical and metaphysical reality that ultimately recedes—in the *Chronology*, in the *Observations on Daniel*, and in the *Opticks*—toward an irreducible complexity. This redefining of order on Newton's part takes a variety of forms: attacks on systematizers, from Leibniz to Athanasius; an obsession with the notions of origins, which he generally equates with a notion of pristine, uncorrupted meaning; and efforts to forestall the kind of closure, the kind of authoritative claims, of which he was deeply suspicious and yet which paradoxically were often made for his work by his eighteenth-century followers.

If Newton can be said to have a concept of "order," it exists only in intimations of a transcendent closure to his open-ended investigations in alchemy, mathematics, optics, history, and the interpretation of prophecy. Because "order," for Newton, always recedes beyond the scope of human investigation, it cannot be expressed in a graphic notational system. It can only be imperfectly glimpsed in what we might call Newton's anticipations of a future revelation, notably mathematics. Paradoxically, precisely because order is a fundamentally religious and mystical concept, it works to authorize or legitimate the logocentric assumptions that underlie his various areas of research and allows him to develop epistemological strategies to celebrate the workings of an "absent" guarantee of transcendental perfection. Yet Newton's efforts to validate the existence of a divinely created order by mathematicizing the universe is hardly a simple process; the quest for an order that forever recedes into greater and greater complexities suggests something of the profound anxiety of the natural philosopher dedicated to bringing the discourses of experimental science within received theological and ideological structures of expectation and knowledge. At the risk of oversimplifying, we might say that the more Newton discovers, the greater his need becomes to devise a theoretical framework to contain and shape the results of his research. As he attempts to demonstrate the order of the universe in a number of related semiotic "systems"—mathematics, theological writing, chronology, history, alchemy, and optics—he becomes increasingly aware of the limitations of each "system" to represent what is ultimately an unrepresentable order.

In an important sense, Newton's mathematics represents an attempt to move the basis of religious belief from faith or revelation—the inward light of the Protestant tradition—to the legibility of an external, objective, self-consistent, and authoritative system of representation. But the authority it claims is guaranteed ultimately not by its internal operations but by its teleological function. Mathematics, therefore, is not Newton's end-product, a final demonstration of celestial mechanics, but a heuristically conceived and provisional attempt to describe a metaphysical order. Although it is a valuable tool, mathematics does not ascend to the status of an ideal semiotics. At best it offers Newton an improved epistemological strategy to pursue the teleological purpose of Baconian and Cartesian philosophy. In effect, Newton's goal—in the *Principia* as well as in the *Chronology*—is to make signifying systems explanatory, to create or discover one-to-one correspondences between mathematical and prophetic signs and thereby to derive as a semiotic as authoritative as humanly possible to provide a method for calculation or interpretation.

At the beginning of Book Three of the *Principia*, Newton calls attention to the fundamental distinction in his work between philosophical and mathematical principles: "In the preceding books I have laid down the principles of philosophy; principles not philosophical but mathematical: such, namely, as we may build our reasonings upon in philosophical inquiries" (397). As Cohen argues, Newton's ability to bracket questions about the ontological status of gravity allows him to derive a mathematical description of gravitational force without having to grapple with the troubling philosophical and theological questions that prevented Huygens, for example, from accepting the reality—and hence mathematical validity—of force operating at a distance (Cohen, *Revolution* 52–154; "Newtonian Style," in Bechler 21–108). But this separation within what Cohen calls the "Newtonian style" between mathematical schemes of representation and the physical reality they describe problematizes their relationship. Mathematics can be a self-consistent, even "universal" system to explain natural phenomena, but its precision does not—and cannot—offer a foolproof assurance that what it describes "really" exists.[13] In an early, "popular" version of Book Three (translated and published separately in 1728 as *Newton's System of the World*), Newton states that his "purpose is only to trace out the quantity and properties of this [gravitational] force from the phenomena, and to apply what we discover in some simple cases as principles, by which, in a mathematical way, we may estimate the effects thereof in more involved cases." Mathematics is essential, he continues, because "it would be endless and impossible to bring every particular to direct and immediate observation" (550). This recognition that "scientific" or mathematical systems of representation do not reflect unproblematically an always and already ordered universe is, as Boyle's writings suggest, not a question of Newton's individual "style" of scientific research (as Cohen would have it) but a cultural and theological problem that lies at the heart of seventeenth-century attempts to construe the universe as ordered and legibile, as "proof" of God's infinite wisdom and boundless authority. In the General Scholium at the conclusion of the *Principia* Newton indicates that the purpose of his mathematics has been, in effect, to offer a more precise description of the argument from design:

> . . . it is not to be conceived that mere mechanical causes could give birth to so many regular motions [of celestial Bodies], since the comets range over all parts of the heavens in very eccentric orbits; for by that kind of motion they pass easily through the orbs of the planets, and with great rapidity; and in their aphelions, where they move the slowest, and are detained the longest, they recede to the greatest distances from each other, and hence suffer the least disturbance from

their mutual attractions. This most beautiful system of the sun, planets, and comets, could only proceed from the counsel and dominion of an intelligent and powerful Being. And if the fixed stars are the centres of other like systems, these being formed by the like wise counsel, must be all subject to the dominion of One . . . . (544)

Newton's emphasis on the "beautiful system" of celestial objects locates mathematical description within a master-discourse of aesthetic and theological order. This order does not proceed from "mere mechanical causes," from a Cartesian or Leibnizian determinism, but from the voluntaristic "counsel and dominion" of God.

Because mathematics, for Newton, is ultimately a means to a theological end, it is grounded in what Knoespel argues persuasively is "a logocentric view of the world." Although mathematics may maintain an "independent logical status" as "an internally consistent system," it cannot be applied to descriptions of the physical universe except through discursive elaborations, through what Newton, many of his contemporaries, and, more recently, postmodern philosophers of language regard as the imperfect medium of language (Knoespel, "Narrative Matter" 35). Put simply, the internal operations of mathematics do not and cannot describe an epistemology; whatever authority, whatever special status mathematics may lay claim to derives from the philosophical, theological, technological, and scientific discourses in which it is embedded and which it helps to constitute. In this respect, Newton's comments in the General Scholium do not represent a theological extension of his mathematical practice but an indication of the terms in which his mathematical demonstrations in the *Principia* take place. In the Scholium, following a series of propositions and theorems on the operation of his "attractive force," Newton observes that he has

now explained the two principal cases of attractions; to wit, when the centripetal forces decrease as the square of the ratio of the distances, or increase in a simple ratio of the distances, causing the bodies in both cases to revolve in conic sections, and composing spherical bodies whose centripetal forces observe the same law of increase or decrease in the recess from the centre as the forces of the particles themselves do; which is very remarkable. It would be tedious to run over the other cases, whose conclusions are less elegant and important, so particularly as I have done these. (202–3)

Newton then adds that he will "comprehend and determine them all by one general method" and offers Lemma XXIX (203). Elegance and mathematical significance are effectively joined. The "attractive force" does not exist simply in self-evident calculations but in the discursive formations—those of elegance and teleological significance—that dispose mathematical proofs and theorems in a particular order. In this respect, aspects of a universal order are reproduced metonymically within the *Principia* on the level of mathematical demonstration. Ernan McMullin suggests that "meaning" in the *Principia* is ultimately dependent on the overall system rather than the individual rule or demonstration, or, to extend this valuable insight, on the complex network of relations that exist among different semiotic systems (McMullin 39–45). In this sense, Newton's mathematics is part of larger relational network defined by his alchemical, theological, and historical concerns. Mathematical order is not a closed system of representation but a provisional inquiry that, by its very nature, can never be finalized.

In his manuscripts, as at the beginning of the *System of the World*, Newton is careful to locate the authority, the ultimate legitimation, of his mathematical description of the universe in the pristine symbolism of ancient knowledge. Mathematics, for Newton, is a means not of discovery but of rediscovery, of re-establishing "the true religion of the children of Noah before it began to be corrupted by the worship of false Gods" (quoted in Westfall, "*Origines*" 30). As Westfall has demonstrated, Newton's theological manuscripts from the 1680s testify to his firm antitrinitarian convictions and to his belief that the coming of Christ was not a "climactic event of human history [but] one repetition of a cyclical pattern" of the corruption, restoration, and renewed corruptions of a primitive, Noachian religion (Westfall, "*Origines*" 29). In this respect, if the mathematics of the *Principia* offers a means to demonstrate order, to enforce order, to "become" order, it nevertheless remains dependent on the theological context of Newton's religious and historical inquiries (Knoespel, "School of Time"). In an important sense, then, the scientific order that Newtonian mathematics creates—the illusion of a "pure" system of describing the natural world—exists only to the extent that Newton in the *Principia* suppresses the semiotic context in which and by which his demonstrations acquire their ideational significance. Given the urgent, even obsessive quality of Newton's work in alchemy, history, and the interpretation of prophecy in the 1670s and 1680s, it would be a mistake to regard his mathematical work up to and including the *Principia* as simply the internally consistent development of new mathematical principles. This body of work testifies not to the independent status of mathematics but to its implication in a variety of semiotic inquiries that have as their ideal—if not, as Newton himself recognized, their practical goal—to accomplish what Wilkins and others aspired to do: bridge or eliminate the gap between sign and signified.

Newton's habits of composition, the obsessive rewriting we find in his mathematical and theological manuscripts, in fact suggests that the seeming unity of mathematics, theology, and sociopolitical concerns can itself become potentially destabilizing: any time his calculations do not "work out," do not yield results that can be readily integrated into an overarching explanatory system, or yield solutions that do not seem to correspond to observable data, the theological order itself is at least implicitly destabilized. His investigations, in this respect, provoke crises that repeatedly lead Newton to attempts to reestablish some sense of equilibrium, to justify in new and ever more complex ways the workings of the universe as evidence of God's authority. Newton's obsession with accuracy, for precise measurement,[14] for reading all available historical and theological sources, stems in part from his awareness of what is at stake in his various semiotic explorations. His reluctance to publish his work (noted by virtually all of his contemporaries) may result from his frustration in never quite being able to make the necessary discoveries that would ideally stabilize the relations among the semiotic systems he studies, that would yield a religiously irreproachable order. In this sense, there is a unified "purpose" to Newton's inquiries, but it is not the sort of mystical order that Castillejo envisions. If Newton is the first "modern" scientist, he is also the first "postmodernist" in the sense that his testing of the limits of his mathematical, historical, and alchemical semiotics questions and destabilizes—even "deconstructs"—the very "systems" he is creating.

3

The problem of conceptualizing the relationship of natural philosophy to mathematics, of theorizing a basis for the practice of eighteenth-century science, is evident throughout the work of Newtonians like John Keill, Henry Pemberton, and Colin McLaurin, to name only three of the many popularizers who sought to explain Newton's achievement. In *A View of Sir Isaac Newton's Philosophy* (London, 1728), Pemberton (the editor of the third edition of the *Principia*, whom Newton called "a man of the greatest skill in these [editorial and mathematical] matters" [*Principia* xxxv]), contrasts the differences between natural philosophy and mathematics:

> The proofs in natural philosophy cannot be so absolutely conclusive, as in mathematics. For the subjects of that science are purely the ideas of our own minds. They may be represented to our senses by material objects, but they are themselves the arbitrary productions of our own thoughts; so that as the mind can have a full and adequate knowledge of its own ideas, the reasoning in geometry can be rendered perfect. But in natural knowledge the subject of our contemplation is without us, not so compleatly to be known: therefore our

method of arguing must fall a little short of absolute perfection. It is only here required to steer a just course between the conjectural method of proceeding, . . . and demanding so rigorous a proof, as will reduce all philosophy to mere scepticism, and exclude all prospect of making any progress in the knowledge of nature. (23)

In attempting to celebrate the "absolute perfection" of mathematical reasoning, Pemberton effectively cuts off "the arbitrary productions of our own thoughts" from the natural world. If mathematics has a theory and a method that natural philosophy lacks, it achieves its "perfection" by presupposing a Cartesian duality of world and intellect, in which the question of what mathematics represents is answered tautologically. Mathematics, for Pemberton, represents an ideal order because it presupposes an order to a "mind" which, at least in theory, "can have a full and adequate knowledge of its own ideas." Like other Newtonians, Pemberton describes this "full and adequate knowledge" in aesthetic and religious terms. His justifications for the study of Newtonian philosophy—which he sees as the judicious application of mathematical reasoning to the problems of natural philosophy—represent a post-Lockean reworking of Boyle's argument from design:

[O]ur desire after knowledge is an effect of that taste for the sublime and the beautiful in things, which chiefly constitutes the difference between the human life, and the life of brutes. . . . [From reason] arises that pursuit of grace and elegance in our thoughts and actions, in all things belonging to us, which principally creates imployment for the active mind of man. . . . [The] taste for the sublime and beautiful directs us to chuse particularly the productions of nature for the subject of our contemplation: our creator having so adapted our minds to the condition, wherein he has placed us, that all his visible works, before we inquire into their make, strike us with the most lively ideas of beauty and magnificence. (3–4)

The mind, for Pemberton, is thus "naturally" predisposed to perceive "beauty and magnificence" in the "visible works" of creation: order, in this respect, is inherent in the act of perception, in the existence of "the human life." It is not something that has to be discovered but is merely apprehended by the observer alert to what is already present in God's creation. More than a century after Galileo, Pemberton has translated the argument from design into the vocabulary of Shaftesburian effusion. He has not, however, questioned the basic premise that the aesthetics of order constitute the "mathematical language" of creation. The origin of the "desire after knowledge" is the sentimental beneficence of the human mind.

Pemberton's *View* is significant precisely because even as it calls attention to the problematic relationship between the natural world and

Newtonian mechanics, it theorizes this gap, this inability of mathematics to describe nature with perfect accuracy, as natural and inevitable. Colin McLaurin, in his influential *An Account of Sir Isaac Newton's Philosophical Discoveries* (1748), remarks that "to all such as have just notions of the great author of the universe and his admirable workmanship, Sir *Isaac Newton*'s caution and modesty will recommend his philosophy; and even the avowed imperfection of some parts of it will, to them, rather appear a consequence of its conformity with nature" (11). Thus, for McLaurin, Newton's "imperfect" philosophy is preferable to "all complete and finished systems" (11), like those of Descartes and Leibniz, all "systems founded on abstract speculations" (24) that, in effect, make God "the object of sense" by refusing to recognize that "his nature and essence are unfathomable" (22). Like Pemberton, McLaurin maintains that Newtonian philosophy accurately reflects an aesthetically and theologically constructed nature: "it is a consequence of [nature's] beauty, that the least part of true [that is, Newtonian] philosophy is incomparably more beautiful than the most complete systems which have been the product of invention" (12). In short, Newton's philosophy is as "true" as it can be because it does not distort nature into invented systems but "excite[s] and animate[s]" the natural philosopher "to correspond with the general harmony of nature" (4). Order, in this regard, becomes identified with a "truth . . . that . . . will be always found consistent with itself" (5), that is both defined by and transcends the operations of Newtonian mathematics.

Pemberton's and McLaurin's accounts of Newton's philosophy illustrate the process of narrativization that the *Principia* and the *Opticks* underwent during the eighteenth century. Newtonianism involves the popularization of a mathematically complex system that comparatively few readers could understand on its own terms; therefore, it encourages the semiotic *translation* of abstruse concepts into a relatively simple language that "everyone" could understand. Significantly, the works of Newtonians like Humphry Ditton and Benjamin Martin tended to flatten out, if not always consciously suppress, the gaps, contradictions, and carefully limited claims in Newton's texts in favor of discourses that offered a thoroughly systematized mechanics, emphasized its religious value, and promoted its actual and potential practical applications.[15] Newton's aloofness from scientific debates in the early eighteenth century, in effect, allowed for both the broad dissemination and distortion of his theories; these were not conscious attempts to revise his work so much as an inevitable part of a complex process of cultural and institutional recontextualizing of his findings. This process, as Pemberton's and

McLaurin's accounts imply, manifested itself in paradoxical efforts to emphasize the dependence of natural philosophy on theology, its role (to paraphrase Boyle) as a mere handmaid to divinity, and to argue for its unique authority as an accurate reflection of nature. This balancing act makes sense if one constantly stresses, as McLaurin does, the Calvinistic notion (earlier promoted by Boyle and Newton) that nature itself was inevitably corrupt and therefore suspect, that it was not immutable but merely a "present frame" that God could and would alter (22–23, 64–90). But once this theological point is rhetorically occluded, or simply forgotten, the internal tensions within Newtonianism—science as a semiadequate means versus science as a new form of authority—become increasingly evident. As early as 1705, Ditton describes the *Principia* as "the Divine Book" and beseeches his reader "to return his Thanks to the great Genius [Newton] to whom all this [the "profound and pleasing Speculations of Nature"] is owing" (Ditton, sigs. B1r, B4r). In effect, Newton and his work have been rhetorically conflated with God and the Bible; William Whiston accomplishes a similar conflation of scientific means and religious ends by praising Newton's "Divine Philosophy" (Whiston, *Mathematical Philosophy* 1). Both Newton and the *Principia* go through a form of beatification that, at least rhetorically, equates the process of scientific investigation with the ends to which science, according to devout Newtonians, should be put.

What we call "Newtonianism," then, must be characterized dialectically by its paradoxical tendencies towards both the championing of mathematics as the most accurate available representation of a mysterious physical creation and towards a fascination with other schemes of interpretation that offer a theological or more broadly religious frame of reference in which to "order" and interpret the universe and—significantly—to harness its powers for practical ends. A number of British Newtonians—particularly, as Force and Rousseau have respectively shown, Whiston and Cheyne—follow Newton into the realms of chronology and the interpretation of biblical prophecies in attempts to locate natural philosophy in relation to other, equally important, epistemological and teleological inquiries. Others, like J. T. Desaguliers and Martin, while praising the religious implications of Newton's work, could emphasize its practical applications. In this respect, by the 1720s Newtonianism had become identified with specific technological programs, with "pure" scientific research, with the institutionalization of scientific inquiry under the auspices of Newton's presidency of the Royal Society, and with the broader sociopolitical implications of Britain's efforts to "apply" its scientific and technological expertise to a variety of areas: navigation, trade, manufacturing, mining, and so on. Newtonianism, therefore, must be seen not simply as a form of empirical science or specialized research but as an ideological structure of beliefs that integrated scientific research into the political and economic operations of eighteenth-century British society. Precisely because the unifying basis of Newtonianism was, as Margaret Jacob notes, ideological, it necessarily operated in a number of complex ways to sanction a variety of political and cultural practices. Newtonianism could be—and was—used to promote and defend the hierarchical ordering of society, the "secularization" of values, and the justification of "order" for its own sake. In this regard, the Newtonian ideology could provide a theoretical vocabularly and an "objective," scientific rationale for challenges to contemporary orthodoxies (Newton's and Whiston's antitrinitarianism, for example) and for the maintenance of the status quo.

Newtonianism marginalizes and complicates notions of "order," of the relation of the inchoate—of nature itself—to attempts to describe the universe. The mathematization of nature that Kline notes can be understood as a series of efforts to render nature legible by idealizing, by imposing heuristic models of coherent systems on what otherwise would have to be taken solely on religious faith. In one sense, the history of science since the eighteenth century can be viewed as a continuing struggle between Cartesian or Leibnizian and Newtonian approaches to the dialectic of chaos and order: the imposition of method, of system, of reifying notions of objectivity, and of deterministic models of cause and effect on the natural world versus the always problematic status of epistemological inquiry, of the observation of discrete phenomena, of the rhetorical nature of scientific explanation, and of voluntaristic assumptions about God's direct and continuing intervention in physical creation. Underlying these seemingly fundamental differences between authoritarian and nondeterministic conceptions of science are different attitudes towards the problematic of representation, of the languages or semiotic systems in which the universe is described. These can and have been defined in a number of ways: realist versus metaphoric conceptions of language, rationalist versus a-rationalist notions of "reality," truth versus rhetoric, and so on. But it is the peculiar legacy of Newtonianism—of the ideational and ideological tensions within seventeenth- and eighteenth-century natural philosophy—that these seemingly antagonistic conceptions of science remain dialectically bound: each can be defined only in relation to the other. Order, in other words, presupposes both a scientific semiotics of some sort and its "other," that which challenges or escapes definition by the system that seeks to define it. In the wake of Newtonianism, there can be no purely objective description of order but neither can there be a "scientific" description of the equivalent of Milton's chaos in *Paradise Lost*—confusion without beginning or end. The mathematics of quantum theory, for example, represents less a rejection of what might be called the deep structure of Newtonian science—less a question about the adequacy of epistemology itself—than a complex development of some of its basic assumptions about the relationship between mathematics and the physical world.

The intersections of religious and scientific descriptions of order in the seventeenth and eighteenth centuries ultimately lead not to the revelation of an absolute truth but to the production of whole series of discourses devoted to explicating, popularizing, and promoting the results of what Newton published in the *Principia* and the *Opticks*. The more remote God becomes, the greater the mediation of the semiotic traces of his presence, the more urgent it becomes—both psychologically and culturally—for Newton and his contemporaries to assert the existence of a benevolent and harmonious nature governed by mathematical laws. In effect, theology is not exorcised from the corpus of science but repressed within it, deferred to the "end" of Bacon's and Boyle's "endless progress." In one respect, the questions that surface in the quasi-mystical or often overtly religious explications of quantum physics—such as in Zee's *Fearful Symmetry* and Barrow and Tipler's *Anthropic Cosmological Principle*—are part of Newton's legacy. The theological imperative—the invocations of religion or mystical teleology that resurface whenever 'normal' scientific practices find themselves straining against the limits of the theories which justify them—is an irrevocable part of scientific revolutions, of the realization that no amount of tinkering with existing systems of representation can make of them an authoritative language to describe an infinite and infinitely mysterious universe. Postmodern science has, in an important sense, inherited not only the values and assumptions of Newtonian method, objectivity, and self-generating progress but its repressed and occasionally half-acknowledged theological imperatives and justifications. The aesthetic vocabulary of contemporary mathematics, the notions of progress that underlie the cultural dissemination of sophisticated technologies, and the very standards of rigor and objectivity that are used to define postmodern science ideationally and institutionally indicate that we are still Newton's heirs, witnessing the return of the theological repressed in our "post-Newtonian" attempts to create and interpret order. Whatever advances have been made in quantum physics, chaos theory, and other contemporary areas of science, we are still, to some extent, struggling within and against the confines of Newtonian notions of order. In Derrida's sense, the "traces" of Newtonianism remain inherent in postmodern science precisely because it is the unanswered—and teleologically unanswerable—questions posed in the seventeenth and eighteenth centuries that contemporary theoretical inquiries must investigate and, in an always more sophisticated and always provisional sense, attempt to resolve.

Notes

1. Galileo's work is often filtered through post-Newtonian conceptions of science. My point is not that Galileo naively equated geometry with some ultimate truth but that as mathematics became more sophisticated during the course of the seventeenth century conceptions of order necessarily became more complex.

2. See Markley ("Objectivity," esp. 364–67). The historical significance of Calvinist assumptions in the development of science is also treated by Easlea (46–87, 154–94).

3. The ideational basis of my argument draws on the work of Mikhail Bakhtin, particularly *The Dialogic Imagination*. I should note that Bakhtin's descriptions of language, identity, and ideology pose significant challenges to the empiricist and realist assumptions that are usually taken for granted by many historians and philosophers of science. Some of the issues that shape contemporary philosophies of science are explored by Rorty and Hesse. My position is more radical than those of either Rorty or Hesse and is influenced by the work of Rossi, Reiss, Feyerabend, and Stepan, among a host of other ideologically oriented historians and critics. I would add, however, that debates about the nature of scientific investigation should not detract from the value of the historical work done by such "internalist" historians of science as Westfall, Dobbs, Cohen, and Whiteside. The significance of their research cannot be underestimated, although I often would emphasize readings of particular texts of Boyle's and Newton's with which they would, I imagine, disagree. My argument would be that my approach, like Knoespel's in the articles cited below, should be read as an elaboration and not a rejection of the work of these historians of seventeenth- and eighteenth-century natural philosophy. Careful explorations of the internal workings of Newton's mathematics and optics are not incompatible with a sophisticated cultural criticism which interrogates the relationships among science, theology, and ideology.

4. On the millenarian aspects of Newtonian thought, see M. C. Jacob, Knoespel ("School of Time"), and Rousseau.

5. The role of metaphor in scientific investigation is a complex topic with a seemingly unending bibliography. For a view of metaphor similar to mine, see Stepan.

6. The ideology of privilege is a recurring theme in late seventeenth-century natural philosophy; it is frequently part of the foundational arguments for the institutionalization of science. See Sprat, Wilkins (especially his preface), Ray, and Glanvill. For valuable discussions of Restoration ideology and the "rise" of

modern science, see Easlea, Keller 33–65, M. C. Jacob, and J. R. Jacob (*Robert Boyle*).

7. A good example of the ways in which scientific and hermeneutic controversies could be, in effect, one and the same is the multisided exchange triggered by Burnet's *Sacred Theory of the Earth* (first published in a more militantly idiosyncratic version in Latin in 1680). Rejoinders to and modifications of Burnet included Whiston (*New Theory* and *Vindication*), Warren, and Keill. See also Rossi (33–40).

8. These radical critics of scriptural authority are discussed by Hill (*World Turned* 259–68, and *Essays* 185–252).

9. On biblical controversies during the period see Reedy, and, specifically on Newton, Markley ("Theological Writings").

10. Boyle's latitudinarian arguments are found throughout his works; see particularly, though, his extended theological discussions in *Style*. See also J. R. Jacob (*Boyle*). Hill deals extensively with problems of religion and ideology in the latter half of the seventeenth century, especially in volume two of *Essays*. On Parker, Stubbe, and the significance of the latter's attack on the Royal Society, see J. R. Jacob (*Stubbe* 78–107). M. C. Jacob also deals extensively with the relations of latitudinarianism, Newtonianism, and upper-class ideology. See also Funkenstein (*Theology*) for valuable historical background.

11. This argument has been developed at length by Slaughter, although my emphases differ significantly from hers, particularly in the relations of Wilkins' project to what will emerge several decades later as Newtonian ideology. See also Reiss (52–53).

12. On Newton's work in these areas see Dobbs, Westfall (*Never at Rest*, esp. 288–308, and "*Origines*" 15–34), Manuel, Castillejo, and Knoespel ("School of Time").

13. The literature on the philosophy of mathematics is too vast to treat intelligently in a note. In addition to the historians cited parenthetically, see Kline, Stolzenberg, Hardison, and Kuhn (31–65).

14. Newton's concern with precise measurements of lunar motion led to his long correspondence and ultimate falling out with John Flamsteed, the Royal Astronomer. See particularly the letters that passed between Newton, Flamsteed, and John Wallis in the early 1690s in volumes three and four of *The Correspondence of Newton*.

15. See, particularly, Desaguliers and Martin. Martin prefaces his work with a list of over fifty previous works which sought to explain Newton's work to the public; he himself had written a number of earlier books on various aspects of Newtonian science. On the scientific lecture circuit and the dissemination of Newtonianism as a form of practical—and profitable—technology, see Stewart.

Works Cited

Bakhtin, M. M. *The Dialogic Imagination*. Trans. Michael Holquist and Caryl Emerson. Austin: University of Texas Press, 1981.

Barrow, John D. and Frank J. Tipler. *The Anthropic Cosmological Principle*. New York: Oxford University Press, *1986*.

Bechler, Zev, ed. *Contemporary Newtonian Research*. Dordrecht, Neth.: D. Reidel, 1982.

Boyle, Robert. *An Examen of Mr. T. Hobbes his Dialogus Physicus de Natura Aeris*. London, 1662.

———. *Certain Physiological Essays*. Oxford, 1661.

———. *New Experiments Physico-Mechanical Touching the Spring and Weight of the Air*. Oxford, 1660.

———. *Some Considerations Touching the Style of the Holy Scriptures*. Oxford, 1661.

———. *The Excellency of Theology as Compar'd to Natural Philosophy*. London, 1674.

Burnet, Thomas. *The Sacred Theory of the Earth*. London, 1686.

Castillejo, David. *The Expanding Force in Newton's Cosmos*. Madrid: Ediciones de Arte y Bibliofilia, 1981.

Cohen, I. Bernard. *The Newtonian Revolution*. Cambridge: Cambridge University Press, 1980.

———. "The *Principia*, Universal Gravitation, and the 'Newtonian Style,' in Relation to the Newtonian Revolution in Science." In Bechler, 1982. 21–108.

Dalgarno, George. *Ars Signorum*. London, 1661.

Desaguliers, J. T. *A Course of Experimental Philosophy*. 2 vols. London, 1734.

Ditton, Humphry. *The General Laws of Nature and Motion; with Their Application to Mechanics*. London, 1705.

Dobbs, B. J. T. *The Foundations of Newton's Alchemy or "The Hunting of the Greene Lyon"*. Cambridge: Cambridge University Press, 1975.

Easlea, Brian. *Witch-hunting, Magic, and the New Philosophy: An Introduction to Debates of the Scientific Revolution 1450–1750*. Sussex: Harvester Press, 1980.

Feyerabend, Paul. *Against Method: Outline of an Anarchistic Theory of Knowledge*. London: New Left Books, 1975.

Force, James E. *William Whiston: Honest Newtonian*. Cambridge: Cambridge University Press, 1985.

Funkenstein, Amos. *Theology and the Scientific Imagination from the Middle Ages to the Seventeenth Century*. Princeton: Princeton University Press, 1986.

Glanvill, Joseph. *Plus Ultra*. London, 1670.

Hardison, O. B., Jr. "A Tree, a Streamlined Fish, and a Self-Squared Dragon: Science as a Form of Culture." *Georgia Review* 40 (1986): 369–415.

Hesse, Mary. "Texts without Lumps and Lumps without Laws." *New Literary History* 17 (1985): 31–48.

Hill, Christopher. *Collected Essays*. 3 vols. Amherst: University of Massachusetts Press, 1986.

———. *The World Turned Upside Down: Radical Ideas During the English Revolution*. Harmondsworth: Penguin, 1985.

Jacob, J. R. *Henry Stubbe, Radical Protestantism, and the Early Enlightenment*. Cambridge: Cambridge University Press, 1983.

———. *Robert Boyle and the English Revolution*. New York: Burt Franklin, 1977.

Jacob, Margaret C. *The Newtonians and the English Revolution, 1689–1720.* Ithaca: Cornell University Press, 1976.
Keill, John. *An Examination of Dr. Burnet's "Theory of the Earth": With Some Remarks on Mr. Whiston's "New Theory of the Earth."* 2d ed. London, 1734.
Keller, Evelyn Fox. *Reflections on Gender and Science.* New Haven: Yale University Press, 1985.
Kline, Morris. *Mathematics and the Search for Knowledge.* New York: Oxford University Press, 1985.
———. *Mathematics: The Loss of Certainty.* New York: Oxford University Press, 1980.
Knoespel, Kenneth J. "Newton in the School of Time: *The Chronology of Ancient Kingdoms Amended* and the Crisis of Seventeenth-Century Historiography." *The Eighteenth Century: Theory and Interpretation* 30 (1989): 19–42.
———. "The Narrative Matter of Mathematics: John Dee's Preface to the *Elements* of Euclid of Megara (1570)." *Philological Quarterly* 66 (1987): 35–54.
Kuhn, Thomas. *The Essential Tension: Selected Studies in Scientific Tradition and Change.* Chicago: University of Chicago Press, 1977.
Manuel, Frank E. *The Religion of Isaac Newton.* Oxford: Clarendon, 1974.
Markley, Robert. "Isaac Newton's Theological Writings: Problems and Prospects." *Restoration* 13 (1989): 35–48.
———. "Objectivity as Ideology: Boyle, Newton, and the Languages of Science." *Genre* 16 (1983): 355–72.
Martin, Benjamin. *Philosophical Britannica: Or, a New and Comprehensive System of the Newtonian Philosophy.* 2d ed. London, 1759.
Mayr, Otto. *Authority, Liberty and Automatic Machinery in Early Modern Europe.* Baltimore: Johns Hopkins University Press, 1986.
McLaurin, Colin. *An Account of Sir Isaac Newton's Philosophical Discoveries.* London, 1748.
McMullin, Ernan. "The Significance of the *Principia* for Empiricism." *Religion, Science, and Worldview: Essays Presented to Ricard S. Westfall.* Eds. Margaret J. Osler and Paul Farber. Cambridge: Cambridge University Press, 1985. 33–59.
Newton, Isaac. *Observations upon the Prophecies of Daniel and the Apocalypse of St. John.* London, 1733.
———. *Principia.* 2 vols. Ed. Florian Cajori. Berkeley and Los Angeles: University of California Press, 1962.
———. *The Chronology of Ancient Kingdoms Amended.* London, 1728.
———. *The Correspondence of Isaac Newton.* Ed. H. W. Turnball, et al. 7 vols. Cambridge: Cambridge University Press, 1959–77.
Pemberton, Henry. *A View of Sir Isaac Newton's Philosophy.* London, 1728.
Ray, John. *Three Physico-Theological Discourses.* London, 1693.
Reedy, Gerard. *The Bible and Reason: Anglicans and Scripture in Late Seventeenth-Century England.* Philadelphia: University of Pennsylvania Press, 1985.
Reiss, Timothy J. *The Discourse of Modernism.* Ithaca: Cornell University Press, 1982.
Rorty, Richard. "Texts and Lumps." *New Literary History* 17 (1985): 1–16.

Rossi. *The Dark Abyss of Time: The History of the Earth and the History of Nations from Hooke to Vico*. Trans. Lydia G. Cochrane. Chicago: University of Chicago Press, 1984.

Rousseau, G. S. "Mysticism and Millenarianism: The 'Immortal Dr. Cheyne'." *Millenarianism and Messianism in Enlightenment Culture*. Ed. Richard Popkin. Leiden: Brill, 1988.

Serres, Michel. *Hermes: Literature, Science, Philosophy*. Ed. Josué V. Harari and David F. Bell. Baltimore: Johns Hopkins University Press, 1982.

Slaughter, M. M. *Universal Languages and Scientific Taxonomy in the Seventeenth Century*. Cambridge: Cambridge University Press, 1982.

Sprat, Thomas. *History of the Royal Society*. London, 1668.

Stepan, Nancy Leys. "Race and Gender: The Role of Analogy in Science." *Isis* 77 (1986): 261–77.

Stewart, Larry. "The Selling of Newton: Science and Technology in Early Eighteenth-Century England." *Journal of British Studies* 25 (1986): 178–92.

Stolzenberg, Charles. "Can an Inquiry into the Foundations of Mathematics Tell Us Anything Interesting about Mind?" *Psychology and Biology of Language and Thought*. New York: Academic Press, 1978. 221–69.

Warren, Erasmus. *Geologia*. London, 1690.

Westfall, Richard S. "Isaac Newton's *Theologiae Gentilis Origines Philosophicae*." *The Secular Mind: Transformations of Faith in Modern Europe*. Ed. W. Warren Wagar. New York: Holmes & Meier, 1984. 15–34.

———. *Never at Rest: A Biography of Isaac Newton*. Cambridge: Cambridge University Press, 1980.

———. "Newton and Order." *The Concept of Order*. Ed. Paul G. Kuntz. Seattle: Univ. of Washington Press, 1968. 77–88.

Whiston, William. *A New Theory of the Earth*. London, 1696.

———. *A Vindication of the New Theory of the Earth*. London, 1698.

———. *Sir Isaac Newton's Mathematical Philosophy More Easily Demonstrated*. London, 1716.

Whiteside, D. T. "Newton the Mathematician." In Bechler, 1982. 109–27.

Wilkins, John. *An Essay Towards a Real Character and a Philosophical Language*. London, 1668.

Zee, A. *Fearful Symmetry: The Search for Beauty in Modern Physics*. New York: Macmillan, 1968.

7

The Authorization of Form: Ruskin and the Science of Chaos

Sheila Emerson

1

Ruskin's relentless discriminations between order and disorder seem to leave no intervening space for what is now named the science of chaos. Yet he would not have been the least bit surprised to learn that in 1984 one of the world's leading physicists would be reported to

> have begun going to museums, to look at how artists handle complicated subjects, especially subjects with interesting texture, like Turner's water, painted with small swirls atop large swirls, and then even smaller swirls atop those. "It's abundantly obvious that one doesn't know the world around us in detail," he says. "What artists have accomplished is realizing that there's only a small amount of stuff that's important, and then seeing what it was. So they can do some of my research for me." (Gleick, "Solving" 72)

In fact Mitchell Feigenbaum's method, as described here and in Gleick's *Chaos: Making a New Science* (1987), is anticipated by Ruskin in the first volume of *Modern Painters* (1843). Defending Turner's drawings against charges that they are chaotic, meaningless, Ruskin discovers in them an order so scientifically accurate that they "afford" viewers

> the capability . . . of reasoning on past and future phenomena, just as if we had the actual rocks before us; for this indicates not that one truth is given, or another . . . but that the whole truth has been given, with all the relations of its parts; so that we can . . . reason upon the whole with the same certainty which we should after having climbed and hammered over the rocks bit by bit. With this drawing before him, a geologist could give a lecture upon the whole system of aqueous erosion, and speculate as safely upon the past and future states of this very spot, as if he were standing and getting wet with the spray. (3:487–88)[1]

Whether for the nineteenth-century geologist or the twentieth-century scientist of chaos, what makes "reasoning on past and future" possible is the representation of "phenomena" in a differently visible medium. The transitions from a subject in nature, to Turner's picture of it, to Ruskin's reasoning back from that account to nature and definitions of composition—these changes might with qualification be compared to those involved in computer experimentation, suggesting the complexity of the work of representation as information becomes pattern, as data become design. Of course Ruskin's modes of representation were developed to contend with problems faced on different disciplinary terms by the scientists of chaos; and a great deal might be said about the inadequacy of any model—whether on canvas or a computer screen—to the flow of a "real system." But there is an important connection to be made between Feigenbaum's dependence on images which display "the way one function could be scaled to match another" (Gleick 179),[2] and the premise in *Modern Painters* that the revelation of "the whole truth" comes only through images expressing "the relation of its parts" (3: 488).[3]

We can approach this connection by recalling that Ruskin studied science as well as art all his life, and that both he and Feigenbaum worked towards order in apparent disorder by rethinking some of the same paradigms—in Plato, in Goethe's holism, in British romantic notions of organic form. In fact some of Ruskin's formulations were probably mixed up in what Feigenbaum rethought. More to the point, like other nineteenth-century scientists and writers, Ruskin was able to move back and forth between the natural world, pictures, and writing by virtue of inherited convictions that the physical creation and art are both inherently languages, the one of God and the other of men. But it is Ruskin's brilliantly innovative substantiation of the familiar assumption that art is a language which forms the crucial link in his work between the order of nature and the order of visual or verbal designs. In *Modern Paintings* I and IV, as George Landow has remarked, Ruskin shows how painting displays "structures of relationships"—whether among forms, colors, or tones—which repeat the proportions, though not the scale or intensity, of "the visual structures of the natural world." These "proportionate relationships" are the "basic element of vocabulary" in Turner's "visual language." Art's capacity to create systems of proportionate relationships parallel to those of nature is what "allows the artist to make statements of visual fact" (Landow, "Nature's Infinite Variety" 377–78).[4]

In the long passage from *Modern Painters* I quoted above, Ruskin seems not only to read what Landow calls a "language of relationships" but to write one as well. By paragraph's end, it is difficult to tell whether the reference to "exquisite and finished marking" is Ruskin's response to the scene or to Turner's painting of it. But by that time, the language of the observer—whether he is Ruskin or a hypothetical geologist whom we might as well call Ruskin—has been segregated from that of the painter and his subject, as if the image could persist in the reader's mind without Ruskin's words.

> But neither [the geologist] nor I could tell you with what exquisite and finished marking of every fragment and particle of soil or rock, both in its own structure and the evidence it bears of these great influences, the whole of this is confirmed and carried out. (3: 488)

Ruskin's disclaimer about his own language has the effect of authenticating Turner's. The sense of a personal, or personified, competition or collaboration between the two individuals was further diminished in the fifth edition of *Modern Painters* I, when Ruskin erased the comparison that had appeared earlier in the same paragraph, in which the geology of the scene is said to be "treated with the same simplicity of light and shade, which a great portrait painter adopts in treating the features of the human face" (3: 487).

But such a face-to-face meeting is exactly what Ruskin requires of his reader in a previous chapter. In this case, though, he is not making claims for a particular artifact but establishing the authority of his own perception of scenes such as those Turner painted.

> Observe your friend's face as he is coming up to you. First it is nothing more than a white spot; now it is a face, but you cannot see the two eyes, nor the mouth, even as spots; you see a confusion of lines, a something which you know from experience to be indicative of a face, and yet you cannot tell how it is so. Now he is nearer, and you can see the spots for the eyes and mouth, but they are not blank spots neither; there is detail in them; . . . there is light and sparkle and expression in them, but nothing distinct. Now he is nearer still, and you can see that he is like your friend, but . . . there is a vagueness and indecision of line still. Now you are sure, but even yet there are a thousand things in his face . . . which you cannot see so as to know what they are. . . . And thus nature is never distinct and never vacant, she is always mysterious, but always abundant; you always see something, but you never see all. (3: 328–29)

From a private perception of the indefinite, Ruskin renders a public definition of nature. The differentiation of order from disorder is a problm posed by and in terms of human perception.

In subsequent volumes Ruskin will repeatedly return to the difficult contradictions about perception that are incipient in *Modern Painters* I: contradictions in criteria whereby distorted representation is distinguished from authoritatively imaginative accuracy, or whereby his own acts of representation are distinguished from Turner's. Some of the resolutions he works towards might be compared to a beginning made by Feigenbaum at about the same age as Ruskin was when he started *Mod-*

ern Painters. Feigenbaum's experience, like those Ruskin uses in his demonstrations, gains credence by being ordinary. But the two observers approach their destinations from opposite directions. Out on one of the long walks he took during graduate school, Feigenbaum passed a group of picnickers and began to puzzle about why their sounds and gestures suddenly seemed incomprehensible as he moved away from them. Questions about how the brain makes sense of all this apparently random noise and movement led him to ask "what sort of mathematical formalisms might correspond to human perception, particularly a perception that sifted the messy multiplicity of experience and found universal qualities" (Gleick 165–66). In the case of Ruskin's experiment, the "resultant truth" is the same whether the object under scrutiny is a simple geometric form or a landscape (3: 329); that the aspect of a person may also be a proof of nature's character suggests that Ruskin, like Feigenbaum, seeks a congruence between the workings of human faculties and the ordering of what those faculties work on.

It was while reminiscing about this early stage in his thinking that Ruskin located, in his autobiography, his own discovery of "the bond between the human mind and all visible things": "the same laws which guided the clouds, divided the light, and balanced the wave" (35: 315). He made these laws out in the face of nature's "palpitating, various infinity" (3: 332). No one has ever loathed chaotic deviation more eloquently than Ruskin, but neither has anyone celebrated more powerfully the variety of nature. And no one, not even Darwin, has so vividly evoked the labor of attending to nature's fluctuant multiplicity. In fact to Ruskin, the capacity to preserve and oversee this multiplicity is an indubitable sign of genius:

> Imagine all that any of these . . . [great inventors . . . Dante, Scott, Turner, and Tintoret] had seen or heard in the whole course of their lives, laid up accurately in their memories as in vast storehouses, extending, with the poets, even to the slightest intonations of syllables heard in the beginning of their lives, and with the painters, down to minute folds of drapery, and shapes of leaves or stones; and over all this unindexed and immeasurable mass of treasure, the imagination brooding and wandering, but dream-gifted, so as to summon at any moment exactly such groups of ideas as shall justly fit each other: this I conceive to be the real nature of the imaginative mind. . . . (6: 42)

Ruskin's insistence on the vast information necessary to the discernment of order may seem less daunting to those who depend on the memory of computers. But this very dependency makes it easier to recognize that Ruskin's great delineations are presented as pictures derived from a disorderly welter of data he has remembered, and made memoranda of, in drawings and in prose. These pictures are not mere imitations but—to use a favorite phrase for chapter titles in *Modern Painters*—"the truth of" the things he recalls. He is concerned not simply with distinguishing finite pattern from infinite variety but also with the definition of each of them in a variety of media. As in chaos theory, where the same pattern of order is discernible in the movements of diverse phenomena, so in Ruskin's renderings of miscellaneous subjects, the crucial "transition is the same in every member . . . and its importance can hardly be understood, unless we take the pains to trace it in [its] universality" (8: 91).

For Ruskin, the perception of "universality" begins with the individual. The problem for physicists—that a system of representation inevitably registers the involvement of the perceiver and may repeat his or her definition of order and disorder—is for Ruskin the advantage of art as an expressive medium. For it requires the expressiveness of the greatest artists to make others see that the present aspect of a thing, whether in nature or on canvas, expresses its own past. Ruskin's diary continually registers his grasp of the relationship between formal design and developmental history. "I was struck in looking over the Shells at [the] Brit[ish] Mus[eum] yesterday," he begins in 1848, going on to present a theory he will soon develop in *The Seven Lamps of Architecture*:

> Now I think that Form, properly so called, may be considered as a function or exponent either of Growth or of Force, inherent or impressed; and that one of the steps to admiring it or understanding it must be a comprehension of the laws of formation and of the forces to be resisted; that all forms are thus either indicative of lines of energy, or pressure, or motion, variously impressed or resisted, and are therefore exquisitely abstract and precise. . . . The same principles apply to the patterns and forms of G[ree]k vases and to mosaics and frescoes, &c. (*Diaries* 2: 370–71)

The lesson of the shells prefigures what he sees among Alpine peaks eight years later in *Modern Painters* IV:

> The hollow in the heart of the aiguille is as smooth and sweeping in curve as the cavity of a vast bivalve shell.
>
> I call these the governing or leading lines, not because they are the first which strike the eye, but because, like those of the grain of the wood in a tree-trunk, they rule the swell and fall and change of all the mass. In Nature, or in a photograph, a careless observer will by no means be struck by them, any more than he would by the curves of the trees; and an ordinary artist would draw rather the cragginess and granulation of the surfaces, just as he would rather draw the bark and moss of the trunk. Nor can any one be more steadfastly averse than I to every substitution of anatomical knowledge for outward and apparent fact; but so it is, that, as an artist increases in acuteness of perception, the facts which

become outward and apparent to him are those which bear upon the growth or make of the thing. (6: 231–32)

The leading line inscribes form in the flux of its own creation. History—whether of nations or architecture or Turner's mastery of the "science of *Aspects*" (5: 387)—is legible in an image. "You need not be in the least afraid of pushing these analogies too far," as Ruskin says elsewhere; "They cannot be pushed too far" (15: 118).

But in fact they *can* be pushed too far for comfort when it comes to Ruskin's realization that the composition not only of his subjects but also of his writing about them is a function or exponent of growth or force, inherent or impressed. Whereas in the 1830s Ruskin was delighted to take his own art as the basis for conclusions about how art takes shape, over the seventeen-year course of *Modern Painters* he struggled to represent himself as a critic rather than an artist, so that it would be his Turner and not himself who would be viewed as "the master of this science of *Aspects*" (5: 387).[5] As we observed in *Modern Painters* I, it is precisely when his own representation of Turner enables him to discern aesthetic and geological patterns that Ruskin insists that he cannot represent what he sees in words. And it is in the midst of his intensely imaginative representation of the imaginative mind that Ruskin distinguishes that mind from his own—again, precisely when he is exercising the power which he implies he lacks. As both of these instances suggest, Ruskin conceals his creativity during, and by means of, his creative renderings from one medium to another.

In the next two parts of this essay, I will be looking at the history of Ruskin's verbal and visual renderings of the physical world, particularly as it develops in *Fors Clavigera* and *Praeterita*. For this history raises telling questions about the discernment of pattern in disorder, and about the relationship of pattern to what Ruskin calls "the movements of his own mind" (5: 365). If thinking about twentieth-century science conduces to thinking about Ruskin, the reverse is also true. Ruskin forces the reader of chaos theory to pursue the connection between definitions of order and definitions of the mind's composition: both the mind's acts of composition, and the way that mind is composed.

2

In *Chaos: Making a New Science*, Gleick repeatedly dramatizes the unconventionality of his chosen characters, whom he finds incongruous or even adversarial in relation to teachers and institutions. As in other disciplines, the formation of new ideas of order is generally redolent of a child's transgression against parental authority. But what Gleick discloses is not finally a series of transgressively isolated accomplishments. A passage from perceptions of disorder to perceptions of order is in a sense personified in the book, as disjunct individuals sometimes form themselves into collectives, and always produce work that eventually coheres in an implication of order. Comparison to Ruskin suggests why it is a "technique" that becomes the key in this transformation from disorder to order (Gleick, 268), and also why the means of representation are generally at the center of a contest between old and new definitions of order.

The fullest account of Ruskin's pictorializing—of his seeing the act of composition in imagery drawn from physical objects, and his finding in such imagery a warrant for personal authority—emerges over the long course of *Fors Clavigera*, which is surely one of the most monumental assertions in English of both coherence and incoherence. In these ninety-six "Letters" addressed between 1871 and 1884 "To the Workmen and Labourers of Great Britain," Ruskin originates a comprehensive design for living in St. George's Company and reluctantly assumes the role of "Master." In fact he writes with such breathtaking mastery of his incapacity to master, that what seems "thrust and compelled" on him—"utterly against my will, utterly to my distress, utterly, in many things, to my shame"—is not authority over the tiny Company but rather his enormous power and ambition as a writer (28: 425). The three massive volumes set themselves against what Ruskin regards as the world's horrifying disorder; and they do it in an "irregular," "desultory," "fragmentary" manner which he repeatedly deplores. Yet this very way of writing—this means of representation—comes to work as a justification both of Ruskin's attack on society and of his designs for its reorganization. So I will concentrate on Ruskin's explanation of how he reads and writes, as that becomes a crucial basis of the authority he claims in *Fors*.

Most people remember that he calls his bible reading with his mother "the one essential part . . . of all my education" (28: 102), but they forget his exclusion of any record of her teaching him to write—although she was his only teacher during those early years when he was regularly sending his prodigious poems and letters to the places where his father was traveling on business. What the history of his reading with his mother does include is a telling disagreement about his memorizing the visible pattern of words on a page.

[T]he mode of my introduction to literature appears to me questionable, and I am not prepared to carry it out in . . . [the schools I propose] without much modification. I absolutely declined to learn to read by syllables; but would get

> an entire sentence by heart with great facility, and point with accuracy to every word in the page as I repeated it. As, however, when the words were once displaced, I had no more to say, my mother gave up, for the time, the endeavour to teach me to read, hoping only that I might consent, in process of years, to adopt the popular system of syllabic study. But I went on, to amuse myself, in my own way, learnt whole words at a time, as I did patterns; and at five years old was sending for my "second volumes" to the circulating library.
>
> This effort to learn the words in their collective aspect, was assisted by my real admiration of the look of printed type, which I began to copy for my pleasure, as other children draw dogs and horses. (28: 274)

The word "aspect" draws attention to his responding to words less as representations of sound or signifiers of ideas, than as visible objects, pictures. Like the reading which accumulates what he calls its "flow" by following the "flow" of what is read, writing assembles its own "aspect" in dutifully retracing the aspect of what was previously written. The self-questioning with which he began the passage has come to coexist with "resolute self-complacency": Ruskin promises that he will "have much to say on some other occasion" about "the advantage, in many respects, of learning to write and read . . . in the above pictorial manner" (28: 275). That writing, which he says he taught himself, precedes reading in this sentence anticipates Ruskin's explicit conviction that children learn best what they teach themselves. Nine years later, he argues that

> nothing could be more conducive to the progress of general scholarship and taste than that the first natural instincts of clever children for the imitation, or often, the invention of picture writing, should be guided and stimulated by perfect models in their own kind. (29: 507)

Here, only paragraphs before he announces that he is closing *Fors* in order to write an autobiography proving that "I had not the slightest power of invention" (35: 608), Ruskin is admiring the "invention" of a manner of writing which he brilliantly mastered in his childhood, and which he considers worthy of use as a model in the education of thousands of other children. When he wants the reader, no less than the children, to understand what he means by "symmetry," "grace," "harmony," he points to picture writing to illustrate that capacity for "composition" which he says he entirely lacked. The most notoriously dutiful of all Victorian sons, Ruskin so describes his being out of order that it issues in a vision of that obedience to higher laws which is the defining characteristic of artists like Turner and Walter Scott. His insistence on learning words "in their collective aspect" evolves into an education for the reader of *Fors*, an education which resolves his double sense of transgression and obedience into a program which implies that his readers must obey their teacher, because he did not.

The importance of his having learned in a "pictorial" manner may begin as a rationale for accounts of his childhood lessons, but it ends as the justification for the lessons in writing, and then reading, which he proceeds to give his reader. Ruskin offers as a model a facsimile of a Greek sentence that begins with an illuminated letter "A." While copying this instance of "pure writing, not painting or drawing," the reader is instructed that "the best writing for practical purposes is that which most resembles print, connected only, for speed, by the current line" (28: 495), and that beautiful writing can be produced only by the hand that is "in the true and virtuous sense, *free*; that is to say, able to move in any direction it is ordered" (28: 494). Writing is "ordered" both by the aspect of print and by the impulse of the person who writes. Ruskin's moralization throughout *Fors* of the implications of handwriting recalls his argument in *The Stones of Venice* and elsewhere that all handwork expresses the moral state of those who produce it. Ruskin's analysis of handwriting is the most fundamental deconstruction of this idea in any of his books. For writing is the work of Ruskin's own hand, and the exfoliation of his self-knowledge substantiates the view that handwriting does not simply register the moral condition of the subject and the self; it also has the "aspect" of them.

The bond between the composition of one's words and one's subject is made graphic in Ruskin's next lesson on the writing of the letter "B." Noting that the model "A" instances no "spring or evidence of nervous force in the hand," Ruskin promises (but never delivers) a "B" from the Northern Schools that has so much "spring and power" that the reader cannot hope to imitate it all at once, but must be prepared "by copying a mere incipient fragment or flourish" (28: 524). What Ruskin does present is the outline of a shell—insisting that "This line has been drawn for you" not by himself but with "wholly consistent energy" by a snail. The "free hand" required to draw this line will thus be retracing not only a portion of a Gothic letter "B" but also, simultaneously, a picture of a living thing: in Ruskin's case, the hand has traced the living thing itself. Or rather, no longer a living thing but the visible creation and record of one—for the line incarnates and memorializes the "strong procession and growth" of the animal. In the next four Letters of *Fors*, Ruskin's fascination with the snail's record of growth—with the implications of the form of its shell—returns amidst discussions of many other historical subjects, including himself. So that the most interesting implication of the fact that Greek writers illuminated "the letter into the picture" after the Egyptians had lost "the picture in the letter" (28: 568), is that the public history of writing is recapitulated in Ruskin's own. His guiding his student's hand as it follows the form, or picture, of its subject is the most basic, and ultimate, demonstration of an identity between his "winding way" of composition and the composition of his favorite subjects. This identity was formed when Ruskin first fused letters and pictures, words and things, in his childhood. Maybe not while he was making the flourishes in his poems and letters to his father, but surely on the basis of them, Ruskin was helped "to understand that the word 'flourish' itself, as applied to writing, means the springing of its lines into floral exuberance,—therefore, strong procession and growth, which must be in a spiral line, for the stems of plants are always spirals" (28: 525).

In fact the leading line in Ruskin's earliest writing is arguably the serpentine or "winding way"—which he celebrates not only as an enactment of a subject that is not himself but also as a sign of his own invention (*Ruskin Family Letters* I: 193).[6] The recurrence of the serpentine in Ruskin's work calls to mind the prominence of curves within recent chaos studies, of spiraling or scrolling forms like sea shells or waves. Ruskin's avowedly ethical preference for such organic forms anticipates the preference of scientists attracted to fractal geometry. Their assessments of the bond between nature and architecture, their valorization of forms that "resonate with the way nature organizes itself or with the way human perception sees the world" (Gleick 116–17), are already resonant with values given currency in *The Seven Lamps of Architecture* (1849) and *The Stones of Venice* (1851–53).

In one of his latest works, a lecture on the motion of snakes called "Living Waves" (1880), Ruskin begins by making a connection between "undulatory" geological movement and his own "curiously serpentine mode of advance towards the fulfillment of my promise"—between his proposed revisions of his text and "one colubrine chain of consistent strength" (26: 295–96). But the energy of this beginning winds up in a venomous passage about paternal ambition, which breaks down the will and breaks up the designs of the sons of England: "fathers love the lads all the time, but yet, in every word they speak to them, prick the poison of the asp into their young blood, and sicken their eyes with blindness to all the true joys, the true aims, and the true praises of science and literature" (26: 329). As this language suggests, the moralization of form allows sinister as well as righteous plots to bear on, or emerge from, representative designs. Thus it is that the shape of Ruskin's writing expresses not only the beauty of organic form and the obedience that issues from perfect freedom, but also the dangers of self-assertion. Like the quintessential Gothic in *The Stones of Venice*—"subtle and flexible like a fiery serpent, but ever attentive to the voice of the charmer" (10: 212)—even the very best handwriting records a sinister association. For the serpent not only has but is a tongue, an "inner language" Ruskin deciphers in *The Queen of the Air* (1869).

> In the Psalter of S. Louis itself, half of its letters are twisted snakes; there is scarcely a wreathed ornament, employed in Christian dress, or architecture, which cannot be traced back to the serpent's coil. (19: 361, 365).

Ruskin catches in this coil the involvement of order with disorder. The continuity between serpents and the devil is spelled out in the disorder of language; so that language is powerless to dispel it. In *The Queen of the Air*, words designed to demystify the creature assemble a mysterious new creation instead: representation meant to redeem the art Ruskin loves is doomed to display its own damnation.

It is because of this taint that he seeks to distinguish different kinds of handwriting in *Fors*. Displayed by Ruskin's editors in a two-page facsimile, the manuscript of Scott's *The Fortunes of Nigel* is said to illustrate "the same heavenly involuntariness in which a bird builds her nest," or in which "the great classic masters" of art produce an "enchanted Design" (29: 265). The instance of "incurably desultory" handwriting which Ruskin facsimiles is, not surprisingly, his own—"distinct evidence . . . of the . . . character which has brought on me the curse of Reuben, 'Unstable as water, thou shalt not excel' " (28: 275). Once again, Ruskin's authority as a teacher develops from his having been a disobedient student. As in the lecture on "Living Waves," it is his own superior capacity to decipher, to follow, and to reproduce the true form of geological forces, of snakes' movements, and of sons' lives that underwrites his assault on fathers, while it is the congruence of these forms with each other which redeems the apparent disjunction of his writing into an approximation of organic design. But in these later works, nature's order is already poisoned by man's disorder, the son's authority by the father's authoritarianism. For all his buttressing of the patriarchal order of things, the aspects of "brightly serpentine perfection" can only lead Ruskin to an act of filial disobedience (26: 328). And filial disobedience of an explicitly phallic kind is just what led Reuben's father to deliver upon his son the biblical curse Ruskin claims as his own blight in *Fors*: " 'Unstable as water, thou shalt not excel' " (28: 275).

Yet the sign of the serpent—"A wave, but without wind! a current, but with no fall" (19. 362)—remains a reminder of the "advantage" of the "pictorial manner." For if writing reflects the "flow" of one's own mental movements, it has a more than personal aspect as it cleaves to the flow of its subject (29: 539–40). Nine months after he writes about the instability of his writing and his character, Ruskin says that he composes *Fors*

> by letting myself follow any thread of thought or point of inquiry that chances to occur first, and writing as the thoughts come,—whatever their disorder; all their connection and cooperation being dependent on the real harmony of my purpose, and the consistency of the ascertainable facts, which are the only ones I teach. . . . (28: 461)

Ruskin's title emblematizes the nexus of values implied or asserted in his writing about writing. Among its other meanings, "Fors" is both the fate that acts upon him from without, and the forces within that work themselves out in his life. In every moment, *Fors* bears the form of those forces in action. It is not simply that Fors will lead to form, but that it already, inherently, has a form, no matter how formless it may seem. The Fors, or force, of the writing reveals a pattern of order in the welter of data, for the interval between flow and form is only the time within which growth and force do their work, the time it takes for them to show themselves as *Fors*—which is the design of a person's life, the picture of a self and writing as "Unstable as water."

What may seem like the book's waywardness with regard to one set of forces turns out to be its lawfulness with regard to another. This is ultimately because Ruskin derives his authority from his denial of having any authority over himself—from his having been absolutely obedient not to his mother or father, but to the forces he could not resist as a boy, including the force exerted by the look of print on a page. Here more than ever the vocabulary of chaos studies seems apt. Like natural objects constrained to take certain shapes, like chaotic activities constrained to repeat certain patterns, Ruskin's writing is, in his account, constrained to assume the forms that it does in *Fors*. The result is a book at once private and public, intimate and estranged. *Fors* is Ruskin's own idiosyncratic signature; and it is a design by whose emergence he was as much astonished as were those twentieth-century scientists, when they first beheld a "strange attractor" taking shape on the screen before them.

3

Ruskin's discovery of a pattern at once within and without receives its last display in *Praeterita* (1885–89), the autobiography which began in passages transferred from *Fors*. Transferred and sometimes transposed, for *Praeterita* "has taken, as I wrote, the nobler aspect of a dutiful offering at the grave of parents" (35:12).[7] It is this changing "aspect" that I want to consider in closing, for it bears the weight of a lifetime's representations of order and disorder.

Coming at the center of the book, the chapter called "Fontainebleau" is centrally about Ruskin's discovery of the laws of composition during the summer before he began *Modern Painters*. The diverse materials assembled in the chapter have elsewhere passed through a number of versions whose evolution constitutes an instance of those laws. The whole story is too long to lay out here. But it should be borne in mind that Ruskin's revisions tend to create a more "dutiful" context for two assertions in particular: that in his sketches of 1842, "Nature herself was composing with" Turner; and that in his own attempts, Ruskin proved, once again, that "I can no more write a story than compose a picture" (35:310, 304). The connection latent between these two discoveries first emerges just after he recalls his recognition that Turner's "sketches were straight impressions from nature,—not artificial designs" (35:310). While thinking this over on a road near his house, Ruskin has a foretaste of what he will describe as a momentous experience.

> . . . "one day on the road to Norwood, I noticed a bit of ivy round a thorn stem, which seemed, even to my critical judgment, not ill "composed"; and proceeded to make a light and shade pencil study of it in my grey paper pocket-book, carefully, as if it had been a bit of sculpture, liking it more and more as I drew. When it was done, I saw that I had virtually lost all my time since I was twelve years old, because no one had ever told me to draw what was really there!" (35: 311)

Ruskin's self-reproach that he "was neither so crushed nor so elated by the discovery as I ought to have been" deflects attention from the way his discovery elevates him above his teachers (who forced him to regard nature in terms of previous paintings and principles), and above his father (whose fault in not buying a supreme Turner landscape is transferred to his son in the *Praeterita* version of the story).

The full force of what has happened is postponed until a few paragraphs (and a month or so) later, when Ruskin finds himself "in an extremely languid and woe-begone condition" in a cart-road at Fontainebleau. He begins only to interrupt himself with miscellaneous quotations and reminiscences of other journeys. But the pieces of the account cohere all the more dramatically for the detours.

> . . . getting into a cart-road among some young trees, where there was nothing to see but the blue sky through thin branches, [I] lay down on the bank by the

roadside to see if I could sleep. But I couldn't, and the branches against the blue sky began to interest me, motionless as the branches of a tree of Jesse on a painted window.

Feeling gradually somewhat livelier . . . I took out my book, and began to draw a little aspen tree, on the other side of the cart-road, carefully. . . .

Languidly, but not idly, I began to draw it; and as I drew, the languor passed away: the beautiful lines insisted on being traced,—without weariness. More and more beautiful they became, as each rose out of the rest, and took its place in the air. With wonder increasing every instant, I saw that they "composed" themselves, by finer laws than any known of men. At last the tree was there, and everything that I had thought before about trees, nowhere. . . . The woods, which I had only looked at as wilderness, fulfilled I then saw, in their beauty, the same laws which guided the clouds, divided the light, and balanced the wave. "He hath made everything beautiful, in his time," became for me thenceforward the interpretation of the bond between the human mind and all visible things . . . . (35:313–15)

Ruskin's critics have made a great deal of the differences between this account and what does or does not appear in his diary of the summer of 1842. But what is more interesting for the purposes of this essay is the fact that Ruskin derives the law he declares from the movements of his own hand. It is entirely characteristic of his prose that in the course of putting his own compositional capacities "nowhere," Ruskin composes one of the most powerful narratives in all his work—a narrative that shows how one can learn what one most inalienably and importantly understands from the process of composition itself. Much as Feigenbaum sees that artists like Turner "can do some of my research for me," Ruskin sees that a person who has devoted years to drawing Turners, and Turner's subjects, can do some of his work for himself.

But Ruskin does not acknowledge that the image of the tree is his own, that it records the movements of his own mind.[8] The translation from tree to image encodes Ruskin's language of relationships—much as Feigenbaum's "inspiration came . . . in the form of a picture," a "wavy image" in the mind (Gleick 175, 179). Ruskin's picture of a tree is in fact a picture of the laws of composition, a picture of relations between mind and its objects. The picture is a statement about, not a mere imitation of, two things at once: about how a tree is composed, and about how an artist composes. Both statements express order as a visual record of movement, change. This order is an abstraction from the growth of his own compositions, verbal or pictorial, through time—it is the comprehensive grasp of form as an inscription of flux.

The drawing of the tree is not there in *Praeterita*. Instead there is an account of a tree that completes the genealogy begun when Ruskin likened the seventeen-year development of *Modern Painters* to the changes "of a tree—not of a cloud" (7:9), and continued in the image of a wind- and flood-swept "birch-tree," that has no more control over the arrangement of its boughs than Ruskin does over the shape of *Fors* (28:254). The account of the aspen in *Praeterita* becomes the evidence that "nature" has broken through the "heavy," "languid" dullness induced explicitly by Ruskin's schooling, and implicitly by parenting that was at once "too formal and too luxurious" (35: 46). Yet for all his exaltation of the natural, Ruskin is interested in the aspen (as he was in the ivy he drew "as if it had been . . . sculpture") because it is composed as an artifact is composed, because it looks like something "painted." Where there is no Turner to bear responsibility, Ruskin comes to n[illegible] as if it already looked like an artifact.

Against the vision of his elders, Ruskin sets not [illegible] but a representation of it—a representation which involves him [illegible]ompetition that could bear as bitter a fruit as the tree forbidden to [illegible] When he wrote about fathers in "Living Waves," it was in sorro[illegible] in fear of those who "prick the poison of the asp" into their sons, [illegible]sicken their eyes with blindness to all the true joys, the true aims, [illegible]he true praises of science and literature" (26:329). But he senses [illegible] for the son who gains his sight in *Praeterita*. This is one of the m[illegible]essing reasons for his affirmations of order in apparent disorderlin[illegible]r his reliance on metaphors of organic form, for his recurrent ass[illegible]s that "I was without power of design" (35: 120). But none of thi[illegible]hough. By the end of the chapter, father, mother, and son are all re[illegible] from what Ruskin regards as the "Eden" of their home—the hom[illegible] shared before the achievement of *Modern Painters*, the home to [illegible] he returned alone to write the preface of *Praeterita*. Not only par[illegible]but also filial ambition is punished in this removal; for "in this ho[illegible]ll our weaknesses," Ruskin succumbed to his "temptation" to occup[illegible]ene where he might make an artifice. His original idea of building a canal was only realized twenty years later in "water-works, on the model of Fontainebleau"—the place where he both experienced and denied the power of his invention (35: 317–18).[9] It is the punishable implication of invention, of a competition with the father, that Ruskin seeks to displace by introducing the tree of Jesse—another inscription of genealogy which composes moving development into "motionless" pattern.[10] The asp may lurk in Ruskin's words about the aspen that opened his eyes, but the root of Jesse is there to identify his composition, and his ideas of composition, with God's design for his own true son.

In disguising the role of the will, the chapter called "Fontainebleau" ironically demonstrates the artist's will to compose himself, to grant himself composure. It takes art to redeem the impulse to make art. At the very moment when a pattern materializes out of Ruskin's own mental and physical movements, he devotes all his force to showing that the pattern certifies a force beyond his own. If there is an "attractor" shaping the design of his "dutiful offering," then it must be God, not the chance of *Fors Clavigera*. And the only thing he finds "strange" in "Fontainebleau" is that he had never before seen the divine design according to which the order of the human mind is fitted to the order of all visible things.

It required years of revisions to work out the will that authorizes form in *Praeterita*. But the self behind the moment of inspiration is not lost as a tree becomes a drawing, and a drawing becomes print on the page. The process by which this is disclosed may seem to be light-years away from the experiments of contemporary physics. But there is a connection between Ruskin's way of thinking about representation and the comments made by the men who made the science of chaos:

> It's an experience like no other experience I can describe, the best thing that can happen to a scientist, realizing that something that's happened in his or her mind exactly corresponds to something that happens in nature. It's startling every time it occurs. One is surprised that a construct of one's own mind can actually be realized in the honest-to-goodness world out there. A great shock, and a great, great joy. (Leo Kadanoff in Gleick 189)

The "shock" and the "joy" are different for Kadanoff and for Ruskin, no doubt. But both of them achieve the exciting correspondence by translating mental "construct[s]" and the "world out there" into other media. Just how "honest-to-goodness," just how far outside the mind, are the operations that produce this correspondence? Like the work of Ruskin, the work of twentieth-century scientists suggests that a shift in the definition of order may be based on an interaction between the thoughts of an individual and patterns that are rendered for others to see. The science of chaos and Ruskin's science of aspects both investigate and finally deny the barrier between private mental movements and the design of the world that is not oneself.

Notes

1. Unless otherwise indicated, all parenthetical references are to *The Complete Works of John Ruskin*, and will be given by volume and page number.
2. Unless otherwise indicated, all parenthetical references to Gleick are taken from *Chaos: Making a New Science*.
3. Cf. Rosenberg's 1969 comparison of Ruskin to contemporary scientists

(44–45), and Rudwick's 1985 observation that "Early nineteenth-century geology may appear at first sight to be poles apart from, say, modern elementary particle physics. . . . Yet it is axiomatic that any one branch of science can be linked to any other, however different, by an indefinitely graded chain of intermediates; an analogous continuity links one historical period to another" (16).

4. The acuteness of Landow's analysis comes of his being a critic both of literature and of visual art. Especially given the connections I am trying to make in this essay, it is suggestive that Landow has since worked with a team of developers at Brown University to create Intermedia, an elaborate hypermedia environment which allows for multiple visual schematizations of literary history—a project that includes, and seems partly inspired by, Ruskin. See Landow, "Hypertext in Literary Education, Criticism, and Scholarship."

5. See especially Helsinger's *Ruskin and the Art of the Beholder.*

6. See my "Ruskin's First Writings: The Genesis of Invention," in Bloom, ed., *Modern Critical Views: John Ruskin* (151–76).

7. See Peterson on Ruskin's struggle in *Praeterita* with the evangelical traditions he inherited (60–90), and Henderson on the book's strategically misleading account of his "vocation as an artist" (65–115).

8. Cf. Levine, who locates in the autobiographies of Mill, Darwin, and Trollope patterns of self-deprecation and "casual and relatively unrhetorical styles" which he associates with the contemporary ascendancy of empirical science. Compare Rudwick: "Charles Darwin's well-known statement that as a young man he had collected facts and worked 'without any theory', referring to a period when he had actually compiled some of the most creatively theoretical notebooks known in the history of science, is but one outstanding example of a pervasive problem" (7).

9. See Landow's lucid account of Ruskin's subordination of art to nature in "Fontainebleau" (*Ruskin* 79–82), and Sawyer's convincing argument that these "water-works" are also redemptive (322).

10. For an approach different from the one I have taken to imagery that combines motion and motionlessness, see Fontaney. And see Helsinger on the "spiral movement" of *Praeterita* ("Ruskin and the Poets" 163).

Works Cited

Harold Bloom, ed. *Modern Critical Views: John Ruskin.* New York: Chelsea House, 1986. 151–76.

Fontaney, Pierre. "Ruskin and Paradise Regained." *Victorian Studies* 12 (1969): 347–56.

Gleick, James. *Chaos: Making a New Science.* New York: Viking Penguin, 1987.

———. "Solving the Mathematical Riddle of Chaos." *The New York Times Magazine,* 10 June 1984:31+.

Helsinger, Elizabeth K. *Ruskin and the Art of the Beholder.* Cambridge: Harvard University Press, 1982.

———. "Ruskin and the Poets: Alterations in Autobiography." *Modern Philology* 74 (1976):142–70.

Henderson, Heather. *The Victorian Self: Autobiography and Biblical Narrative.* Ithaca: Cornell University Press, 1989.

Landow, George P. "Hypertext in Literary Education, Criticism, and Scholarship." *Computers and the Humanities* 23 (1989): 173–98.

———. "J. D. Harding and John Ruskin on Nature's Infinite Variety." *The Journal of Aesthetics and Art Criticism* 28 (1970):369–80.

———. *Ruskin.* Oxford: Oxford University Press, 1985.

Levine, George. "Science and Victorian Autobiography: The Arrogance of Humility," MLA Convention, San Francisco, 29 December 1987.

Peterson, Linda H. *Victorian Autobiography: The Tradition of Self-Interpretation.* New Haven: Yale University Press, 1986.

Rosenberg, John D. "The Geopoetry of John Ruskin." *Etudes Anglaises,* January-March 1969, 42–48.

Rudwick, Martin J. S. *The Great Devonian Controversy: The Shaping of Scientific Knowledge among Gentlemanly Specialists.* Chicago: University of Chicago Press, 1985.

Ruskin, John. *The Complete Works of John Ruskin.* Ed. E. T. Cook and Alexander Wedderburn. 39 vols. London: George Allen, 1903–12.

———. *The Diaries of John Ruskin.* Ed. Joan Evans and John Howard Whitehouse. 3 vols. Oxford: Clarendon Press, 1958.

———. *The Ruskin Family Letters: The Correspondence of John James Ruskin, His Wife, and Their Son, John, 1801–1843.* Ed. Van Akin Burd. 2 vols. Ithaca: Cornell University Press, 1973.

Sawyer, Paul L. *Ruskin's Poetic Argument: The Design of the Major Works.* Ithaca: Cornell University Press, 1985.

8

Linear Stories and Circular Visions: The Decline of the Victorian Serial

Linda K. Hughes and Michael Lund

Progress has been much more general than retrogression.

Charles Darwin, *Descent of Man* (1871)

We have no right to assume that any physical laws exist, or if they have existed up to now, that they will continue to exist in a similar manner in the future.

Max Planck, *The Universe in the Light of Modern Physics* (1931)

As many of the essays in this volume suggest, classic notions of stability which permit the inference of a whole from a part are increasingly problematical. It is dangerous work to infer the doctrine of an individual from an excerpt (as Markley demonstrates in the distortions wrought by Newton's followers), much more so the tenor of an entire age from an individual exponent or ideology. Still, while Darwin's evolutionary theory and Planck's quantum theory are complex, highly detailed concepts that elude rendering into a simpler, uniform substance, the quotations which head this essay can at least point to important reference points in the two scientists' respective eras. Biology might be said to be the premier science of the nineteenth century (Jordanova 28; Eichner 10; Shuttleworth 2–4), physics of the twentieth century. Darwin rejected simple notions of development or linear progression, but his theory could easily be absorbed into such a simple framework by others (Beer 5; Cosslett 11); the work of twentieth-century physics actively resists such a framework and calls into question not only linearity but also simplified notions of causality.

Another possible reference point (among many) to distinguish the nineteenth and twentieth centuries lies in a literary model. A dominant publishing format in the Victorian era was the serial, a work appearing in parts over an extended period of time. Louis James in fact suggests that, as much as it was the age of the novel, the Victorian era was the age of the periodical (352), in which so much installment literature appeared. During the Victorian era serialization was such a pervasive practice that major texts of poetry and nonfiction prose as well as fiction first appeared in installments.[1] Dickens' novels, George Eliot's *Middlemarch* and *Daniel Deronda*, Robert Browning's *The Ring and the Book*, William Morris's *Pilgrims of Hope* and *News from Nowhere*, and most of Ruskin's works, including *Fors Clavigera*, all were first published serially. Even less expected titles such as Flaubert's *Madame Bovary*, Dostoevsky's *The Brothers Karamazov*, and James's *Portrait of a Lady* were first known to their nineteenth-century audiences as installment publications. But while serialization continued as a medium for popular literature (e.g., detective stories) in the twentieth century—even though Pound's first *Cantos* were published in three monthly parts and Joyce's *Ulysses* was serialized for more than two years before obscenity charges halted this process—the serial form lost its place as a primary medium for "high" literature in the twentieth century. Instead, the modernist tradition privileged the work grasped all at once as an autonomous, seamless aesthetic whole, and for this reception the work published as a whole volume was essential.[2] We argue that the displacing of biology by physics as dominant science, and of serialization by whole-volume publishing as premier publication format, are related events underwritten by shifting notions of order.

Serialization, though first introduced as an economic incentive for purchase and tied to increasing literacy rates (see, e.g., Altick 259–80; Brooks 147; Feltes), shares fundamental assumptions of nineteenth-century life sciences and historicism. All three conceptual frameworks endorse developmental approaches to life patterns and their significance. For example, nineteenth-century embryological theory (first articulated by C. F. Wolff in 1759) displaced older notions of the fetus as a miniature human being and insisted instead on viewing the origin of human life as a linear series of developmental phases leading inevitably from simpler to more complex organization of life (Chapple 2; Shuttleworth 11–12). A serial literary work also grew from simpler to more complex order, from a single initial fragment to an accreting and diversifying collocation of characters and plot lines, and likewise grew in a strict chronological sequence (over a span often lasting longer than fetal development) determined by a publishing schedule.

True, developmental concepts did not exclude retrogression or reversals, so that their paths of progression were not simple straight lines. As Darwin commented, natural selection "in some cases will even degrade or simplify the organisation, yet leaving such degraded beings better fitted for their new walks of life" (330). Serial plots, as we have discussed elsewhere, often took two steps back for each step forward in the story of a character's fortunes or emotional experiences (Hughes and Lund, chap. 5). But the issuing of installments per se insured an inherently forward thrust in any serial work, just as evolutionary patterns, viewed globally rather than locally, suggested forward movement. As Darwin asserted, "Although we have no good evidence of the existence in organic beings of an innate tendency towards progressive development, yet this necessarily follows, as I have attempted to show in the fourth chapter, through the continued action of natural selection" (200; see his similar remarks on embryonic development, 410–11). A serial's forward thrust was also reinforced by the linear narrative technique sanctioned by literary realism. As Peter Brooks remarks, "the time it takes, to get from the beginning to the end—particularly in . . . nineteenth-century novels—is very much part of our sense of the narrative. . . . if we think of the effects of serialization . . . we can perhaps grasp more nearly how time in the representing is felt to be a necessary analogue of time represented" (20–21).

Again, just as biblical accounts of creation as an instantaneous and full-blown event were challenged by the Darwinian theory of evolution, which viewed creation as a slow unfolding of life forms over vast amounts of time with pauses (or at least gaps in the fossil record) between developments, so serialization fostered an approach to narrative as a gradually developing story and pattern of significance, with pauses between parts for additional reflection and speculation, rather than as a finished aesthetic product to be read and considered as a whole all at once. Victorian historicism, like serial emplotment, emphasized nonreversible sequences of events essential to cultural development, and history was viewed as an analogue to the developmental process of nature. As a result of the work by Darwin, the geologist Charles Lyell, and others, nature itself was historicized (Jordanova 28; Toulmin and Goodfield 232), while history consolidated its prestige by linking its work to that of science, since both examined the processes of change and development (Jann xi-xii). We should not forget that serialized works like *David Copperfield* carried the term "History" in their titles,[3] a convention inherited from the eighteenth century but which acquired greater force because serialization imparted literal history to fiction during the many months of serial publication. As Jordanova observes of the era, "Analogies abounded: individual life histories, the history of the earth, of nations, tribes, races and empires, cities and familiies, the development of a foetus, of languages themselves" (28–29)—and, we would add, of serialized literary works.

This developmental notion of order, however, was countered in the early twentieth century by what N. Katherine Hayles (15–59) has termed "the cosmic web," an idea of order as emergent interrelations rather than predictable sequence, and embraced under the general term of field theory. Work on scientific field models received special impetus from Einstein's 1915 papers on the general theory of relativity and continued into the early decades of the twentieth century with the solidification of quantum mechanics around 1930 (43–45). The distinguishing traits of a field view of reality, according to Hayles, "are its fluid, dynamic nature, the inclusion of the observer, the absence of detachable parts, and the mutuality of component interactions" (15). Within this view linearity is not necessarily banished but is complicated by and certainly subordinated to notions of simultaneity, since according to the field concept everything (including the observer) is connected to everything else and parts cannot be teased out for classic linear analysis—or for the kinds of plot lines associated with Victorian serial narratives.

In this chapter we are especially interested in literary works published in the late nineteenth and early twentieth centuries which reflect the convergence of the two different ordering systems identified above. At this time works were still being published serially, and so their installment issue of necessity enforced the importance of linear progression; yet the content and imagery of works by several major authors gestured toward the newer paradigm embodied in field theory. Three such works, Thomas Hardy's *The Woodlanders* (1886–87), Joseph Conrad's *Lord Jim: A Sketch* (1899–1900), and Hardy's *The Dynasts* (1904–8), illustrate the coexistence of these divergent conceptual frameworks and suggest not only that the decline of the serial for "high" literature was partly prompted by a shifting paradigm[4] but also that what may at first appear as "chaotic," disordered, or contradictory literary techniques are the result of competing ordering strategies.

Hardy's *The Woodlanders*, for instance, appeared in the monthly *Macmillan's Magazine* in England from May 1886 to April 1887 and was first received within expectations established by the tradition of Victorian serial fiction, that is, with the assumption that its parts would move the story forward in a customary linear fashion. The novel also appeared as a complete work in one volume on March 15, 1887; but that later single-volume format emphasized Modernist assumptions quite different from those defined by the serial tradition. Concern with aesthetic design was perhaps most frequently applied to a novel's conclusion, as Modernists insisted on a resolution consistent with the whole. Serial issue, however, distanced a work's ending from the rest of the text, allowing the author more freedom to leave loose ends and to explore byways before concluding the audience's journey. Tension between the middle and the ending of Hardy's late-1880s text, between the linear serial form and the instantaneously perceived field of the single volume, reflects an opposition between theories of order in Western culture at large.

Editors and publishers knew that creating a readership for a novel published serially in the 1880s involved generating favorable early reviews and making use of promotional techniques. The beginning and the middle of an installment text would establish its fundamental success and sales, delaying and perhaps reducing any disappointment the ending might later create. The single-volume issue, of course, made the novel's ending immediately available, and no action could be taken to counter unfavorable reviews of the whole. Thus, Hardy composed much of *The Woodlanders*' serial version to avoid offending his editor or his public; but the ending of the novel, and the shape it retroactively imposed on the whole after April 1887, inspired an adverse reaction from Victorian readers.

Critics have frequently measured *The Woodlanders* by the strength (or weakness) of its ending, which presents an ancient, agricultural, preindustrial England threatened by an analytical, urban mentality. Hardy himself, according to Robert Gittings, "set out [in this novel] to demolish the 'doll' of English fiction, which demanded a happy ending" (44–45). Michael Millgate states that Hardy also later claimed "the 'hinted' ending of the story—'that the heroine is doomed to an unhappy life with an inconstant husband'—would have been made much stronger and clearer if current conventions had permitted" (274). At the end of the novel the two urban, socially refined characters, Edred Fitzpiers and Grace Melbury, move away from the traditional rhythms of woodland life, attempting, probably unsuccessfully, to come to terms with their modern sensibilities in the changing environment of the Midlands. Marty South, who represents a harmonious identification with the cycles of orchard and season, remains at the novel's conclusion in rural Wessex, but with no companion after Giles Winterbourne's death and with no future, given the inevitable advance of the outside world.

This ending posed a challenge to Victorian notions of causality, especially moral consequence. R. H. Hutton's 26 March 1887 review in the *Spectator* articulated Victorian dissatisfaction with the novel's central character, Fitzpiers, and with his place at the conclusion of the action.

Calling the doctor an "unworthy and godless creature," Hutton found it incomprehensible that he should "find favor, as he evidently does, in the mind of the artist" (included in Cox 143). That Hardy in the end spared Fitzpiers from being caught in Tim Tangs' man-trap only exacerbated what Hutton called the novel's "moral drift" (Cox 142), for, rather than suffering, this "sensual and selfish liar" is unjustly "received back into his wife's favour and made happy" (Cox 143–44). For Hutton and many Victorian readers Fitzpiers should have been punished by the consequences of his actions in the novel's conclusion. Readers also admired Marty South and generally regretted her lonely figure at the novel's conclusion, despite the elegiac tone Hardy invested in her description. A plot built on firm notions of cause and effect should have shown an inevitable moral advance, rewarding the worthy and punishing the guilty.

During its serial run, however, *The Woodlanders* was popular enough to inspire several editors from other magazines to make offers for Hardy's next work, and Robert Gittings reports that *Macmillan's* had "done well" with *The Woodlanders* (58).[5] The design of individual installments suggests how the serial text fit in with accepted nineteenth-century notions of history, progress, and personal growth. Further, Hardy's technique derived from the traditional linear structure of installment fiction and adhered to Wilkie Collins's standard formula of making an audience laugh, weep, and wait.

For instance, a repeated situation of individual installments in *The Woodlanders* involves a reversal, which reinspires readers' interest in the plot: one character, who had just felt strongly for another, suddenly reverses him- or herself. For example, in chapter 15, in the August 1886 installment, Grace begins to change her mind about Giles Winterborne, the orchard farmer whom she had just rejected because her father wishes, and she concurs, that she pursue a higher place in society. When she learns that Winterborne is losing his home, her attitude shifts: "And yet at that very moment the impracticability to which poor Winterborne's suit had been reduced was touching Grace's heart to a warmer sentiment on his behalf than she had felt for years concerning him" (304). The "pendulum of the clock" (304) at Giles's house described in the very next paragraph represents the continuing alternation of feelings characteristic of the narrative. Such alternations encouraged readers' hopes that, despite many obstacles, Hardy's characters would ultimately achieve some stability and happiness.

This fact of reversal, a fundamental structural element in every number of the novel, is also frequently reflected in the environment which surrounds the characters.[6] When Winterborne cuts down the huge tree outside John South's cottage in chapter 14, for instance, the narrator concentrates on how things formerly distant and indistinct are now brought close and seen clearly: "the elm of the same birth-year as the woodman's [South's] lay stretched upon the ground. The weakest idler that passed could now set foot on marks formerly made in the upper forks by the shoes of adventurous climbers only; once inaccessible nests could be examined microscopically; and on swaying extremities where birds alone had perched the bystanders sat down" (August 1886: 303). The reversals here—the weak with the adventurous, the upper with the lower, birth with death, sitting with standing—suggest that no character is forever chained to one place in this universe; the last will be first, if enough time passes. While the passage itself anticipates to some extent a field theory of order (in Hayles's terms, a "fluid dynamic nature, . . . the absence of detachable parts, and the mutuality of component interactions"), its place in a sequence of similar passages in other installments also insists on a linear development over time.

The sense of reversal of feelings and exchange of place is also evident in the crucial scene (chapter 26, November 1886) in which Mrs. Charmond's carriage, with "the figure of a woman high up on the driving-seat" (1), overturns at the spot where Giles Winterborne's cottage once stood: "Winterborne went across the spot, and found the phaeton half overturned, its driver sitting on the heap of rubbish which had once been his dwelling, and the man seizing the horses' heads. The equipage was Mrs. Charmond's, and the unseated charioteer that lady herself" (1–2). Immediately after this fall, almost precisely at the midpoint of the novel's serial run, Mrs. Charmond meets Edred Fitzpiers. Their attraction to each other initiates the most dramatic reversals in the novel, as it also inspires Grace to recover her feeling for Giles.

These two fundamental elements of the middle of Hardy's novel, the reversal of characters' relationship and the reversals implied in landscape, of course underscore the novel's thematic emphasis on transition, Little Hintock threatened by the outer world. In each installment of the novel, characters and landscape vacillate between the old order (still with its power in this remote corner of Wessex) and the new order (surely coming with the railroad). No matter which force seems ascendant in any part of the novel, a meaning of all parts is that soon will come a reversal.

This context of reversals affects how we might view one of the most famous passages from *The Woodlanders*, Hardy's reference to "Unfulfilled Intention." Describing the woodland scene in chapter 7, June 1886, the narrator observes: "On older trees still than these huge lobes of fungi grew like lungs. Here, as everywhere, the Unfulfilled Intention, which makes life what it is, was as obvious as it could be among the depraved crowds of a city slum. The leaf was deformed, the curve was crippled, the taper was interrupted; the lichen ate the vigour of the stalk, and the ivy slowly strangled to death the promising sapling" (92). Linking society to nature here, Hardy suggests a universal crippling in creation, a condition apparently inescapable for both the old world of Little Hintock and the new world of Sherton and beyond. Yet this passage is itself another of Hardy's reversals.

The description of the natural scene that precedes this one about unfulfilled intention presents a more harmonious and fulfilled environment: Melbury and his daughter Grace "went noiselessly over mats of starry moss, rustled through interspersed tracks of leaves, skirted trunks with spreading roots, whose mossed rinds made them like hands wearing green gloves; elbowed old elms and ashes with great forks, in which stood pools of water that overflowed on rainy days, and ran down their stems in green cascades" (91–92). This passage mixes human and plant, hand and rinds, elbow and fork, feet and roots, to suggest a connected scene, a fulfilled intention. At one moment, then, growth characterizes nature in *The Woodlanders*; at another, decay reigns.

When Hardy continues in this particular passage, he pursues the vision most frequently associated with his work, where fate or destiny in an imperfect world leads certain unfortunate individuals to doom. But even that doom in much of his fiction is delayed by reversals, is, in fact, most realized in the novels' endings. Because single-volume readers can read literature in reverse, looking from the ending back to signs or foreshadowings of a final disposition, the middles of many Hardy texts have been thought to be drifting ultimately toward disappointment. Reading the middle, however, without knowing the end, subscribers to *Macmillan's* in 1887 encountered a principle of reversal, from one emotion to its opposite, from bad to good, happy to sad. Generally optimistic Victorian readers were able to believe that a happy or at least just resolution of the novel's problems lay in front of them, just beyond one final reversal. The appearance of installment after installment of *The Woodlanders* in *Macmillan's Magazine* permitted traditional Victorian expectations of progress toward a distant, if difficult, fulfillment.

This conformity to basic Victorian ideas of progress in *The Woodlanders'* serial form, however, appeared less visible once the novel had ended. Hardy's novel was judged at its conclusion by literary principles of artistic unity. Those principles dictated that the novel's last page be its final word indeed, that the author come down on one side or the other of the fiction's many reversals to create a single model of the universe not dependent on time. Readers, however, found Fitzpiers's reform unconvincing and Marty South's fate unearned; the ending did not match their sense of the story thus far. The single-volume issue of *The Woodlanders* placed this ending closer to the rest of the entire text than had serial issue, giving it added force. That resolution retroactively privileged all similar earlier moments in the text, endorsing a permanently changed protagonist and a doomed heroine. To satisfy himself and all his readers, perhaps Hardy actually needed to depict unending reversals; but the forms available to him at this point in literary history made that representation impossible.

Thus, the novel puzzled readers as it seemed somehow both to endorse and challenge values central to its culture. Its serial form fit in well with the notions of linear progress that dominated the nineteenth century; but its conclusion, given special emphasis in a single-volume edition, failed to satisfy readers' desire for a unified aesthetic whole. Endings, in fact, might be seen as formally opposed to installment publication: a serial is inherently linear, whereas an ending, once appearing and closing off a text, creates the field of an autonomous work of art. The same contradictory or even "chaotic" state, in which a literary work bifurcates into two versions of order, which then vie for dominance, surrounds the initial publication and reception of a novel with one foot in each century, Conrad's *Lord Jim*.

Beginning in October 1899, Conrad's *Lord Jim: A Sketch* appeared in fourteen monthly issues of *Blackwood's Magazine*, concluding in November 1900.[7] The strain within this novel's reception did not derive from an opposition between middle and end, as was the case with *The Woodlanders*. Instead, the Modernist emphasis on simultaneity, on a single, autonomous model of the world, dominated *Lord Jim: A Sketch*.[8] A major manifestation of new ideas about order in Conrad's story is the fact that each part enacts in miniature the entire novel.[9] Thus, there are no separate, isolated pieces of this Modernist text; the whole is visible in the part. The August installment contains, as we will show here, references to all the basic elements of Jim's story, tracing the full course of his life from youthful innocence to final tragedy. Yet once again, the serial publication of Conrad's tale retained a linear order beneath the new vision. Thus, readers' experience at the turn of the century continued to include a tension between two kinds of order, the developmental patterns of Victorian installment fiction and the simultaneity of Modernist single-volume novels.

The unusual narrative structure of *Lord Jim: A Sketch* has been commented on from its first publication,[10] but Conrad's complex use of point of view and a narrative style of extended flashbacks and frequent foreshadowings can be linked to the larger cultural concern with ideas of order. In installment 11 (August 1900), for instance, Conrad's text challenged the established conventions of linear narrative, particularly a belief in the temporal relationship of cause and effect. His circling, repetitive style flirted with negating one of the most basic principles of storytelling, that more narration will ensure greater understanding. In the course of that installment Conrad's first readers in *Blackwood's Magazine* confronted a "chaotic" situation: Modernist literary technique in a traditional Victorian format.

Number 11 began with adventure, apparently promising Victorian readers familiar material in an accepted mode: "The defeated Sherif Ali fled the country without making another stand . . . and [Jim] became the virtual ruler of the land"(251).[11] This quick summary of events concluded with Marlow's saying, "This brings me to the story of [Jim's] love" (252). Readers would have expected in an installment that began this way a fuller account of Jim's history as leader and lover, narrated by an old friend who once visited there. That narrator, however, soon finds himself in great difficulty:

> I suppose you think it is a story that you can imagine for yourselves. We have heard so many such stories, and the majority of us don't believe them to be stories of love at all. For the most part we look upon them as stories of opportunities: episodes of passion at best, or perhaps only of youth and temptation, doomed to forgetfulness in the end, even if they pass through the reality of tenderness and regret. This view mostly is right, and perhaps in this case, too. . . . Yet I don't know. To tell this story is by no means so easy as it should be—were the ordinary standpoint adequate. (252)

Both audience and speaker here are apparently so familiar with love stories that none needs to be told, but some new framework is necessary to explain fully what is involved in Jim's situation. Marlow, like his creator Conrad, is forced in the August installment to move beyond "the ordinary standpoint," beyond the standard Victorian modes of narration, in an effort to satisfy himself and his listeners. The truth Marlow hopes to convey, "a cruel wisdom" (252) about disappointed human aspiration, however, remains locked behind the "sealed lips" of Jewel's mother in "a lonely grave" (252).[12] All the pieces of Jim's history can be identified, but Marlow discovers that he cannot shape them into a traditional linear story of cause and effect.

In a place "three hundred miles beyond the end of telegraph cables" (256), Marlow realizes that "the haggard utilitarian lies of our civilisation" are unreliable; the referential language of urban, industrial England will not suffice. Here "pure exercises of imagination, that have the futility, often the charm, and sometimes the deep hidden truthfulness, of works of art" (256) hold sway. Marlow's narrative has such a "charm" or aesthetic appeal, but its "truthfulness" is strained in this installment. What the narrator achieves is a self-sufficient, autonomous literary creation, artistically satisfying but never clearly conveying the linear experience it was intended to represent. Like its central figure, Jim's one love, this installment might be called "The Jewel."

The motif of the gem or valued thing is introduced in "The Jewel" by Marlow at the end of number 11's exposition: "Jim called her by a word that means precious, in the sense of a precious gem—jewel. Pretty, isn't it?" (254). Marlow learns that others have interpreted this woman's identity from their own perspective, projecting what they value into the word "jewel." For instance, a representative of debased colonialism believes Jim "has really got hold of something fairly good—none of your bits of green glass—understand?" (255). Citizens of both East and West are attracted to the story of "an extraordinary gem—namely an emerald of an enormous size, and altogether priceless" (255). Marlow learns from "a sort of scribe to the wretched little Rajah of this place [230 miles south of Patusan River]" (255) that such jewels are "best preserved by being concealed about the person of a woman. Yet it is not every woman that would do. She must be young—he sighed deeply—and insensible to the seductions of love" (256).

Both Rajah and the deputy-assistant resident are imagining what they desire rather than perceiving what exists. Confronted with "the mystery" (253) of mother's and daughter's trials, Marlow works in this installment to understand that reality, to name or rename Jewel; but his language too often wanders off into meaningless abstractions, such as "a region of unreasonable sublimities seething with the excitement of [women's] adventurous souls" (253). Marlow is committed to the value of stories, to their traditional function in tracing linear development: "Indeed the story of a fabulously large emerald is as old as the arrival of the first white men in the Archipelago; and the belief in it is so persistent that less that forty years ago there had been an official Dutch inquiry into the truth of it" (255). Like others seeking the true history of the jewel, however, Marlow will find it elusive; he will find no traditional cause and effect explanation of its past and present.

Jim, however, apparently found an effective language to represent Jewel and her world, at least for a time. Jim's success in Patusan is linked in this part to his ability to picture or name elements of the realm he has entered and to place them in temporal and causal relationships. Jim's sudden vision of a "plan for overcoming Sherif Ali," for instance, leads to immediate successful action: "His brain was in a whirl; but, nevertheless, it was on that very night that he matured his plan for overcoming Sherif Ali. . . . He could see, as it were, the guns mounted on the top of the hill. . . . He jumped up, and went out barefooted on the verandah" (262). When Cornelius soon appears, clearly advancing some scheme against Jim, the young hero surprises himself with an outburst, opposing that plot with his own order: "He let himself go—his nerves had been over-wrought for days—and called him many pretty names,—swindler, liar, sorry rascal: in fact he carried on in an extraordinary way. He admits he passed all bounds, that he was quite beside himself—declared he would make them all dance to his own tune yet, and so on, in a menacing, boasting strain" (263). Despite the "ridiculous" (263) nature of this explosion, this naming and ordering does seem to "make them all dance to his own tune," to conquer Cornelius: "What stopped him at last [Jim] said, was the silence, the complete deathlike silence, of the distinct figure far over there, that seemed to hang collapsed, doubled over the rail in a weird immobility. . . . 'Exactly as if the chap had died while I had been making all that noise,' he said" (263). Since Marlow's audience knows of Jim's ultimate success as "virtual ruler of the land" (251) and his romantic fulfillment, "marital, homelike, peaceful" (254), they must conclude that his strategy, his language, was for a time effective.

Reflecting on Jim's life in Patusan, Marlow also desires to arrange the elements of that scene in language, to lay out in a traditional, linear, causal analysis an explanation of the past. However, he seems incapable in the end of matching word and deed, of explaining "the unanswerable why of Jim's fate" (252). For instance, he accounts for the girl's satisfaction with her life initially through a naive faith in words that resembles Jim's "romantic conscience" (253): "she would listen to our talk; her big clear eyes would remain fastened on our lips, as though each pronounced word had a visible shape" (256). Like Jim's her innocence had been assaulted by an ugly reality. Where Jim faced a "misfortune" (254) that brought him to Patusan and "crossed the river" (258) to start up a new life, Jewel overcame her father to achieve a new self: "'Your mother was a devil, a deceitful devil—and you, too, are a devil,' he would shriek in a final outburst, pick up a bit of dry earth or a handful of mud (there was plenty of mud around the house), and fling it into her hair" (260). Cornelius's efforts seem to involve keeping Jewel tied to the earth, to his reality, but she escapes this trap: "Sometimes, though, she would hold out full of scorn, confronting him in silence, her face sombre and contracted, and only now and then uttering a word or two that would make the other jump and writhe with the sting" (260). Her "word or two" here controls this reality so that she does not need Jim, who says, "I can stop his game. . . . Just say the word" (261). When she explains that she would "kill [Cornelius] with her own hands" if she did not know he was already "intensely wretched," Jim is shocked at her "talk" (261), at her command of words, things, and time.

In the end, Marlow admits that the innermost identity of Jewel, like that of her mother, is beyond his knowledge and experience: "it is only women who manage to put at times into their love an element just palpable enough to give one a fright—an extra-terrestrial touch. I ask myself with wonder—how the world can look to them—whether it has the shape and substance WE know, the air WE breathe!" (253). Woman's "extra-terrestrial" vision remains illusively out in front of this observer no matter how hard he races after it, a facet of Jim's fate beyond Marlow's narrative and his masculine audience.

Admitting the failure of linear narrative, then, Marlow in this August 1900 installment circles back again and again, as the novel as a whole does, to a central subject, here the feminine consciousness, the "jewel" of Jim's desire. Although Jim's triumphant mood at the end of this episode is understandable ("he went to sleep for the rest of the night like a baby" [263]), the image of the girl turning toward him haunts Marlow in the last words of the number: "'But I didn't sleep,' struck in the girl, one elbow on the table and nursing her cheek. 'I watched.' Her big eyes flashed, rolling a little, and then she fixed them on my face intently'" (262; this sentence concludes the installment). In addition to the danger she correctly foresees in Cornelius, other messages in this fixed look must be traced back to the "lingering torment" (253) Marlow admits women endure "with a peculiar cruelty," the sense of "having been deserted in the fullness of possession by some one or something more precious than life" (253). Her questioning look at Marlow recalls Doramin's wife at the beginning of the installment, who, without "removing her eyes from the vast prospect of forests stretching as far as the hills," raised the possibility that Jim might return to a home country: "Had he no household there, no kinsmen in his own country?" (252). Thus, Jewel's tense appearance at the end of the installment stresses the theme that disappointment and failure are inherent in life and, recalling a scene from the beginning of the number, is a resonant conclusion to the entire installment.

Marlow, however, is unable to understand such willingness to accept a flawed world, to endure Cornelius and to suspect Jim. He cannot put in satisfactory order the facts of this couple's history, and at the end of the August 1900 installment he seems to have added almost nothing to his own understanding of Jim's story. As narrator he has turned the question over brilliantly in this number, looking at mother, daughter, and lover directly and through analogues. "The Jewel" is, thus, an autonomous whole, all parts fitting together into an artistic success; but the evolution of Jewel's consciousness and the history of Jim's life cannot be traced from distant origin to final condition: "Apparently it is a story very much like the others: for me, however, there is visible in its background the melancholy figure of a woman, the shadow of a cruel wisdom buried in a lonely grave, looking on wistfully, helplessly, with sealed lips" (252).[13]

Despite the fact that each installment of *Lord Jim: A Sketch* created the entire novel in miniature to a certain extent, the succession of parts did, of course, have a cumulative effect, a linear movement.[14] Victorian readers then and modern readers now see Jim in the fullest light after having considered the subject from many different vantage points. But the reading experience of 1899–1900 was characterized by a tension between the traditional expectations of added knowledge in a serial and the repetitive nature of all parts of *Lord Jim: A Sketch*. The serial form encouraged readers to find progress in the narrative because this had been the case with installment literature for sixty years. Conrad's more modern style, ultimately designed for the single volume rather than many installments, contradicted and frustrated those expectations.

Thomas Hardy's *The Dynasts* also (like *The Woodlanders* and *Lord Jim: A Sketch*) manifests a strain between linearity and simultaneity, serial issue and single-volume publication, but it suggests even more clearly that this strain was linked to a major shift in paradigms. Hardy's poetic drama is largely unread today, though many have pronounced it his most ambitious and greatest work. The work's first readers, and those who later declined to read it at all, were often struck by what seemed anomalies or sheer quirkiness. The full title of the work when first published was *The Dynasts. A Drama of the Napoleonic Wars, In Three Parts, Nineteen Acts, & One Hundred and Thirty Scenes*—not the sort of twentieth-century title to lure masses of readers into the text. Hardy presented the work as a drama, moreover, but stated in his Preface of 1903 that it was "intended simply for mental performance, and not for the stage" (1:x). The apparent anomalousness of *The Dynasts* extends to its publication history. It appeared in three parts: Part I in 1904, Part II in 1906, and Part III in 1908. Its sequencing—not only its appearance in successive parts but also its mostly linear plot line—affirmed the continuing vitality of the serial tradition, yet *The Dynasts* also articulates a Modernist perspective and aesthetic.

The drama's presentation of an Immanent Will blindly determining the events of the cosmos has caused *The Dynasts* to be linked to the work of Schopenhauer and Eduard von Hartmann, but Hardy himself asserted that "My pages show harmony of view with Darwin, Huxley, Spencer, Comte, Hume, Mill, and others, all of whom I used to read more than Schopenhauer" (quoted in Orel 23). If Hardy thoroughly absorbed the major Victorian works that presented evolutionary theory, however, his drama was appearing within a few years of Einstein's special and general theories that helped effect such a marked change in twentieth-century conceptual frameworks. Within *The Dynasts* it is possible to locate perspectives that accord with evolutionary theory and with field theory, and the resistance of these divergent perspectives to absorption in a single synthesizing framework illuminates Hardy's positioning in a time of shifting paradigms of order.

These two converging ordering systems are already apparent in the preface and end matter of the drama's first part, published in 1904. The preface on the one hand invokes the familiar nineteenth-century notions of chronicity and causality:

> the provokingly slight regard paid to English influence and action throughout the struggle by those Continental writers who had dealt imaginatively with Napoleon's career, seemed always to leave room for a new handling of the theme which should re-embody the features of this influence in their true proportion. . . . [*The Dynasts*] may, I think, claim at least a tolerable fidelity to the facts of its date as they are given in ordinary records. Whenever any evidence of the words really spoken or written by the characters in their various situations was attainable, as close a paraphrase has been aimed at as was compatible with the form chosen. (1:vi-vii)

A projected table of contents for the second and third parts appearing at the end of the 1904 volume places the drama in the same stream of linear time as its subject matter, since Hardy labels the contents as "Subject to revision" and asserts at the end of his outline, "The Second and Third Parts are in hand, but their publication is not guaranteed." When Hardy insists that future parts are subject to revision with the passage of time and are not even guaranteed to appear, he appeals to serialization's suspense, anticipation, and provisional judgment. When he invokes fidelity to historical realism and the causal role of the English he seems to endorse the developmental views associated with both historicism and serialization.

But even as he places his work within an ongoing linear perspective he also appeals to a simultaneity of vision consonant with a Modernist framework:

> in devising this chronicle-piece no attempt has been made to create that completely organic structure of action, and closely-webbed development of character and motive, which are demanded in a drama strictly self-contained. A panoramic

show like the present is a series of historical "ordinates" (to use a term in geometry): the subject is familiar to all; and foreknowledge is assumed to fill in the curves required to combine the whole gaunt framework into an artistic unity. (1.ix)

Insofar as Hardy directs readers to the "artistic unity" of his drama and asks readers to bring "foreknowledge" of historical outcomes to the play and "fill in the curves" (or "gaps," as Hardy elsewhere calls them) in the drama, he appeals to the simultaneous perception endorsed by Modernist aesthetic principles. Indeed, readers who remember the futile life and early death of Napoleon's son as they approach the scene presenting the birth of the king of Rome, or who steadily keep in mind the outcome of Waterloo as they read of Napoleon's escape from Elba, not only activate the drama's acerbic ironies but also weaken the forward thrust of linear plot because they superimpose later on earlier events and view both at once. Hardy's appended table of contents for the succeeding parts is an attempt to "compensate" for the drama's serialization and encourage a simultaneous perception of the whole that likewise circumvents the play's linear unfolding.[15]

The same multiplicity of perspectives is evident in the matter of the drama itself. Hardy explicitly adopted the materials of history and chose as his subject a story central to the nineteenth century and to Victorian England's status as a world power. In notable respects Hardy's treatment of history was also Victorian. There are, in effect, two dramas embedded in *The Dynasts*: the events in the European theater of war, and the reactions embodied in the "Phantom Intelligences"—the Spirit of the Years, Spirit of the Pities, Spirits Sinister and Ironic, Spirit of Rumour, and Recording Angels—who watch events in the human realm. Most often the human drama, the conventional materials of history, are also situated within the conventional assumptions (especially linear progression and causality) of Victorian historicism.

The drama's central character is of course Napoleon, and the drama, as has often been observed, presents a clear beginning, middle, and end in the story of his empire. In the first act of Part I Napoleon crowns himself in the Milan cathedral and so (within Hardy's work) inaugurates his reign as emperor. The remainder of the drama shows Napoleon's attempt to entrench and extend his reign, first through territorial expansion (Part I), then through dynastic succession (Part II, in which Napoleon divorces Josephine and marries Marie Louise), followed by the demise of his empire (Part III). Lest we miss it, we are even alerted to the exact middle point of his career, which comes halfway through the second part and hence, in the original issuing of the poem, exactly halfway through the poem's appearance (in 1906). Napoleon's preoccupation with matters to the east causes him to leave Spain (which the English have invaded) in the hands of others and return to Paris, at which point the Spirit of the Years remarks, "More turning may be here than he designs. / In this small, sudden, swift turn backward, he / Suggests one turning from his apogee!" (2.122). *The Dynasts*' serialization in three parts, in chronological order, would have given added force to the linear unfolding of the drama, since each volume was literally a beginning, a middle, and an end and was embedded in the flow of a four-years' publishing time span.

The Dynasts, moreover, carefully develops lines of causality, partly through plotting which upholds a strict connection between initial events and results which issue from them. Even the Intelligences (which, as we will see, are generally associated with perspectives more consistent with field theory) cleave to a view of causality and developmental phases.[16] In the Fore Scene which opens the play, the Spirit of the Years, looking down upon the earth, notes that despite superficial changes "phase and phase / Of men's dynastic and imperial moils / Shape on accustomed lines" (1.5). In act 6 of the second part, similarly, the Spirit of the Years watches the fierce battle between the French and English at Albuera and discerns the active forces of causality behind all:

A hot ado goes forward here to-day,
If I may read the Immanent Intent
From signs and tokens blent
With weird unrest along the firmament
Of causal coils in passionate display.

(2:264)

Hardy's canvas is so large partly because he attempts to dramatize manifold causes behind Napoleon's rise and fall, from fever to weather to human character and the intricate dances of diplomacy. Even within parts Hardy establishes causal sequences, as when George III's refusal to bring the opposition into Pitt's cabinet in act 4 hurries along Pitt's death, which occurs at the end of the first part with the close of act 6.

As the drama approaches its end, however, the pacing of events speeds up. In the third part the initiation and the results of the disastrous Russian campaign are presented in a single act. Hardy is still working within a linear framework, but the appalling contrast between Napoleon's setting forth with some half-million soldiers and his returning to Paris alone encourages the perception of the Russian campaign more as an instantaneous act of rash and cruel despotism than as a slowly occurring sequence (as it would have been for those who survived late into the campaign). The Russian campaign can also serve to remind us of another factor in Napoleon's rise and fall, one that interrogates simple notions of causality.

The drama's most famous figure is not Napoleon but the Immanent Will, and this makes a mockery of sequence and even the unidirectional flow of time. The Immanent Will, Hardy's personification of the nonteleological force which impels the universe, absorbs and transforms what may appear (to limited human vision) as linear cause and effect into the instantaneous, unified workings of an unconscious mind. The first glimpse of the Immanent Will comes in the Fore Scene, when the Spirit of the Years makes visible the understructure of the turmoil in Europe: "A new and penetrating light descends on the spectacle, enduing men and things with a seeming transparency, and exhibiting as one organism the anatomy of life and movement in all humanity and vitalized matter included in the display" (1.10). The Immanent Will is not a straight line but a web, as the Spirit of the Years remarks at Borodino (it is "*the will of all conjunctively; / A fabric of excitement, web of rage, / That permeates as one stuff the weltering whole*" [3:31]), and again at the drama's end:

Yet but one flimsy riband of Its web
Have we here watched in weaving—web Enorme,
Whose furthest hem and selvage may extend
To where the roars and plashings of the flames
Of earth-invisible suns swell noisily,
And onwards into ghastly gulfs of sky,
Where hideous presences churn through the dark—
Monsters of magnitude without a shape,
Hanging amid deep wells of nothingness.

(3.349–50)

Hardy's concept of the Immanent Will as one enormous web pervading the cosmos bears a striking relation to models of twentieth-century field theory. As Hayles comments,

> Characteristic metaphors [of the field model] are a "cosmic dance,: a "network of events," and an "energy field." A dance, a network, a field—the phrases imply a reality that has no detachable parts, indeed no enduring, unchanging parts at all. Composed not of particles but of "events," it is in constant motion, rendered dynamic by interactions that are simultaneously affecting each other. . . . (15)

She later elaborates this metaphor or model as a web, and notes that in particle theory individual parts subject to strict causality disappear in lieu of particles as an expression of "the field's conformation at a given instant, appearing as the field becomes concentrated at one point and disappearing as it thins out at another. Particles are not to be regarded as discrete entities, then, but rather . . . as 'energy knots' " (16).

This is very close to the model or metaphor Hardy constructs of human beings within the grasp or "field" of the Immanent Will, who are likewise never detachable parts operating according to local, individualized intentions, and who cluster in multitudes (especially on battlefields) or thin out into small knots or groups according to the dynamics of history's "events."[17] This field view of the Immanent Will is especially apparent in the heat of battle at Austerlitz, when the Spirit of the Years again causes the Immanent Will to become visible: "the scene becomes anatomized and the living masses of humanity transparent. The controlling Immanent Will appears therein, as a brain-like network of currents and ejections, twitching, interpenetrating, entangling, and thrusting hither and thither the human forms" (1:197).[18] It is even possible to view the field of the Immanent Will as including the observer as well as the participants. Noting that the various spirits are termed the "flower of Man's intelligence" at the end of Part I and that the unconscious Immanent Will is figured as a brain, Susan Dean argues that, together, the spirits and the will are an analogue of the human brain, which also has a conscious and unconscious element. Thus, when Hardy states that the drama is intended for mental performance, Dean suggests that the mind is both substance and setting for the play, the Overworld and earthly domains corresponding respectively to the sentient observer and the unconscious body (and layer of mind) (55–60). In this reading, the field of the play encompasses not only the events of the play but also the reader and playwright.

But if the linear, developmental perspective within *The Dynasts* is contested by the simultaneous, all-encompassing web of the Immanent Will, the perspective consistent with field theory is itself pulled into the domain of linear development at a crucial point in each of the three acts. In act 5 of Part I the Spirit of the Pities laments that the impercipient will occasioned the accidental growth of consciousness in humanity, which is burdened with perceiving and feeling the pain of events which it cannot determine or control. The Spirit of the Years then replies,

> *Your hasty judgments stay,*
> *Until the topmost cyme*
> *Have crowned the last entablature of Time.*
> *O heap not blame on that in-brooding Will;*
> *O pause, till all things all their days fulfil!*
>
> (1:166)

Pausing and looking to see the fulfillment wrought by distant time are actions deeply connected with the developmental perspectives of evolution, historicism, and serial reading. And the notion of absorbing the will back into a linear and developmental framework intensifies in Part II because similar lines occur at the rhetorically powerful end point of the volume, which in its serial issuance then also paused and looked ahead to the fulfillment of the drama to be brought with time. In response to the Years' assertion of the will's impercipience, the Chorus of the Pities invokes the absence of closure and the potential for development:

Yet It may wake and understand
Ere Earth unshape, know all things, and
With knowledge use a painless hand,
A painless hand!

(2:302)

The jostling of views associated with evolutionary and field theories receives its most emphatic presentation at the close of *The Dynasts*. In a passage we have already quoted, the Spirit of the Years describes the cosmic "web Enorme" which pervades the universe. The Pities then respond with a vision derived from an evolutionary framework:

Thou arguest still the Inadvertent Mind.—
But even so, shall blankness be for aye?
Men gained cognition with the flux of time,
And wherefore not the Force informing them,
When far-ranged aions past all fathoming
Shall have swung by, and stand as backward years?

(3:350)

The Spirit of the Years concedes evolutionary development in its own viewpoint ("*You almost charm my long philosophy / Out of my strong-built thought, and bear me back / To when I thanksgave thus*" [3:353]), and the play ends with the Chorus of the Pities anticipating evolutionary development in the Will:

But—a stirring thrills the air
Like to sounds of joyance there
That the rages
Of the ages
Shall be cancelled, and deliverance offered from the darts that were,
Consciousness the Will informing, till It fashion all things fair!

(3:355)

Hardy appears to resolve two competing models of existence by absorbing the field view of the universe into an evolutionary model, a resolution to which the drama's serialization would have given added force since the possibility for change within his text developed slowly over an expanse of years. But the inability of the developmental model to account for all of *The Dynasts*—especially its most famous image of the Immanent Will as a network simultaneously connecting and directing all of humanity—suggests the rise of a new paradigm which would soon enter twentieth-century culture at large and effectively label major assumptions of the nineteenth century, including those which helped make serialization a plausible mode for literary works, as naive and reductive.

The theory of evolution did not "cause" serialization, nor did field theory "cause" serialization's demise. The serial form was already a fixture of the Victorian literary scene when Darwin published the *Origin* in 1859—though as Peter Morton reminds us, the most readily absorbed elements of Darwin's theory, "organic evolution as a historical process," was unoriginal with Darwin and was widely accepted in the 1840s and already circulating in the 1830s (22),[19] the same decade in which Charles Dickens made the serial form wildly popular with the appearance of *The Pickwick Papers* (1836–37). Moreover, thousands of novels and poems were published in their entirety, as "wholes," before Einstein or Planck ever began their researches. But the general prevalence of developmental and linear views in the nineteenth century would have worked to make the serial form an acceptable format for the era's best literary works.

Some of the same economic conditions which generated serialization still exist today. Every bank has its installment-loan department, and twentieth-century serials, both written and broadcast, abound. The first film serials in 1912 and 1913 even worked on the same basis as fiction serialized in Victorian periodicals, since the serial films, each installment called a "chapter," were issued in conjunction with stories serialized in magazines and newspapers and were designed to boost circulation (Stedman 3–18). But for twentieth-century audiences the serial is acceptable as a vehicle only for popular culture: comic books, television programs, detective stories, and so on. Tom Wolfe may have first serialized *The Bonfire of the Vanities* in *Rolling Stone Magazine* in 1984 and 1985, but as a "serious" author who received serious, and enthusiastic, critical success when *Bonfire* appeared as a whole volume in 1987, Wolfe in his publicity interviews (like reviewers in their columns) was studiously quiet about the work's first appearance in parts. We suggest that one reason for the serial's demise among the twentieth-century intelligentsia is that the intellectual framework which once helped endorse the serial is no longer intact. A field view of reality is the common property of contemporary intellectuals; it is not surprising that twentieth-century literary critics prefer to approach works as autonomous, whole entities in which all elements can be perceived at once.

Yet obviously one need not approach serialization and whole volumes, evolution and field theory, as opposing choices, though our own linear narrative has made it most efficient to treat these ideas of order in this way. We need to remember that Darwin's world of evolution is also susceptible to being viewed as a field, a point indicated not only by the emergence of ecology studies in the twentieth century but also by Darwin's famous "tangled bank" passage which closes *Origin of the Species*:

> It is interesting to contemplate a tangled bank, clothed with many plants of many kinds, with birds singing on the bushes, with various insects flitting about, and with worms crawling through the damp earth, and to reflect that these elaborately constructed forms, so different from each other, and dependent upon each other in so complex a manner, have all been produced by laws acting around us.[20] (450)

While Darwin goes on to evoke linear development—"whilst this planet has gone cycling on according to the fixed law of gravity, from so simple a beginning endless forms most beautiful and most wonderful have been, and are being evolved" (450)—it is clear that he endorsed a notion of the earth's living forms as one interconnected web (or tangled bank) acted upon by consistent laws but also acting on one another within the field that contains them. Hence many followers of Darwin emphasized linearity less and interconnectedness more, as Shuttleworth argues is the case for George Eliot's *Middlemarch*, with its famous image of the web (though hers is a decidedly earthly and social, not cosmic, web).[21] Hardy himself wrote that "The discovery of the law of evolution, which revealed that all organic creatures are of one family, shifted the centre of altruism from humanity to the whole conscious world collectively" (F. Hardy 138).[22]

Linearity and divisible parts are often set in opposition to simultaneity and indivisible wholes, but the instances provided by Hardy and Conrad, even by Darwin, suggest that this particular bifurcation is another construct by which we attempt to order our universe, an attempt shaped by historical context and dominant ideology. It is probably impossible to reconcile the two views, just as it seems impossible fully to reconcile the convergent ordering systems in *The Dynasts*. The essays in this volume, however, suggest that it is not necessary to reconcile orders. We in turn propose that serial and whole-volume reading patterns be seen not as rival systems but as coexistent reading strategies that offer equally viable, if not equivalent, results.

Notes

The writing of this essay was supported by an Interpretive Research grant from the National Endowment for the Humanities, a federal agency which supports the study of such fields as history, philosophy, literature, and languages.

1. For a fuller account of serialization and of the impact of publication format on audiences' construction of literary meaning, see our *The Victorian Serial*.

2. We are not suggesting that Victorian readers did not wish to look at literary works as wholes. Coleridge insisted that poems be viewed as organic wholes: "ALL the parts of an organized whole must be assimilated to the more *important* and *essential* parts" (*Biographia* 2:72); "the Spirit of Poetry like all other living Powers, must of necessity circumscribe itself by Rules, were it only to unite Power with Beauty. . . . It must embody in order to reveal itself; but a living Body is of necessity an organized one,—& what is organization, but the connection of Parts to a whole, so that each Part is at once End & Means!" (*Lectures* 1.494; Coleridge's definition of part and whole is derived from Kant [see Shuttleworth 3]). In fact, reviewers occasionally complained when a work was serialized because this format precluded an immediate view of the whole. The pre-Jamesian novel, of course, was less likely to be rigorously subjected to organicist aesthetics. Even so, this nineteenth-century articulation of "wholes" differs from the Modernist tradition in seeing parts as at once an end, if also a means. In twentieth-century Modernist aesthetics, especially insofar as they are congruent with field theory as outlined by N. Katherine Hayles, there can be no detachable parts—though as Hayles points out (17–19) field theory shares selected features of Romanticism.

3. Beer, interestingly, notes that in the late eighteenth century "evolution" was a term tied to the history of the individual, since it referred to "the stages through which a living being passes in the course of its development from egg to adult" (15). Not until the 1830s, Beer states, was "evolution" used to apply to species' development.

4. For another recent account, see Hobsbawm 188, 244, 256, or Kern, who details the many conceptual and technological innovations (e.g., the cinema) which authorized a perception of time as reversible. Kern also notes that the "modern electric clock with the sweeping fluid movement of its second hand . . . invented in 1916" encouraged the conception of time as flux rather than as periodic progression (20).

5. The two-stage reception of Hardy's *The Woodlanders* was also bound up with the state of the publishing industry at the time. In *Modes of Production of Victorian Novels*, Norman Feltes has recently explained how the system of book production was changing in the 1880s and 1890s. For much of the nineteenth century, the three-volume novel, selling for a high initial price to lending librar-

ies, particularly Mudie's, allowed publishers to control the market in fiction (57–75). Novels that failed to meet the standards of Victorian circulating libraries had little chance of turning a profit, and Mudie's limits were, as Guinevere L. Griest writes (87–101), well-known. Very much the same values governed the major literary periodicals of the day, applying the limits of conventional middle-class belief to authors writing their fiction in installments. This, of course, was a major reason writers in the 90s like Gissing, Hardy, and James rebelled against serialization and endorsed the lower priced, single-volume novel, which skipped over the magazine editors and circulating libraries to reach the public more directly.

6. Hardy was using new techniques in his descriptions of nature in *The Woodlanders*, moving toward a style that scholars have seen as impressionistic. Gregor explains that Hardy was specifically studying "the late Turner and the new Impressionists" during the composition of *The Woodlanders*, writing in his journal in January 1887, "I don't want to see landscapes, i.e. scenic paintings of them, because I don't want to see the original realities—as optical effects that is. I want to see the deeper reality underlying the scenic, the expression of what are sometimes called abstract imaginings" (13).

7. The familiar title of Conrad's classic is *Lord Jim: A Tale*, but the serial version in *Blackwood's* ran under the title of *Lord Jim: A Sketch*.

8. Many readers complained of the repetitive nature of Conrad's tale. The reviewer for *Sketch* wrote, for instance, on 14 November 1900: "It is a short character-sketch, written and rewritten to infinity, dissected into shreds, masticated into tastelessness" (included in Sherry 118).

9. This argument is made in more detail in chapter 6 of *The Victorian Serial*. All references here to *Lord Jim: A Sketch* are to the August 1900 installment of *Blackwood's Magazine*, pp. 251–63.

10. Ian Watt provides a thorough analysis of Conrad's style and its place in the narrative conventions of the nineteenth century (263–320).

11. Fleishman has discussed Jim's success in the Patusan portion of the novel (106–11).

12. Cox has noted of the narrative in general: "Marlow is not sure in what genre he is composing his story, and so his images are constantly changing, like patterns in a kaleidoscope" (21).

13. Roussel makes a similar point about the failure of language to represent reality in Conrad: "Language, like all of man's creations, is a part of this surface, and to bring it into direct contact with the darkness is simply to confront it with its own negation" (106).

14. In *Fiction and Repetition*, J. Hillis Miller argues that because Marlow cannot end his narrative with a final definitive explanation, the work itself cannot have a determinate meaning: "The overabundance of possible explanations only inveigles the reader to share in the self-sustaining motion of a process of interpretation which cannot reach an unequivocal conclusion" (39). While such an assertion is possible when we read the text as a static whole, the original serial reading of *Lord Jim: A Sketch* added a temporal frame of reference to the novel's

meaning, suggesting that there is value in the process of Marlow's sustained questioning excluded from Miller's analysis.

15. In fact the projected table of contents of 1904 and the actual contents of the 1906 and 1908 volumes did not entirely correspond. For Hardy's first readers the drama literally embodied the principles of chance and the flouting of human plans which form part of the subject matter of the Napoleonic Wars (see our *The Victorian Serial* for an extended discussion of this point).

16. James Gleick asserts, "an adaptationist explanation for the shape of an organism or the function of an organ always looks to its CAUSE. . . . Final cause survives in science wherever Darwinian thinking has become habitual" (201).

17. Susan Dean argues that the "dumb show" at Leipzig (3:III.ii) embodies "what in this poem lives are at their base—temporarily charged bits of matter: 'So massive is the contest that we soon fail to individualize the combatants as beings, and can only observe them as amorphous drifts, clouds, and waves of conscious atoms, surging and rolling together' " (81).

18. In *The Linguistic Moment* J. Hillis Miller asserts the derivation of Hardy's Immanent Will from "nineteenth-century formulations of the wave theory of light and of electromagnetism," and, like Hayles, Miller notes the implication of field or particle theory for language and concepts of the self (307–11).

19. Beer (25) and Gould (99) comment on the (erroneous) tendency of popular accounts to absorb Darwin's theory into a progressive, optimistic, intentionalist framework.

20. As Beer points out, he also refers to "an 'inextricable web of affinities' " within the natural world in chapter 13 of the *Origin* (22, 167). Beer also notes that Darwin's theory is susceptible to more than one intellectual perspective: "His theory could be extrapolated to suggest a random and disordered play of forces, or it could be made to yield the assurance of irreversible upward growth (his own image of the TREE emphasized verticality)" (114).

21. J. A. V. Chapple makes a similar point: "If what Darwin called 'a severe struggle for life' was not so obvious in Victorian Britain as it might be in more savage societies, he nevertheless provided a model of special value for prose fiction when he emphasised 'the infinite complexity of the relations of all organic beings to each other and to their conditions of existence, causing an infinite diversity in structure, constitution, and habits" (80–81). He also sees *Middlemarch* as embodying evolutionary "fictional laws," which include "complexity of relations, repetition with variation, progression and continued divergence" (87). Eliot herself made it clear that her web was "not," in Beer's terms, "co-extensive with the universe" when her narrator asserts in Book I of *Middlemarch* that she must focus " 'on this particular web' " rather than on " 'that tempting range of relevancies called the universe' " (Beer 171).

22. On the other hand, Hardy implied a teasing out of part from whole in his 21 December 1914 letter to Dr. Caleb Saleeby about the Immanent Will: "The nature of the determination embraced in the theory is that of a collective will; so that there is a proportion of the total will in each part of the whole, and each part has therefore, in strictness, *some* freedom, which would, in fact, be operative

as such whenever the remaining great mass of will in the universe should happen to be in equilibrium" (F. Hardy 269–70). Appropriately, Hardy elsewhere enunciated views which accord with the major focus of the present volume, as evident in this journal entry from 8 September 1896: "Why true conclusions are not reached, notwithstanding everlasting palaver: Men endeavour to hold to a mathematical consistency in things, instead of recognizing that certain things may both be good and mutually antagonistic: E.G. patriotism and universal humanity; unbelief and happiness" (F. Hardy 54). In 1901 he again wrote, "My own interest lies largely in non-rationalistic subjects, since non-rationality seems, so far as I can perceive, to be the principle of the Universe. By which I do not mean foolishness, but rather a principle for which there is no exact name, lying at the indifference point between rationality and irrationality" (F. Hardy 90).

Works Cited

Altick, Richard D. *Victorian People and Ideas*. New York: W. W. Norton, 1973.

Beer, Gillian. *Darwin's Plots: Evolutionary Narrative in Darwin, George Eliot and Nineteenth-Century Fiction*. London: Routledge & Kegan Paul, 1983.

Brooks, Peter. *Reading for the Plot: Design and Intention in Narrative*. New York: Vintage, 1984.

Chapple, J. A. V. *Science and Literature in the Nineteenth Century*. Houndsmill, Basingstoke, Hampshire and London: Macmillan Education, 1986.

Coleridge, Samuel Taylor. *Biographia Literaria or Biographical Sketches of My Literary Life and Opinions*. Ed. James Engell and W. Jackson Bate. Collected Works of Samuel Taylor Coleridge, vol. 7. Princeton, NJ: Princeton University Press, 1983.

———. *Lectures 1808–1819 On Literature*. Ed. R. A. Foakes. Collected Works of Samuel Taylor Coleridge 5. Princeton, NJ: Princeton University Press, 1987.

Conrad, Joseph. *Lord Jim: A Sketch*. *Blackwood's Magazine* October 1899-November 1900.

Cosslett, Tess. *The "Scientific Movement" and Victorian Literature*. Sussex: Harvester Press, 1982; New York: St. Martin's Press.

Cox, C. B. *Joseph Conrad: The Modern Imagination*. London: J. M. Dent, 1974.

Cox, R. G., ed. *Thomas Hardy: The Critical Heritage*. London: Routledge & Kegan Paul, 1970.

Darwin, Charles. *The Origin of Species*. New York and Toronto: New American Library, 1958.

Dean, Susan. *Hardy's Poetic Vision in* The Dynasts*: The Diorama of a Dream*. Princeton: Princeton University Press, 1977.

Eichner, Hans. "The Rise of Modern Science and the Genesis of Romanticism." *PMLA* 97 (1982): 8–30.

Feltes, Norman N. *Modes of Production of Victorian Novels*. Chicago: University of Chicago Press, 1986.

Fleishman, Avrom. *Conrad's Politics: Community and Anarchy in the Fiction of Joseph Conrad*. Baltimore: Johns Hopkins University Press, 1967.

Gittings, Robert. *Thomas Hardy's Later Years*. Boston: Little, Brown, 1978.
Gleick, James. *Chaos: Making a New Science*. New York: Viking, 1987.
Gould, Stephen Jay. Appendix A: "Stephen Jay Gould's Extemporaneous Comments on Evolutionary Hope and Realities." In *Darwin's Legacy*. Ed. Charles L. Hamrum. San Francisco: Harper & Row, 1983. 97–103.
Gregor, Ian. Introduction. *The Woodlanders*. By Thomas Hardy. New York: Penguin Books, 1981.
Griest, Guinevere L. *Mudie's Circulating Library and the Victorian Novel*. Bloomington: University of Indiana Press, 1970.
Hardy, Florence E. *The Later Years of Thomas Hardy, 1892–1928*. New York: Macmillan, 1930.
Hardy, Thomas. *The Dynasts. A Drama of the Napoleonic Wars, in Three Parts, Nineteen Acts, & One Hundred and Thirty Scenes*. London: Macmillan, 1904–8.
———. *The Woodlanders*. *Macmillan's Magazine*, May 1886-April 1887.
Hayles, N. Katherine. *The Cosmic Web: Scientific Field Models and Literary Strategies in the Twentieth Century*. Ithaca: Cornell University Press, 1984.
Hobsbawm, E. J. *The Age of Empire: 1875–1914*. New York: Pantheon, 1987.
Hughes, Linda K., and Michael Lund. *The Victorian Serial*. Charlottesville: University Press of Virginia, 1991.
James, Louis. "The Trouble with Betsy: Periodicals and the Common Reader in Mid-Nineteenth-Century England." *The Victorian Periodical Press: Samplings and Soundings*. Ed. Joanne Shattock and Michael Wolff. Toronto: Leicester University Press and University of Toronto Press, 1982. 349–66.
Jann, Rosemary. *The Art and Science of Victorian History*. Columbus: Ohio State University Press, 3985.
Jordanova, L. J. Introduction. *Languages of Nature: Critical Essays on Science and Literature*. Ed. L. J. Jordanova. New Brunswick: Rutgers University Press, 1986. 15–47.
Kern, Stephen. *The Culture of Time and Space, 1880–1918*. Cambridge: Harvard University Press, 1983.
Miller, J. Hillis. *Fiction and Repetition: Seven English Novels*. Cambridge: Harvard University Press, 1982.
———. *The Linguistic Moment*. Princeton: Princeton University Press, 1985.
Millgate, Michael. *Thomas Hardy: A Biography*. New York: Random House, 1982.
Morton, Peter. *The Vital Science: Biology and the Literary Imagination, 1860–1900*. London: George Allen & Unwin, 1984.
Orel, Harold. *Thomas Hardy's Epic-Drama: A Study of The Dynasts*. Lawrence: University of Kansas Press, 1963.
Roussel, Royal. *The Metaphysics of Darkness: A Study in the Unity and Development of Conrad's Fiction*. Baltimore: Johns Hopkins University Press, 1971.
Sherry, Norman, ed. *Conrad: The Critical Heritage*. London: Routledge & Kegan Paul, 1973.
Shuttleworth, Sally. *George Eliot and Nineteenth-Century Science: The Make-Believe of a Beginning*. Cambridge: Cambridge University Press, 1984.

Stedman, Raymond W. *The Serials: Suspense and Drama by Installment.* 2d ed. Norman: University of Oklahoma Press, 1977.

Toulmin, Stephen, and June Goodfield. *The Discovery of Time.* Chicago: University of Chicago Press, 1977.

Watt, Ian. *Conrad in the Nineteenth Century.* Berkeley: University of California Press, 1979.

9

Science and the Mythopoeic Mind: The Case of H.D.

Adalaide Morris

"Mythopoeic mind (mine) will disprove science and biological-mathematical definition," proclaimed the American poet H.D. in an autobiographical novel about a girl who fails conic sections, flunks out of college, and takes on the task of undermining the formulations of her scientist father (*HERmione* 76). Throughout her long career, in the questions she asked, in the answers she attempted, H.D. aspired to reconcile science and poetry by becoming what one of her characters calls "a sort of scientific lyrist" (*Nights* 24). Her poems bristle with terms from biology, chemistry, physics, and astronomy, with geometrical angles and shapes, with mathematical signs and equations, all of which she uses to adjust small details to larger patterns and formulate ever more exact, enduring, and inclusive laws. Although her work does not "disprove" her father's science, it achieves something more important and remarkable: it displaces his classical Newtonian physics with observations which parallel the revolutionary advances of twentieth-century science. The existence in H.D.'s work of ideas that also animate the theories of relativity, quantum mechanics, and the science of chaos is one of its most powerful and intricate features, a conjunction that tells us much not only about her writings but also about the larger relations between science, poetry, and the "mythopoeic mind" of the culture that contains them.

Born in 1886, H.D. grew up in a household of patriarchs who spent their days or nights peering through powerful magnifying lenses: her maternal grandfather, the Reverend Francis Wolle, an internationally known microbiologist, examined algae; her father, Professor Charles Leander Doolittle, charted the orbit of the earth around the sun; and her much older half-brother, Eric Doolittle, catalogued double and multiple star systems. Throughout the first years of H.D.'s life, her grandfather worked in the sitting room of the house next door and her father in a small transit house behind their home; after 1896, when her father became director of the University of Pennsylvania's Flower Observatory, her family lived on the grounds of the observatory where her father and half-brother conducted their research and H.D. helped introduce large companies of visitors to the wonders of the night sky.[1] Throughout her early life, scientific activities were as ordinary to H.D. as the bustle in the kitchen or the garden, a family tradition which entered her own professional life to widen her speculations, deepen her exploratory and experimental spirit, and encourage her to keep abreast of and appropriate for her work some of science's most intricate conceptual advances.

In these ways, for H.D., science was a direct influence and source of inspiration, "a new Muse," as one of her characters puts it, "one not to be treated lightly" (*Hedylus* 25). For many of the most interesting intersections between science and H.D.'s mythopoeic mind, however, direct influence is unlikely or even impossible. In considering these instances the critic's task is not to specify a Muse but to comprehend a shared musing, a meditative absorption in a cultural matrix. My assumption in this essay is that there are deep parallels between the literature and science of any given period, parallels that emerge as shared terms, concepts, rhetorical strategies, interpretive paths, inferences from events and behaviors, and unstated premises, the stuff, in short, of cultural history. That H.D.'s poetry plays out patterns formalized by chaos theorists more than a decade after her death is only one instance of an always complicated and compelling interplay between manifestations of a culture's myth-making mind.

H.D.'s life spanned the most important conceptual revolutions since Copernicus contended that the earth was not the center of the universe. When she was born in the late nineteenth century, physics was widely considered a closing book. Newton had explained gravitation, Maxwell and Boltzmann had explained heat, Maxwell and Faraday had explained electromagnetism, and scientists were advising their students to abandon hope of further breakthroughs and devote themselves to the labor of refining existing laws. By the time H.D. reached mid-career, however, almost every known "law" had been disputed, diminished, or discarded. "Nothing," Alfred North Whitehead exclaimed in the 1930s, "absolutely nothing was left that had not been challenged, if not shaken; not a single major concept. This I consider to have been one of the supreme facts of my experience."[2]

The breakthroughs that created the new physics coincided with H.D.'s formative years. In 1900, while she was listening to her father and half-brother discuss astronomy across the dinner table, Max Planck took the first crucial step toward quantum theory by noting that energy is absorbed and emitted not smoothly or continuously, as classical physicists had assumed, but in chunks, packets, or "quanta." Einstein published his theory of special relativity in 1905 when H.D. was pursuing her studies at Bryn Mawr, and in 1916, the year H.D. published her first volume of poems, he completed his greatest work, the theory of general relativity. In 1927, the year H.D. completed the bildungsroman in which her heroine pits her "mythopoeic mind" against her father's Newtonian certitudes, Niels Bohr and Werner Heisenberg formulated the principles of complementarity and uncertainty, principles which became known as the "Copenhagen Interpretation" and convinced most physicists of the coherence and correctness of quantum theory. And finally, in the early 1970s, some ten years after H.D.'s death, isolated mathematicians, physicists, chemists, biologists, and meteorologists began the investigations into the nature of irregular phenomena that have resulted in chaos theory, a theory some scientists consider the third major scientific revolution of the twentieth century.

The heroes of H.D.'s meditations on creativity are thinkers like Leonardo da Vinci and Sigmund Freud, figures who unsettle any easy opposition between science and art. In a crisscross she executes with some glee, H.D. habitually defines great artists as those who push toward truth and great scientists as those who devote themselves to a pure and poignant beauty. As she describes them, creators like da Vinci, Einstein, Freud, and Sappho sound remarkably alike, for each combines "artistic wisdom" and "scientific precision" and enacts the Greek notion of the "physicist" as the student of the nature of things. These artistic scientists and scientific artists ask the kind of questions that compel children—what is the universe made of, what are dreams, how does love work—and discover their answers not only through sophisticated intellection but also through hunches, flashes, and imaginative flights.[3]

Just as she aligned great scientists with extraordinary artists, H.D. understood ordinary creative work in terms of the investigations she witnessed in the households organized around her grandfather the microscopist and her astronomer father and half-brother. With few exceptions, artists in H.D.'s poetry and fiction are not lone, introspective geniuses but diligent workers in a collective enterprise, members of a team or group addressing a shared problem. When H.D.'s friend William Carlos Williams reported that the Doolittle household was dominated by the spirit of "a life of scientific research" (2) he meant precisely this kind of work: not the grand shifting of paradigms but a day-by-day, year-by-year, slow, sustained, exact and exacting observation of a multitude of similar instances. This, the life of the "research worker," was the vocation H.D. claimed for herself, a vocation she played out not as a microbiologist or an astronomer but as a poet.[4]

The activity that drove the men in H.D.'s family to spend long hours at the eyepieces of powerful magnifying lenses was the collection and classification of data characteristic of late Victorian science.[5] The lenses they looked through took them toward the edges of Newton's three-dimensional, linear, deterministic model of the world. Beyond the reach of her grandfather's microscope lay the subatomic units of matter whose behavior provoked the theories of quantum mechanics. Within the range of her father's and half-brother's telescopes but beyond their understanding lay the phenomena that verified Einstein's theories: the deflection of starlight near the sun which supports the hypothesis of curved space, the furious and enduring stellar burning that demonstrates the famous formula $E = mc^2$, and the double stars which illustrate a key postulate of special relativity, the notion that light, whether it comes from the side of the star rotating towards us or the side of the star rotating away, always moves with a constant speed.

Despite their approach toward the very small and the very large, however, the researchers in H.D.'s household stayed well within Newtonian parameters, and it was here, within these parameters, that H.D.'s "scientific lyricism" began. Before examining her drive to "disprove" Newtonian science, then, it will be useful first to look at her opposite and earlier desire to align herself with it. Although traces of this desire linger throughout H.D.'s life, it is dominant in the first phases of her career during which she developed three strategies to harmonize her poetry with her forebears' science. Each of these strategies is linked with a powerful figure from her youth: the two simplest, imitation and supplementation, are associated with her grandfather and father; the most ambitious, assimilation, is associated with her companion and suitor, the poet Ezra Pound.

I

H.D.'s grandfather, the Reverend Francis Wolle, was a passionate amateur. An inventor, schoolteacher, educational administrator, and Moravian minister, he retired in 1881 at the age of sixty-four to devote himself to his long-standing hobby: the collection, microscopic observation, and classification of cryptogamous plants. For at least four hours each day he peered through a J. Zentmayer "Army Hospital" microscope at slides smeared with pond scum and recorded his observations in a series of notebooks that provided the basis for his internationally recognized publications: *Desmids of the United States* (1884), the two-volume *Fresh-Water Algae of the United States* (1887), and *Diatomaceae of North America* (1890).[6] These catalogues not only name, classify, and describe thousands of varieties of algae, desmids, and diatoms but contain, in addition, some 6,000 precisely scaled and meticulously rendered front, lateral, and vertical views of cryptogamous plants, two-thirds of them hand-tinted to reproduce the colors that so delighted Wolle, first "green," he tells us, "then pink grading off into all the shades of purple, and finally olive, from golden green and bright tawny to black" (Wolle, *Desmids* xii).[7]

At a time when aspiring poets wandered the twilight landscapes of the *symbolistes* or paraded the heroic terrain of William Morris, compendia of pond scum might seem an unlikely model, but Charlotte Mandel is surely correct in contending that the Reverend Wolle's botanical passion, eye for structure, and attentiveness to detail enter H.D.'s first volume of verse (307–8). The terrain of *Sea Garden* is the coastland, salt marsh, and stream bank, places where earth and water mix to generate small, stubborn, resourceful vegetation. Carefully spaced through the volume are poems that vindicate its title by identifying and describing tough watery plants that most of us overlook. Together with their companion poems in *Sea Garden*, H.D.'s "Sea Rose," "Sea Lily," "Sea Poppies," "Sea Violet," and "Sea Iris" reproduce not only the Reverend Wolle's insistence on the "singular beauty" of his discoveries (*Diatomaceae* ix) but also his delicate language, his transfixed, interrogating gaze, and his push for taxonomic precision.

Wolle's descriptions of the varieties of algae are succinct and luscious. Prefaced by a name created from Greek or Latin components, each entry is first grounded in a particular locale—"attached to rocks," for example, "in rapids below water-fall, Pike County, Pennsylvania"—and then catalogued in precise, rhythmic, sinewy prose. "Oscillaria Froelichii," Wolle writes: "Stratum dark steel blue, or dark olive green, often elongated, radiating, opaque, shining"; "Floridiae": "plants rosy red or purple, dark reddish brown or blackish; multicellular, various in form; crustaceous, filamentous, fasciculate, verticillately branched" (*Fresh-Water Algae* I: 77, 315, 51). Unlike the overblown, vague, or sentimental language of his poetic contemporaries, Wolle's wording here is spare, precise, concrete, and vivid, an excellent model for the imagist poems which would bring his granddaughter acclaim.

H.D.'s *Sea Garden* poems are taut, restrained, and exact. Like Wolle's Greek and Latin nomenclature, their classical settings and mythological references give them an air of inscribing universal truths and enduring laws. Each of the poems is carefully situated in a specific landscape where, like a microscopist's, H.D.'s gaze isolates, suspends, and magnifies one small element of the scene. In the almost hallucinatory clarity of this gaze, each fleck, each streak and leaf-spine of the plant becomes visible. Many of the poems move with mesmerized precision toward a definitive description or name: "Amber husk," one begins, "fluted with gold, / fruit on the sand / marked with a rich grain"; "Weed, moss-weed, / root tangled in sand," another declares, "sea-iris" (*Collected Poems* 21, 36). These are investigatory poems, poems plotted and paced by a series of procedures meant to uncover a truth: in them the poet watches as wind, water, sand, and sun fling, lift, slash, stain, or shatter the plants, releasing them from her gaze only when they disclose some identifying mark of their vitality.

Although a bold break from prevailing poetic norms, H.D.'s imitation of the Reverend Wolle's science was facilitated by the fact that her grandfather recognized no disharmony between science and art and in his catalogues felt equally comfortable classifying his little forms, exclaiming over their beauty, admiring their resourcefulness, and comparing them to mythological beings.[8] By contrast, Professor C. L. Doolittle's *Treatise on Practical Astronomy as Applied to Geodesy and Navigation* seems stark and severe, the work of a professional tackling a dryly formulated problem within his discipline. It has, that is, the narrow focus and cool intellectual distance that Stephen Toulmin identifies as a mark of increasing scientific specialization and Evelyn Fox Keller sees as the result of the division of labor that has traditionally maintained science as a male preserve (Toulmin 231; Keller 178 and passim).

Although H.D. tells us that the professor wanted his daughter to become a "scientist like (he even said so) Madame Curie,"[9] she chose not to imitate what she understood as the masculine, intellectual, or critical cast of his science but instead to supplement it by positioning her poetry as its feminine, emotional, and creative complement. Such a configuration repeated what she understood to be the blending of science and art in her parents' marriage, but with a crucial difference: unlike that marriage, this was to be a relationship of equals and collaborators.[10] Its neatest figuration is encoded in the name H.D. selected for her fictional counterpart in a story set in classical Rome. This story portrays a writer whose poems are to be inset in and thereby complete a rare scientific manuscript. Her name—and the story's title—is "Hipparchia," female complement of the name of Hipparchus, the classical astronomer who made his fame by calculating the positions of the stars in their nightly transit across the skies (*Palimpsest*, 71).

H.D.'s strategy of supplementing her father's science was at once conservative and critical, for although it left conventional gender divisions in place, it was designed to rebalance a system dangerously tilted toward the "masculine." Professor Doolittle's dry clarity of mind, his distance from the objects of his scrutiny, the rigor of his mathematics, and the austerity of his work made him an almost paradigmatic embodiment of the cultural conflation of the words "scientific," "objective," and "masculine." Leading the arduous life of an academic astronomer, Doolittle conducted regular classes in astronomy and mathematics, directed an observatory, and made observations by night that he turned into complex calculations by day. Acclaimed for its exactitude, his work was to mark, again and again, the instant of time stars crossed the threads in the eyepiece of his zenith telescope and to use those observations to formulate mathematical laws for variations in latitude.[11] "My father studied or observed the variable orbit of the track of the earth round the sun," H.D. explained. "He spent thirty years on this problem, adding a graph on a map started by Ptolemy in Egypt" (*Tribute to Freud* 142).

H.D.'s historical reference here is accurate but not innocent, for she prefers Ptolemy's map to her father's diagram. Ptolemy formulated laws for determining variation in latitude, struggled, like Doolittle, with the problem of atmospheric refraction, and developed a predictive model to understand planetary motions, but he also codified the ancient Babylonian astrological tradition and became the founder of modern popular astrology (Sagan 50–53). For Ptolemy there was no clear line between Mars the astronomical notion and Mars the astrological presence. It is this amalgam of science, poetry, and religion that H.D. wills back into being in *Tribute to Freud* when she chooses to describe her father's work from the point of view of the child for whom the "hieroglyph" at the top of his "columns and columns of numbers . . . may stand for one of the Houses or Signs of the Zodiac, or it may be a planet simply: Jupiter or Mars or Venus" (25).

As if to correct the dry, formalized, and instrumental cast of her father's science and thereby remedy what was, in her view, a diminished enterprise, throughout her life H.D. habitually studied astrological star-maps, star-lore, and star-catalogues. In her thinking, astrology seems to stand as a kind of synecdoche for the hermetic traditions left behind at the birth of modern science, traditions which insisted that matter is everywhere suffused with spirit.[12] As the twentieth century progressed through two world wars toward the atomic and hydrogen bombs, H.D.'s desire to temper modern mechanical science with the old hermetic wisdom became more and more urgent. In 1957, long after she had found more effective means of displacing her father's science, she recorded a dream which revived and reenacted her old supplementary strategy, a dream in which she escorted her distinguished father up the steps of a cathedral to meet "the Queen, the Mother" and "reconcile [his] purely formal, rational, scientific mathematics + astronomy with the inner mystery of the letters and numbers + the astrology + star lore + 'myth' of the *Kabballe*."[13]

Each time H.D. scanned, recorded, and dreamed about the great drift of stars overhead, wore her zodiac ring, signed her letters with the sigil of a star, or conducted the reading, chart-deciphering, and meditation she called her "star research," she was engaged in the strategy of supplementation. For the most part, however, this strategy failed to nourish her art. When astrology enters her poetry and novels, it most often results in a forced and elaborate rhetoric of starry puns, allusions, and parallelisms.[14] Although the rhetoric can be interpreted as a hermetic and elusive metaphysics, it does not function as a counterpart or corrective to modern science. However yoked through their common ancestry in Ptolemy's maps, astrology and astronomy have diverged too radically to be effectively reconciled: as twentieth-century practices they inhabit separate spheres, speak different languages, and satisfy disparate human needs. Although supplementation persisted in the rituals of H.D.'s life, in her writing it soon yielded to other, more ambitious and dynamic strategies.

Where supplementation makes science and poetry adjacent and complementary disciplines, assimilation, H.D.'s third strategy, like imitation, her first, overlaps them. Instead of turning poets into perpetual apprentices, however, assimilation makes them full partners. In imitation the poet follows a scientific pattern, model, or sample, as H.D. did in copying her grandfather's botany; in assimilation, the poet takes up a position as a practitioner within the scientific community and makes independent contributions to knowledge. The claim of this strategy is not that poetry is *like* science but that it *is* science. Here poets aspire to do precisely what the lensmen in H.D.'s family did: to bring into focus regions hitherto invisible, to record what they see accurately and disinterestedly, and to turn their data into laws and equations.

The man who most vigorously formulated this aspiration was a poet who wooed the astronomer's daughter with such panache that he convinced her, she tells us, to choose, "because my life depends upon it, between the artist and the scientist" (Hirslanden Notebooks II: 27).[15] H.D.'s anecdote has the resonance of a fable, for like a princess selecting a suitor she is also at this moment choosing her fate. But the story has a deconstructive twist: having forced her to choose between two apparently opposite vocations, H.D.'s suitor then went on to elaborate a program that collapsed and conflated them. Between 1910 and 1920, the years during which he most had H.D.'s ear, Ezra Pound built his aesthetics on the conviction that the proper method of poetry is the method of science, a conviction he developed in a discourse that drew heavily on the terminology of chemistry, biology, mathematics, electromagnetism, radiology, and telecommunications.[16] Although, as we will see, H.D. eventually abandoned the Newtonian premises of Pound's argument, she never rejected its gist: the insistence that poets, like scientists, can discover and report on fundamental, enduring, and universal laws.

"Consider the way of the scientists rather than the way of an advertising agent for a new soap," Pound advised the aspiring imagist. "The scientist does not expect to be acclaimed as a great scientist until he has *discovered* something. He begins by learning what has been discovered already. He goes from that point onward. He does not bank on being a charming fellow personally" ("Retrospect" 6). Soap salesmen, charming fellows, and romantic poets persuade through force of personality; scientists and imagists persuade by accuracy of access, by a sensitivity so keen that it generates new data which can, in turn, produce new equations, formulas, or laws. For Pound, art has the two-phase rhythm of Doolittle's nightly observations and daily calculations: artists first observe, collate, and record occurrences, then formulate laws to explain them, laws that have the precision and authority of mathematics. "We learn that the equation $(x-a)^2 + (y-b)^2 = r^2$ governs the circle," Pound explains. "It is the circle. It is not a particular circle, it is any circle and all circles. It is nothing that is not a circle. It is the circle free of space and time limits. It is the universal, existing in perfection. . . . It is in this way that art handles life" (Gaudier-Brzeska 91).[17]

Like the botany and astronomy practiced in H.D.'s family, imagism relies on an assumption so fundamental as to seem self-evident: the belief that there is a "real" world separate from the observer, a world we can assess without in any way altering. When Einstein demonstrated that measurements depend on the observer's particular clock or ruler, he pegged relativity to the observer's location and in this way maintained his belief in an objective world that exists apart from the observer, but when Heisenberg demonstrated that our choice of what we observe makes an irretrievable difference in what we find, the foundations of Newtonian science and imagist poetry began to erode. When she turned from imitating, supplementing, or assimilating classical science and moved instead to "disprove" it, H.D. did so by formulating in her own way insights that also animate the sciences of relativity, quantum mechanics, and chaos. Her first sustained explorations of this cultural matrix occur in *HERmione*, a comic masterpiece in which she blasts free of all her mentors—the Reverend Wolle, Professor Doolittle, and Ezra Pound alike—and commits herself to the "mythopoeic mind (mine)."

II

"Her Gart went round in circles," H.D.'s novel begins. "'I am Her,' she said to herself; she repeated, 'Her, Her, Her'" (3). Her is in a hurry; Her has been hurt; Her is hurtling home through the woods carrying letters written to her (Her). The name of this heroine is a kind of mass-energy that manifests now as object, now as subject, now as possessive pronoun, now in some indeterminate form that seems to contain all the others. Her name appears always in a tangle of self-referentiality and usually with a grammatical wrench. Her doesn't fit. Rachel Blau DuPlessis and Susan Stanford Friedman have explored the way in which this "object case, used in subject place, exactly locates the thematics of the self-as-woman" (DuPlessis 61).[18] I am interested in the way in which this heroine named Her also challenges the thematics of classical science, among them its clear distinction between subjectivity and objectivity, its positing of absolute space and absolute time, and its faith in the possibility of precise linear statements.

At the opening of the book, our heroine, having flunked conic sections, returns home from college in disgrace. "Science . . . failed her" (6), the narrator explains in a formula that at first seems clear but soon begins to shimmer and spin. What does it mean to "fail?" Who has failed whom? Science in the guise of the academy has defined Her as lacking, which means Her has disappointed Science in the guise of her father and brother, the brilliant Carl and Bertrand Gart, formulators of "the Gart theorum of mathematical biological intention" (4).[19] Her has been declared inadequate; she has not passed. During the course of the novel, however, stimulated by the poet George Lowndes and the visionary Fayne Rabb, her mythopoeic mind examines science and finds it wanting. "Words may be my heritage," she declares, "and with words I will prove conic sections a falsity" (76). When words undermine Euclid's lucid schema, the tables turn and Her dismisses Science. Because Science has failed Her, that is, Her now fails Science.

In its purposeful, even perverse slipperiness, a phrase like "science failed her" foregrounds language. The oscillation between opposite meanings and the difficulty of establishing an absolute or "proper" interpretation mark the divide between Newtonian physics and the physics of relativity and quantum mechanics. Like the Reverend Wolle and Professor Doolittle, classical scientists strove to make true statements about a world they assumed to exist independently of the scientists who observe it; Einstein's relativity theory and the quantum mechanics of Heisenberg and Bohr make statements about a world they assume to be intimately entangled with the observer. Both relativity theory and quantum mechanics began in a critical evaluation of the process by which scientists observe the world they inhabit, but there is an important difference: while relativity theorists aim to extend the possibility of making "true" statements about the world, quantum theorists argue that all the mind can ponder is its own cognitive constructions. Written in 1927, H.D.'s novel is situated at a crossroads between these three visions of the world. The science of the Gart theorum looks backward toward Newton; the mythopoetics of Hermione Gart looks forward toward Einstein, Heisenberg, and Bohr.

The Garts are the Doolittles squared. Charles becomes the rigid and Germanic Carl, Hilda becomes the resonant Her, and the Doolittle house, which was merely attached to an observatory, becomes Gart Grange, where all the outbuildings have been converted into laboratories and every closet, linen cupboard, and butler's pantry has been crammed with bottles, zinc tanks, aquariums, and little dishes of sizzling acids. In the center of it all, smack in the midst of the family living room, perched under a screw of bright light at a desk stacked high with professional journals, is the unmoved mover of this enterprise: Carl Gart with his eye affixed to a microscope. Last and also—in this context—least, Hermione's mother, Eugenia, knits in an obscure corner: "He can work better," she explains, "if I'm sitting in the dark" (79).

Gart is a comic composite: part Wolle (his microscope slide is smeared with algae), part Doolittle (he explains the variations in planetary orbits), he is microbiologist, biologist, chemist, meteorologist, and astronomer. Like Newton's law of gravity, which joined the fall of an apple with the motions of the moon, the Gart theorum applies across all scales and explains everything from molecules in solution to earthquakes in Peru. Like Newton's, its theater is absolute time and absolute space, its geometry is Euclidean, and its claim is universal. With his tall skinny frame and piercing abstract eyes, Gart is the naive imperialist, "some sort of Uncle Sam, Carl-Bertrand-Gart God" who claims hegemony over everything: "There is only one solution," Her explains (96), and we know she means "Gart, Gart, Gart and the Gart theorum of mathematical biological intention" (4).

The only "science" Gart shuns is psychology (4), a lapse H.D. accentuates by making his primary preoccupation not physical science but the study of living organisms. For him vital processes are explicable in terms of matter alone; there is no need to posit a psyche. The materialism of his theory clashes merrily with its name, each term of which embeds some aspect of spirit: the "art" in "Gart," the "bios" in "biology," the learning (*mathema*) in "mathematics," and, most specifically, the resolve or intent in "intention." All these terms point to the subject, to the scientist who nonetheless believes his work to be "objective" and thus wholly uncontaminated by animus, design, or purpose. Gart is the paradigmatic classical scientist, so dispassionate and detached that when his daughter comes to announce *her* "intention," her plan to marry the poet George Lowndes, he cannot bring his abstracted eyes to focus on her (99–100). The Gart theorum, which kept Mrs. Gart in the dark, also "dropped out Hermione" (4).

H.D.'s depiction of the Gart family exposes the gendered substructure of a scientific ideology that pretends to neutrality while casting objectivity, reason, and mind as male, and subjectivity, feeling, and nature as female (see Keller, 6–7). In the Gart marriage the bifurcation is absolute: Gart is Athenian, his wife Eleusinian; he inhabits light, she dwells in darkness; he acts, she reacts. When fifteen years of his experimental work is destroyed in a flood, Gart's wife is heartbroken but he remains impassive. "Your mother takes these things too seriously," Gart explains to Hermione (92). If biology is in part the study of groups of organisms, and mathematics is in part the study of numerical operations, interrelations, and combinations, this novel's "mathematical biological intention" is its sly critique of the rigid and reductive operations of late Victorian familial and social life.

The problem is set in H.D.'s description of Gart Grange: "The house is columns of figures," she writes, "double column and the path at right angles to the porch steps is the line beneath numbers and the lawn step is the tentative beginning of a number and the little toolshed and the springhouse at the far corner of the opposite side is bits of jotted-down calculations that will be rubbed out presently" (83). As the novel begins, Bertrand's new wife, the washed-out, whiny Minnie Hurloe, must be factored in "like some fraction to which everything had to be reduced," a "common denominator" by which the family is minnie-mized, hurled low (15). The addition of Minnie divides the family into heterosexual couples, so that at dinner "Gart and Gart sat facing Gart and Gart" (35) and, once again, Hermione is dropped out, rendered an odd number until she is "normalized" by her engagement to George. The end of the novel undoes this equation, however, as Hermione attaches herself to Fayne Rabb, attenuates her commitment to George, is dropped out by their coupling, and then breaks down. She has flunked Normal Life as definitively as she failed Conic Sections.

"'Normal . . . ,'" muses Hermione toward the end of the novel, "their vocabulary gets more meagre" (177). The advantage of being thrust out of institutions like the family or the academy is a viewpoint from which sanctioned paradigms appear thin and insufficient and alternate forms of thought become visible. Hermione's "subtle form of courage" (4)—and, we surmise, her potential as a writer—is her ability to sustain this perspective, a perspective that renders all forms of "mathematical biological intention" suspect. The novel's mathematics poses late Victorian social expectations as a sort of mechanics of the heart analogous to Newton's celestial mechanics: a rigid set of rules specifying predictable patterns of attraction and permissible orbits. What the novel puts in question is the belief that such rules are inherent in nature rather than part of the framework we use to describe nature. Surrounding the cleared space of Gart Grange, outside the "numbers fencing us in" (97), is a "wild zone,"[20] a place where normal rules do not apply. This forest is Her's laboratory, the place where she conducts her own investigation into conic sections and discovers several principles that displace her father's Newtonian theories, principles analogous to those that animate the theories of relativity and quantum mechanics.

To "prove conic sections a falsity" would be to pull a linchpin in Newton's celestial mechanics. The stage of the Newtonian universe is the world of classical Euclidean geometry, a world coextensive with Western "common sense." Here material particles move through absolute space, which is empty, at rest, and unchangeable, and absolute time, which flows smoothly from past to present to future. What Newton's equations for gravitational force demonstrate is that on this stage the only possible orbit for a particle moving under the influence of another particle is a conic section: an ellipse, parabola, or hyperbola made by passing a two-dimensional plane through a three-dimensional cone. The consequences of these equations are profound, for they suggest that a hidden, preordained universal order drives everything from atoms to planets. Because orbits are fixed, the argument runs, if we knew any particle's initial conditions in sufficient detail, we would be able to calculate its motion into the indefinite future. The course of the universe, therefore, is both determined and, at least in principle, knowable.[21]

Newton's observations about orbits still have enough validity to get a man to the moon. What, then, would it mean to "prove conic sections a falsity"? Like Einstein, who surpassed Newton by pondering the vast reaches of outer space, and the quantum theorists, who displaced him by exploring the minute world of subatomic particles, H.D. displaced Newton by examining a region to which his assumptions were clearly insufficient. In the forest that surrounds the cleared space of Gart Grange, conic sections dissolve in a riot of vortices: here, in "cones of green set within green cones" (71), trees "swung and fell and rose" (64), birds whirled up and whirred down, and "heat seeped up, swept down, swirled about . . . with the green of branches that was torrid tropic water" (70). In the absolute space of Newton's world, particles in orbit are as hard, massy, and discrete as billiard balls, but here in the forest surrounding Gart Grange what goes round is a dynamic, oscillating mass-energy, a matter-wave manifesting now as color, now as heat, now as the flutter of bird-wings, the swirl of leaves. There is no privileged position apart from or above the whirl of this world, no universal clock or ruler to provide "true" or "objective" measurements of time and space. In the forest the angle from which any particular observer views the universe is only one among many, all of them partial, contingent, and relativistic.

Once these Einsteinian postulates are brought into play, the kingdom of Carl Gart tumbles and Hermione can claim her prerogative as an observer with her own acute and peculiar instruments. "Her Gart peered far," the narrator tells us, "adjusting, so to speak, some psychic lens" (5). Like an exuberant lab notebook, the novel records her relativistic measurements: a mosquito near the ear on a hot night is as huge as a chicken hawk (85), a peony petal near the eye covers the whole house (71), and a bee buzzing close almost "blot[s] out the sun itself with its magnified magnificent underbelly and the roar of its sort of booming" (13). Given the hegemony of "Gart, Gart, Gart, and the Gart theorum," it takes courage to affirm these perceptions as valid statements about the world, but to do so confirms two key postulates of relativity theory: the contention that measurements of size, distance, and duration vary from one frame of reference to another and the insistence that no one frame is inherently superior to any other.

When the poet George Lowndes dances into the novel wearing a conical hat, he seems at first to be the one who can free Her from the reductive formulas of Gart Grange. Like Pound, however, Lowndes is a classical scientist in poet's clothes. For him the observer retains both objectivity and mastery because he possesses a universally valid system of measurement. When Hermione enters the forest, she allows its dynamic swirl to dissolve all her preconceptions; when Lowdnes enters the forest, he forces it into alignment with his stock of literary tags. His measurements make Gart Grange's oak trees into "*the forest primeval, the murmuring pines and the hemlocks*" (65) or, alternatively, the forest of Arden (66). Longfellow's formula releases Lowndes's parody of Yankee literature, Shakespeare's formula sanctions his courtship of Hermione: together these two tags turn the forest into a stage for the rehearsal of Lowndes's literary preconceptions and the pursuit of his "biological intentions." Far from liberating Hermione, this way of thinking reduces the forest to yet another cleared space, brings back "the heated scrape of slate pencil across slate surface," and once again makes "numbers jog and dance . . . in [Hermione's] brain" (66).

At the end of the novel Hermione dismisses both the Gart formula and the Lowndes formula and embraces the intricate mode of perception she calls "mythopoeic mind (mine)." The mythopoeic or myth-building mind is the agent of the poet's powerful impulse toward pattern. Alert to the dynamics of representation, on the one hand, it is also firmly engaged with the world on the other. Without making exclusive claims to validity, the mythopoeic mind attempts to locate structures that will be the complex story of all the stories it feels to be true. Unlike materialist, mechanical, or behavioristic modes of thought, the mythopoeic impulse springs from a respect for "the mysterious," an awe Einstein located "at the cradle of true art and true science" (*Ideas and Opinions* 22).[22] Once committed to the workings of the mythopoeic mind, H.D. never again turned her art to the project of imitating, supplementing, or assimilating the science of her father and grandfather.

In her writings after 1927, H.D.'s mythopoeic way of seeing coincides in important particulars with both relativity theory and the theory of quantum mechanics. Although they developed in two quite different directions, these two bodies of thought are frequently lumped together as part of the revolution in physics that occurred during the first three decades of the twentieth century. Both of these theories stem from their founders' intuitions about the processes through which we observe the world, both deal largely with phenomena outside our daily experience, and both contain concepts that seem at first absurd or paradoxical, concepts that contradict our commonsense notions of space and time, cause and effect. Where quantum theory diverges sharply from relativity, however, is in its insistence on an unavoidable uncertainty in our descriptions of the world, an uncertainty H.D.'s work seems paradoxically at once to endorse and deny.

A number of exciting practices in her poetry suggest that like Heisenberg and Bohr, H.D. found uncertainty to be intrinsic to our apprehensions of the world. These practices, present throughout her work, surface most clearly in her late epic, *Helen in Egypt*: they include a habit of making probabilistic claims, an insistence on the simultaneous validity of mutually exclusive statements, and a suggestion that all the mind can ponder is its own enticing but insubstantial eidolons. Although many twentieth-century intellectuals confused this sort of thinking with "relativity," in actuality it was on this point that Einstein broke with the quantum theorists. For Einstein, relativity did not mean "everything is relative"; in harmony with a second and contradictory strain in H.D.'s work, he insisted on the possibility of making true statements about the world.

As N. Katherine Hayles reminds us, before Einstein settled on the term "relativity," he considered calling his hypothesis the "Theory of Invariance" (*Cosmic Web* 45). Although it argues that no one frame of reference can be privileged, his theory allows us to translate from one frame of reference to another. In this sense, Einstein's equations are a kind of Rosetta stone that lets us agree on the facts of a situation even though we each observe it from a separate point of view. The belief that there is an objective world these equations describe, a world that possesses a "radiant beauty" (22) which we can at least begin to comprehend, led Einstein to distance himself from the quantum theorists. "The theory accomplishes a lot," he wrote Max Born in 1926, "but it does not bring us close to the secrets of the Old One. . . . I am convinced that He does not play dice" (quoted in Bernstein 192). Some thirty years later in *Helen in Egypt*, H.D. echoed Einstein's formulation: "*God does not weave a loose web*," she wrote, "*no*" (82). In understanding H.D.'s attempts to trace the articulations of this cosmic web, it is helpful to leave the friction between relativity and quantum physics behind and read her work side by side with yet one more kind of science: contemporary chaos theory.

III

Like the cleared space of Gart Grange, classical science stopped at the edge of chaos. The green swirl of the forest, the storm clouds massed over the house, and the flood that washed out Gart's experiments were exceptions to the rule of ordered systems and, as Gart warned Hermione, only women took them "seriously." Gart's remark was offhand but not incidental: as unpredictable and embarrassing as a hysteric, turbulence was the unrepresented, unarticulated, unthought "other" of classical science.[23] The terms used to describe it, words like "ir-regularity" or "dis-order," merely reasserted the rule of the regular and orderly, a rule epitomized by the spare, lucid abstractions of Euclidean geometry. From our earliest schooling, this geometry has taught us to think of mountains as cones and coastlines as curves and thus to ignore the jaggedness and complexity of the world we inhabit. Scientific investigation of this complexity awaited the intuition that what we have termed "chaos" is in fact richly organized pattern.

The chaos theorists who began their work in the late 1960s and early 1970s were mathematicians, meteorologists, astronomers, biologists, chemists, and physicists, many of them mavericks who worked on eccentric problems with improvised equipment and considered their results idiosyncratic because they had no systematic way of understanding them.[24] What links their disparate work is the hunch that the phenomena they studied displayed a regular irregularity. In pursuit of this hunch, working at first independently and then, tentatively, in alliance, they developed a sophisticated language, an intricate set of mathematical procedures, and an array of astonishingly beautiful computer graphics.[25] In their studies of such phenomena as the shifts of weather, the ups and downs of animal populations, the rise and fall of the Nile, the fluctuation of cotton prices, and the fibrillations of the heart, they have been able to demonstrate that apparently random and accidental events have a structure that is in fact complex, flexible, and rich with information.

Like its precursors, chaos theory undoes the certainties of classical science: if relativity eliminated the Newtonian illusion of absolute space and time and quantum theory ended the Newtonian dream of precise, controllable measurements, chaos puts a definitive stop to the idea that the course of the universe is both determined and predictable. Chaos returns science from the very large and the very small to the dynamics of everyday life, to the "flapping, shaking, beating, [and] swaying" that surround us (Gleick 262). In its aim and exuberance, the science of chaos is closer to the practice of Carl Gart and the Reverend Wolle than it is to the fragmented, highly professional, scrupulously specialized discourse of contemporary science: like Gart, chaos theorists seek the harmonies of the whole, the patterns that link earthquakes in Peru to fish in tanks and molecules in solution, and, like Wolle, they unabashedly admire the "wondrous promise" and beauty of their findings (Mitchell Feigenbaum, quoted in Gleick 187).

Like chaos theory, H.D.'s mythopoeic writings pull seemingly random or disorganized phenomena into dynamic relation by discovering patterns which repeat across scales or recur one inside the next. Chaologists call these patterns "scaling" and "recursion," locate them in price charts and gene development, and give them mathematical formulation

(Gleick, 115–16, 179); H.D. locates them in the events of autobiography, history, and mythology and gives them mythopoeic expression. For her, the presence of such patterns makes an event "come true," a phrase she uses again and again to claim universality and validity for her perceptions. Things "come true" for H.D. not when they can be empirically verified or logically deduced but when they display symmetries between scales or within interlocking levels of a system.[26] Examples of scaling in H.D.'s writing range from the repetition of universal mythologies in local events to the reiterated structures in a seashell's whorl, the swirl of sparks from a bonfire, and the whirling of stars in a galaxy; examples of recursion include the many symmetries that link a poem's sounds, rhythms, images, and poetic or narrative structures to the "laws" toward which they drive.[27] For H.D., such patterns are not metaphors substituting for something similar, nor are they metonymies associated with something analogous: they are templates for particular shapes of matter or forms of behavior, the mythopoeic equivalents of the laws of chaos theory.

Scientists have developed several ways of representing the rich coherence of chaotic phenomena. The most haunting of these is the "strange attractor," an odd variety of a familiar scientific abstraction. An "attractor" is any point in an orbit that seems to pull the system toward it. Classical science recognized two kinds of attractors—fixed points and limit cycles—both of which can be illustrated by the orderly behavior of pendulums. Fixed-point attractors are characteristic of systems whose behavior reaches a steady state, like free-swinging pendulums that eventually stop at the midpoint of their arc; limit cycle attractors are characteristic of systems that repeat themselves continuously, like motor-driven pendulums that oscillate from one side of an arc to the other. Both fixed-point and limit-cycle attractors are simple and predictable; neither is "strange."

Strange attractors occur in the orderly disorder of chaos and are more complicated and difficult to understand. A pendulum's swing depends on only two variables, velocity and momentum, and can thus be represented on a two-dimensional graph, but chaotic data like stock-market prices or weather shifts depend on a vast number of variables and are therefore best charted in what physicists call "phase space." Phase space can have as many dimensions as a system has variables: in it the state of a system at any given moment is represented as a point that moves as the system shifts, tracing a continuous trajectory across a computer screen. What phase space diagrams show is that chaos too has an "attractor," a pattern that is neither a fixed point nor a limit cycle but an orbit that always stays within certain bounds without ever crossing over or repeating itself.[28]

"Strange attractors" are the forest that surrounds the Gart Grange of fixed points and limit cycles: they are everywhere and everything else. Whether the data charted in phase space come from measles epidemics or lynx trapping, whether it spans a week or a month or a millennium, whether it is local or global, the same trajectory appears again and again: a line that never doubles over itself loops round and round the computer screen in an infinitely deep and complex demonstration of the fine structure that constrains what we have thought of as disorder. The most famous strange attractor is a pattern first discerned in a phase space picture the meteorologist Edward Lorenz made from a set of nonlinear equations for the chaotic rotation of heated fluid.[29] Like all strange attractors, its trajectory is a continuous path of infinite complexity that never runs off the page and yet never exactly replicates itself. The shape it traces is a shape H.D. returned to again and again in her writings: a double spiral in three dimensions that looks like a butterfly with two large wings or an owl's mask with two deeply set ringed eyes.[30]

The coincidence between H.D.'s mythopoeic mind and the science of chaos offers rich and resonant access to aspects of her work that have been difficult to capture through conventional literary analysis. One particularly powerful demonstration of the overlap is the cascade of images that structures H.D.'s long poem *Trilogy*. Written in London amidst the turbulence of World War II, this is a poem about forms in motion: in it air thickens, wind tears, rain falls, bombs descend, and roofs tumble into ruins. Like the chaos theorists to come, the poet searches the borderland between order and disorder for the pattern that pulls all else toward it, a pattern that prevails across scales, through time and over space, a pattern whose every recurrence mixes characteristics that are "the same—different—the same attributes, / different yet the same as before" (105). The poem, though lyrical, is presented as the result of "research" and constructed hypotactically using connectives like "so," "moreover," "but," "however," and "for example": (51, 19, 54). Like Lorenz who sought the patterns that structure atmospheric turbulence, H.D. aimed to discover the "true-rune[s]" which she believed to be "indelibly stamped / on the atmosphere somewhere, // forever" (5, 17).

Like Lorenz's attractor, the shape H.D. traces in her poem is a looping spiral which is bounded and therefore finite but also unending and therefore, mysteriously, infinite.[31] The pattern appears again and again in a pulsing of hollow spaces within which degeneration turns into regeneration. The mollusc flutters its fans shut and open, the worm spins a shroud, the heart contracts and expands: from the first comes a pearl; from the second a butterfly; from the third, mysteriously, a "tree / whose roots bind the heart-husk // to earth" (8, 53, 35). A tree in a bombed courtyard, "burnt and stricken to the heart" (82), bursts suddenly into bloom; the city, fallen to ruins, becomes a rune of regeneration (3–4). The germ of new life comes from outside the system and throws it into turbulence. Sand cast into the mollusc shell, a grain cast into the heart, a bomb cast into the city, the philosopher's stone cast into a crucible, a seed cast into the womb: each occurrence propels the system from steadiness through turbulence into richly reorganized life. As in a strange attractor, so in the poem there is no end to the loop: the mummified pharoah in the ruined tomb of the poem's opening becomes the swaddled infant from the womb at its close, and the beginning, which had seemed to be an end, turns into an end which is also a beginning.

Like many chaotic patterns, H.D.'s rune is self-embedded: it repeats itself on finer and finer scales not only in the world but also within the poem. From its largest narrative, argumentative, and imagistic structures through the smallest details of its rhythm and phrasing, every level of *Trilogy* repeats the cycle of disturbance, disintegration, and reintegration. Even individual words are shattered and reconstituted so that "here" emerges from "there" (3), "mother" from "smother" (30), "word" from "sword" (18), and "ember" from "dismembered" (4). Like all chaotic patterns, the patterns in H.D.'s poem do not replicate each other exactly, but neither are they a jumble. They have a disorderly order that emerges slowly but surely, so slowly and so surely that the reader experiences something like the eerie inevitability one observer remembers feeling when he first watched a strange attractor form on the computer screen: the pattern "appears," he tells us, "like a ghost out of the mist. New points scatter so randomly across the screen that it seems incredible that any structure is there, let alone a structure so intricate and fine" (Gleick 150). What is crucial to remember, what differentiates this pattern from ornament, is the insistence of Pound's early poetics: for a poet like Pound or H.D. these intricate structures are not decorations but discoveries, laws discerned by the mythopoeic imagination.

In the constant exchange between the imaginative paradigms of physics and metaphysics, many writers have elaborated the implications of classical science's two attractors: the fixed-point downslide into entropy and the endless binary push-pull of limit cycles. H.D.'s mythopoeic mind does not so much disprove these classical paradigms as open another pattern to our imagination, one that suggests why, despite the second law of thermodynamics, nature continues to generate patterns of immense complexity and why, despite the seeming tyranny of limit cycles, most phenomena remain richly unpredictable. Weather changes, prices rise and fall, and we all go on thinking thoughts which repeat but never exactly replicate each other. Forced repetition of pattern may be totalizing, but the flowing geometry that builds coral banks, snowflakes, arterial paths, or tripartite poems is not: "my mind (yours)," H.D. insists, ". . . differs from every other / in minute particulars, // as the vein-paths on any leaf / differ from those of every other leaf // in the forest, as every snow-flake / has its particular star, coral or prism shape" (*Triology* 51–352). What chaos theory gives us is a way to think of order as a delicate interplay between the forces of stability and the forces of instability, between the pull toward similarity and the generation of difference, between determinism and unpredictability.

A theory like the Gart formula assumes a world in which effect follows cause in a rational and regimented fashion. For Gart, as for many Newtonians, events were both determined and, at least in theory, predictable: should sufficient data be available, they believed, our equations would be fully capable of forecasting the future. Einstein demonstrated that we cannot always tell cause from effect, however, and Heisenberg disputed our ability to gather any data beyond "probabilities" or "tendencies to occur." One of the most exciting aspects of chaos theory is its formulation of yet another way to conceive of the relationship between determinism and predictability. In the geometry of chaos, small scales interlock with large in such a way that minute changes in initial conditions can lead to large changes overall. Small perturbations cascade upward through a system in a manner that is determined but not predictable. Edward Lorenz gave this "sensitive dependence on initial conditions" its most graphic example by suggesting that a butterfly stirring the air today in Peking can cause storm systems next month in New York. For the scientists who led the revolution, the Butterfly Effect was the beginning of chaos theory (Gleick 8).

Born at least a generation earlier than most of the prominent chaos theorists, H.D. suffered not only the terrors of World War I and World War II but also a persistent nightmare of the impending holocaust of World War III. For this reason, I believe, the urgency behind her work is not just intellectual and aesthetic but also moral. In looking for the fine structure hidden within disorderly streams of data, she is also looking for the ways in which a poet's local intervention might avert global disaster. This ambitious project is a consequence of her understanding of the manner in which small scales intermesh with large. "Sensitive dependence on initial conditions" suggests that poets in wartime are not "useless" or "pathetic" (*Trilogy* 14), for the stroke of a poet's pen might change the course of the world. If to a scientist like Lorenz the Butterfly

Effect meant that even the most informed forecast is speculative, to an artist like H.D. it means the possibility of metamorphosis. This is the nonscientific but nonetheless fervent hope that gives her search for recursive structures its distinctive energy and edge.

H.D.'s early attempts to imitate, supplement, or assimilate classical science are consistent with the received view of the relationship between science and literature as a hierarchy in which the writer's task is to help transmit scientific insight. What H.D. managed to achieve in her writings, however, was something much more bold and interesting. In 1927, when she cut loose from her father and grandfather, committed herself to her own peculiarities, and boasted that "mythopoeic mind (mine) will disprove science," she pitted her poetic intelligence against classical science. Paradoxically, this turn away from science allowed her not just to keep up with it but to anticipate it: she made good on her claim of disproving classical science by following her intuitions toward the revolutionary insights of relativity, quantum mechanics, and chaos. Her marginality as a woman and a mystic, her knowledge of mythology and pre-Newtonian science, and the history that situated her again and again in the midst of turbulence no doubt contributed to these intuitions, but the main factor was surely the one H.D. herself identified: the intricate will-to-pattern in the mythopoeic mind of poet and scientist alike.

What H.D.'s work suggests is that the relationship between science and poetry is not finally hierarchical or oppositional. Artists and scientists participate in the same mythopoeic matrix, a matrix that shapes them as definitively as they in turn shape it. When chaos theorist Mitchell Feigenbaum maintains that "[artists] can do some of my research for me" (quoted in Gleick 186), he is making the same point H.D. makes by identifying herself as a "researcher" and arguing that "we are each one of us a Galileo, a Newton. We may make discoveries that the human mind has not yet, so far, been in a position to make, about the human mind" (*Within the Walls*, "Blue Lights"). There is little evidence that H.D. read scientific theory[32] and none that Feigenbaum reads modernist poetry, but direct influence is beside the point: if H.D. and Edward Lorenz are right, sensitive dependence on initial conditions is at work in the world and we are all affected by even the slightest and most delicate movements of the mythopoeic mind.

Notes

1. H.D.'s participation in observatory tours is noted by Norman Holmes Pearson in his Foreword to H.D.'s *Hermetic Definition*, n.p. Although I have been unable to locate them, Pearson reports that H.D. also at this time published

articles on astronomy for children in a Presbyterian paper ("Norman Holmes Pearson on H.D.: An Interview," *Contemporary Literature*, 10 [1969]: 437). For the Doolittle family's move to the grounds of the Flower Observatory, see Barbara Guest, *Herself Defined: The Poet H.D. and Her World* (Garden City: Doubleday, 1984), 15–16.

2. The summary statements and quotation from Whitehead come from Timothy Ferris's excellent survey of twentieth-century astrophysics, *The Red Limit: The Search for the Edge of the Universe* (New York: Quill, 1983) 220. I am also indebted to Robert H. March's formulations in his *Physics for Poets*.

3. For H.D.'s meditations on creativity, see especially *Thought and Vision* 64 and passim. For H.D.'s account of Freud's "flash of inspiration," see *Tribute to Freud* 77; for an account of one of the imaginative flights that sparked Einstein's theories of relativity, see March 110–12.

4. To take seriously H.D.'s statement that "[my father] did make a research worker of me but in another dimension" is to significantly realign our reading of her work. In light of this self-definition, for example, H.D.'s repetition of plot, incident, and character cannot be regarded simply as a mark of obsession or blockage, for it is through just such reiteration of instances that researchers confirm a pattern or establish a law. H.D.'s statement appears in the Hirslanden Notebooks and is quoted in Norman Holmes Pearson's Foreword to *Tribute to Freud* x.

5. For a helpful summary of H.D.'s scientific forefathers, see Mandel.

6. H.D.'s first cousin, also named Francis Wolle, gives a detailed account of the Reverend Wolle's career in his memoir, *A Moravian Heritage*. See especially chapter 4, "Francis Wolle: Teacher, Inventor, Minister, Scientist (1817–1893)," 21–27. For more information on the Reverend Wolle, see H.D.'s "Notes" to *The Gift* 4–5, in typescript at the Beinecke Rare Book and Manuscript Library, Yale University, and an essay by Francis Drouet, Curator of Cryptogamic Botany at the Field Museum, "Francis Wolle's Filamentous Myxophyceae," *Field Museum of Natural History*, Botanical Series, 20.2 (1939): 17–64.

7. H.D.'s cousin tells us that as a child she earned pennies by carrying the loosely rolled bundles of black-and-white key sheets to the "needy female relatives" whose task it was to tint them (Wolle, *Moravian Heritage* 25).

8. See, for example, the description of *Chlamydococcus pluvialis* in Wolle, *Fresh-Water Algae* I: 164.

9. From the Hirslanden Notebooks, quoted by Pearson in his Foreword to H.D., *Tribute to Freud* x.

10. For two of H.D.'s many figurations of her parents' marriage, see *Tribute to Freud* 145, and "The Sword Went Out to Sea," typescript at the Beinecke Library 144.

11. For Doolittle's professional standing, see the obituaries in *Publications of the Astronomical Society of the Pacific* 31 (1919): 103–4, and *The Observatory* 42 (1919): 219–20. For a description of the mathematical calculations involved in his work, see his *Practical Astronomy* 283–339.

12. See Keller's description of the advent of modern science in *Reflections on*

Gender and Science 43–65. For a thorough and important discussion of H.D.'s occult studies, see Friedman 157–296.

13. This dream, recorded in the Hirslanden Notebooks, is quoted by Friedman 188.

14. For a typical stellar dream, see "The Last Day," in "Within the Walls" 3, at the Beinecke; for the zodiac ring which she considered a token of her "marriage to the 'STARS,'" see her letter to Bryher dated "Halloween 1930," at the Beinecke; for her "star research," mentioned in a letter to Viola Jordan, July 2, 1941, see her letters to Silvia Dobson, Robert Herring, and Elizabeth Ashby, many of them signed with her star glyph, at the Beinecke. Finally, for a sustained example of H.D.'s rhetoric of starry puns and parallels, see her letter to Bryher, May 28, 1933, also at the Beinecke. All unpublished material from the Beinecke is cited with the kind permission of Perdita Schaffner and New Directions Publishing Corporation.

15. The two most sustained versions of the courtship are contained in H.D.'s novel *HERmione* and her *End to Torment: A Memoir of Ezra Pound* (New York: New Directions, 1979).

16. For two informative studies of the scientific strain in Pound's aesthetics, see Ian F. A. Bell's *Critic as Scientist: The Modernist Poetics of Ezra Pound* (London: Methuen, 1981) and Max Nanny's *Ezra Pound: Poetics for an Electric Age* (Bern: Francke, 1973).

17. See also Pound's statement that "poetry is a sort of inspired mathematics, which gives us equations, not for abstract figures, triangles, spheres, and the like, but equations for the human emotions" (*Spirit of Romance* [Norfolk: New Directions, 1952] 5). For a few of the many examples of this mathematical vocabulary in H.D.'s work, see *Compassionate Friendship*, typescript at the Beinecke, 114; *Palimpsest* 155; "Pontikonisi (Mouse Island)," by "Rhonda Peter," *Pagany* 3.3 (1932): 1; and *Tribute to Freud* 146.

18. For a full elaboration of this argument, see Susan Stanford Friedman and Rachel Blau DuPlessis, "'I Had Two Loves Separate': The Sexualities of H.D.'s *Her*," *Montemora* 8 (1981): 12.

19. "Theorum" is H.D.'s own, perhaps comical spelling of the proper "theorem." Because it seems to reproduce Carl Gart's rough-cut mumbling, however, I have retained H.D.'s spelling. For an account of H.D.'s failure in "Analytical Conics and Theory of Equations," which precipitated her withdrawal from Bryn Mawr, see Emily Mitchell Wallace, "Athene's Owl," *Poesis* 6, no. 3–4 (1985): 116.

20. The term "wild zone" comes from anthropologists Shirley and Edwin Ardener and is applied to women's experience by Elaine Showalter, "Feminist Criticism in the Wilderness," in *The New Feminist Criticism: Essays on Women, Literature and Theory*, ed. Elaine Showalter (New York: Pantheon, 1985) 262.

21. For a brief summary of Newton's equations and the Marquis Pierre Laplace's formulation of their deterministic consequences, see Bernstein 29–31.

22. In understanding what H.D. means by the term "mythopoeic," it is helpful to keep in mind Robert Duncan's work, especially his *The Truth and Life of Myth: An Essay in Essential Autobiography* (Fremont: The Sumac Press, 1968).

23. For an elaboration of this argument in the context of contemporary literary theory and the science of chaos, see N. Katherine Hayles, "Turbulence in Literature and Science: Questions of Influence," manuscript.

24. For my understanding of the science of chaos, I am indebted to the work of James Gleick and N. Katherine Hayles, especially to Gleick's *Chaos* and to Hayles's *Chaos Bound*.

25. For three of the most compelling of these graphics—the "Lorenz attractor," "Koch curve," and "Mandelbrot set"—see the illustrations in Gleick, inset after p. 114.

26. For an example of H.D.'s use of the phrase "come true," see *The Gift* 17.

27. See H.D.'s *Tribute to Freud* and *The Gift* for the coincidence of global mythology and local events; see *Trilogy* and *Helen in Egypt* for the overlapping patterning of shells, sparks, and stars. For recursive patterns in H.D.'s poems, see the analysis of *Trilogy* below.

28. For more detailed descriptions of strange attractors, see Gleick, 49–51, 132–37; Hayles, *Chaos Bound*, chapter 6, "Strange Attractors: The Appeal of Chaos"; Robert Shaw, "Strange Attractors, Chaotic Behavior, and Information Flow," *Zeitschrift für Naturforschung*, 36A (January 1981): 79–112.

29. For a description of Lorenz's process, see Gleick, 23–30.

30. The butterfly and the owl are two of H.D.'s sigils. For the butterfly, see especially "The Walls Do Not Fall," the first poem of *Trilogy*; for the owl, see especially "Sagesse" in *Hermetic Definition*.

31. For an elaboration of the paradox of infinite forms in finite space, see Gleick 98ff.

32. The catalogue of H.D.'s personal library at the Beinecke includes no scientific texts beyond *The Stars in Their Course* (Cambridge: The University Press, 1931) and *The Mysterious Universe* (Cambridge: The University Press, 1932), two explications of contemporary science by the respected mathematician, theoretical physicist, and astronomer Sir James Jeans.

Works Cited

Bernstein, Jeremy. *Einstein*. New York: Penguin, 1976.

Doolittle, C. L. *A Treatise on Practical Astronomy, as Applied to Geodesy and Navigation*. New York: John Wiley and Sons, 1885.

DuPlessis, Rachel Blau. *H.D.: The Career of That Struggle*. Brighton: Harvester Press, 1986.

Einstein, Albert. *Ideas and Opinions*. New York: Dell, 1973.

Friedman, Susan Stanford. *Psyche Reborn: The Emergence of H.D.*. Part II. Bloomington: Indiana University Press, 1981. 157–296.

Gleick, James. *Chaos: Making a New Science*. New York: Viking, 1987.

Hayles, N. Katherine. *Chaos Bound: Orderly Disorder in Contemporary Literature and Science*. Ithaca: Cornell University Press, 1990.

———. *The Cosmic Web: Scientific Field Models and Literary Strategies in the Twentieth Century*. Ithaca: Cornell University Press, 1984.

H.D. *Collected Poems 1912–1944*. Ed. Louis L. Martz. New York: New Directions, 1983.

———. *The Gift*. New York: New Directions, 1982.

———. *Hedylus*. Redding Ridge, Conn.: Black Swan Books, 1980.

———. *Helen in Egypt*. New York: New Directions, 1961.

———. *HERmione*. New York: New Directions, 1981.

———. *Hermetic Definition*. Foreword by Norman Holmes Pearson. New York: New Directions, 1972.

———. Hirslanden Notebooks, manuscript. Beinecke Rare Book and Manuscript Library, Yale University.

———. *Nights*. New York: New Directions, 1986.

———. *Notes on Thought and Vision & The Wise Sappho*. San Francisco: City Lights, 1982.

———. "Notes" to *The Gift*, manuscript. Beinecke Rare Book and Manuscript Library, Yale University.

———. *Palimpsest*. Carbondale: Southern Illinois University Press, 1968.

———. *Tribute to Freud*. Foreword by Norman Holmes Pearson. New York: New Directions, 1984.

———. *Trilogy*. New York: New Directions, 1973.

———. "Within the Walls," manuscript. Beinecke Rare Book and Manuscript Library, Yale University.

Keller, Evelyn Fox. *Reflections on Gender and Science*. New Haven: Yale University Press, 1985.

Mandel, Charlotte. "Magical Lenses: Poet's Vision Beyond the Naked Eye." In *H.D.: Woman and Poet*. Ed. Michael King. Orono: National Poetry Foundation and the University of Maine at Orono, 1986. 301–17.

March, Robert H. *Physics for Poets*. New York: McGraw-Hill, 1978.

Pound, Ezra. "A Retrospect." *Literary Essays of Ezra Pound*. Ed. with an introduction by T. S. Eliot. New York: New Directions, 1935. 3–14.

———. *Gaudier-Brzeska: A Memoir*. New York: New Directions, 1970.

Sagan, Carl. *Cosmos*. New York: Random House, 1980.

Toulmin, Stephen. *The Return to Cosmology: Postmodern Science and the Theology of Nature*. Berkeley: University of California Press, 1982.

Williams, William Carlos. "A Letter from William Carlos Williams to Norman Holmes Pearson Concerning Hilda Doolittle and Her Mother and Father (11 July 1955)." *William Carlos Williams Newsletter* 2, no. 2 (1976): 2–3.

Wolle, Francis. *A Moravian Heritage*. Boulder: Empire Reproduction & Printing Company, 1972.

Wolle, Rev. Francis. *Desmids of the United States*. Revised and enlarged ed. Bethlehem: Moravian Publication Office, 1892.

———. *Diatomaceae of North America*. Bethlehem: The Comenius Press, 1890.

———. *Fresh-Water Algae of the United States*. Bethlehem: The Comenius Press, 1887.

III

CHAOS AND ORDER: PROBING THE LIMITS

10

Representation and Bifurcation: Borges's Garden of Chaos Dynamics

Thomas P. Weissert

Turbantibus aequora ventis: pockets of turbulence scattered in flowing fluid, be it air or salt water, breaking up the parallelism of its repetitive waves. The sweet vortices of the physics of Venus. How can your heart not rejoice as the flood waters abate (*décliner*) and the primordial waters begin to form, since in the same lofty position you escape from Mars and from his armies that are readied in perfect battle formation? In these lofty heights that have been strengthened by the wisdom of the sages, one must choose between these two sorts of physics.

Michel Serres, *Hermes: Literature, Science, Philosophy*

The Plurality of times presupposed by the special theory of relativity must be construed as illusion or, rather, as the effects of perspective.

Henri Bergson, *Durée et Simultanéité*[1]

To each partial system, relatively independent of the environment, we assign a local model that accounts qualitatively and, in the best cases, quantitatively for its behavior. But we cannot hope, a priori, to integrate all these local models into a global system. . . . The era of grand cosmic synthesis ended, very probably, with general relativity, and it is most doubtful that anybody will restart it, nor would it seem to be useful to attempt to do so.

René Thom, *Structural Stability and Morphogenesis*

Michel Serres argues that great literature often discovers scientific truth long before the scientists get around to it (*Feux et signaux*).[2] In "The Garden of Forking Paths," Jorge Luis Borges discovered the essence of bifurcation theory thirty years before chaos scientists mathematically formalized it.

If we think of all the various disciplines of creative thought in this culture as unaffected by one another, we would have serious difficulty making sense of Serres's argument and regarding Borges's story as anything more than an inspired fluke. But if we can conceptualize the complex dynamics of culture as a fluid system in which each of the disciplines is a current of information, we could easily understand how the situation described by Serres could arise. These currents are not isolated but are constantly intermixing their ideas in a process which could only be described as stochastic. Each current carries a quantity of information for a while, processes it, changes it, and then returns it to the central flow. Because people are influenced by the representations of their society, no discipline can remain isolated. Thus an idea or structure may arise in literature first before it makes its way into scientific formalism. But it might also be the case that a scientific theory influences a work of literature. With this model, all the interesting dynamic structures of fluids come into metaphorical play: eddies, flows, bifurcations, feedback loops, mixing, and, of course, the most interesting feature of all, turbulence. In cultural dynamics, as in hydrodynamics, linearity must be abandoned because the flow of ideas is clearly nonlinear. As the culture flows from movement to movement, or from paradigm to paradigm, currents become complex. In this century, the most significant example of such a passage is the transition from modernity to postmodernity. As disciplines move beyond their modern boundaries, they generate ripples in the cultural flow, and these ripples influence the other currents in a nonlinear, random way. I will discuss this transition in terms of the reciprocity between two of these currents—literature and physics.

Jorge Luis Borges stands as a transitional figure between modern and postmodern literature. Whereas determinism, manifesting itself in his work as closure and textual unity, connects him to literary modernism, his self-conscious narrative convolutions point toward postmodern literary techniques. Saturated with scientific thought, his work makes an excellent medium for a discussion of modernism and postmodernism in physics. The first revolution in physics in this century followed the publication of Einstein's theories of relativity. The multiplicity of perspectives implied by these theories, in conjunction with their global formulation, defines them as exemplars of modernist physics. Einstein himself was a determinist, and his project of trying to find the unified field theory exemplifies the character of modernist physics. Quantum mechanics was the second revolution, confuting determinism on the microscopic level but leaving macroscopic predictability intact. Chaos theory is the third, and I would argue that it is the first, move toward a postmodern physics in practice.[3] In nonlinear dynamical systems, islands of order arise from the sea of chaos. The interspersed order, comprehensible where chaos is not, implies abandoning the modernist pursuit of a global theory and attempting to compile a postmodern catalogue of local theories.

In his "Garden," Borges makes references to Einstein, and his theories locate his construction of several levels of narrative reality within a relativistic universe. Further, he presents a narrative labyrinth which involves an infinity of relative perspectives. Thus we see the influence of modernist physics in his work. But his narrative also involves nonlinearity and a theory of bifurcations remarkably similar to a formalized theory devised by chaos theorists some thirty years after the publication of "Garden." This nonlinearity implies the defeat of an entirely comprehensive global theory. As representations, both scientific models and literary texts employ levels of reality. Correlating these levels of reality in model and text for a specific case will show how science and literature work as consensual constructions.

Levels of Reality: Modeling and Narrative

Mathematical modeling and computer simulation have become invaluable tools for a theorist studying dynamics. A typical physical system studied this way is the atmosphere—a large, fluid system which cannot be solved exactly with the deterministic laws of Newtonian dynamics. Because it is chaotic and therefore never repeats itself exactly, its dynamics can never be wholly predictable. However, certain recognizable, large-scale dissipative structures do tend to reappear randomly from this chaos—for example, "streets" of clouds, which are formed when roll-like convection cells, heated at the surface of the Earth, rise to dissipate that heat into the cooler air of the upper atmosphere. These are the spontaneous local islands of order which a dynamicist wishes to understand and categorize. With mathematics and computer text, he constructs a three-tiered system of representations, each of which, although intrinsically different in character, reveals some essential quality of the original physical system. Each additional level adds more information to the whole system. Upon the completion of his experiment, the dynamicist returns his processed information to the cultural stream by submitting an article to a refereed journal.

In "Garden," Borges also creates levels of reality, which are arranged in concentric succession like Chinese boxes (Barrenechea 38). Borges uses four levels of reality in "Garden," three of which parallel the three levels generated by the dynamics theorist, and one of which serves to represent man's subjectivity in his attempts to understand his world. The closest analogue to this fourth level in the scientific journal article is the affirmation of the journal's referee.

In dynamics, the first level of reality is often an experimental, laboratory-scale, physical model of the larger, natural system. Typically, the experimenter will build a small-scale device which simulates some of the essential conditions of the natural system. Obviously, simplifications must be made to render the analysis wieldly. The scientist makes these choices using his acquired scientific intuition. This level of representation has tangibility as its obvious main feature; however, its originary position among the sequence of representations also implies a certain Platonic purity. The combination of tangibility and Platonic purity generates a special reader-confidence in the data reported in this phase. This assumed reader-confidence has led to the existing protocol requirement that these measurements be held up as a standard to which findings from other, more removed representations must consistently conform. The reader perceives this representation, via pictures of the apparatus and the measurements reported, as a first mediated reality within the text (the article).

In the analogous first level of "Garden"—the first level of representation of a fictive reality—the central character, Dr. Yu Tsun, is a Chinese native working as a German spy in England during World War I. He pays a visit to the home of Mr. Albert, a sinologist who has studied the writings of one of Tsun's ancestors. The entire narrative present takes place within a one-day interval in the lives of these two characters. The reader obtains direct knowledge of the activity of these characters through the first-person narrative of Dr. Tsun, which is analogous to the scientific reader's acquisition of knowledge of the tangible, physical model via the measurements and description included in the journal article.

The second level of representation in dynamics is a mathematical abstraction. The theorist constructs a set of differential equations that mathematically represents the forces experienced by the dynamic elements of the physical model, thereby creating an abstract conceptualized model of the physical mechanism. These are called *differential* equations because they contain derivatives, which describe how the dynamic elements respond to changes in the independent variables.[4] In chaos theory, theorists employ *nonlinear* differential equations in their models.

Nonlinearity is chaos theory's point of departure from classical dynamics, because Newtonian dynamics is only equipped to handle linear phenomena. A system is linear if, given some specific change in an independent variable such as temperature, the system's response is a proportional change in the dependent variable such as particle velocity. However, most real physical systems do not behave in this linear manner. Quite often, small changes in independent variables can cause very large changes in the dependent variables. Some nonlinear systems can be described by linear differential equations—exponential growth is an example—but many require equations containing products of derivatives, which in general have no explicit solutions. With intuitive insight, the chaos theorist attempts to find the best set of nonlinear equations to represent the physical system—one that most closely describes and predicts the measurements obtained in the first representation.

Whereas the first level is essentially tangible, this second level is inaccessible to direct sensory experience. Although we can never directly experience this representation, if we ask the right questions of the equations (i.e., manipulate them in a suitable way), they will give us information about the system. One of the activities of a chaotician is the creation of an abstract "space" in which he can study the behavior of the theoretical entities that represent physical systems. In his article, he includes his equations, their solutions, his method of finding those solutions, and various graphs, but he cannot include the representation itself because it exists only in a mathematical, nonphysical space that is directly inaccessible to his audience. Even though equations describe a model, they in themselves are not the model.

Just as a mathematical model can be known only through equations, a historical person can be known only through historical writings. The second narrative level in "Garden" is the story of Ts'ui Pên, Dr. Tsun's great-grandfather—and it too has an ephemeral quality. Our knowledge of Ts'ui Pên and his life is made doubly uncertain, first by the mediation of that information through Tsun and Albert, and, second, by the epistemological barrier of historical conjecture itself. This information has been passed down orally and, we must assume, through some scraps of Chinese historical documentation. The reader—as well as Albert and Tsun—can never know Ts'ui Pên directly but only know about him through hearsay.

Even though the first two narrative levels are qualitatively different, they are connected, just as are the levels of modeling. This second narrative level conforms to the pattern of the first in that there are definite parallels between Mr. Albert, a character from the first level, and Ts'ui Pên.[5] Albert had been a missionary in Tientsin before he aspired to become a sinologist. He gave up this vocation in life to study the life and work of Ts'ui Pên. Ts'ui Pên had also given up his past life to create "The Garden of Forking Paths": "A strange destiny . . . that of Ts'ui Pên—

Governor of his native province, learned in astronomy, in astrology and tireless in the interpretation of the canonical books, a chess player, a famous poet and a calligrapher. Yet he abandoned all to make a book and a labyrinth" (95). Albert lives in solitude, and his house is surrounded by a garden with zigzag paths (95). Analogously, Ts'ui Pên had "shut himself up in the Pavilion of the Limpid Sun" which was "set in the middle of an intricate garden" (96). Finally, at the end of the story Albert is killed by Tsun, a stranger. So, the reader is told, was Ts'ui Pên also killed by a stranger. Therefore this second, directly inaccessible level of the story represents a life which conforms to the model created in the first level of representation. In both cases conformity is maintained between levels to provide the reader with continuity, although each additional level also adds new information to the system as a whole.

The third level of reality in dynamics is a visual, graphical representation of the possible behavior of the physical system over a specified period of time. To obtain this diachronic demonstration, the experimenter uses the equations describing the variables to plot them on a set of axes. The graphs show how the system's variables—usually position and velocity—change over time. These plots conform to plots of the measurements taken in the first representation, so that we again obtain continuity. This final representation offers the scientific reader the opportunity to spatialize the system's flow in time, an important step in the development of his dynamical intuition.

The third and innermost level of Borges's story within a story within a story is, as we might expect, a diachronic demonstration of the dynamic flow of the variables—the actions of its characters. Ts'ui Pên's manuscript is a graphic representation of the flow of its characters, just as the short story itself is a graphic representation of the flow of Albert and Tsun during one day. Whereas the flow of Borges's narrative is linear and easy to follow, however, the flow of Ts'ui Pên's novel is nonlinear, which is the reason that no one until Albert could understand it. We can attribute Albert's solution to his ability to spatialize the nonlinear flow; only then could he recognize its hidden, ordered structure. Thus these three levels of "Garden" are structured analogously to the three levels of modeling, each adding additional complexity and information to the whole system while maintaining internal consistency.

Subjectivity: The Other Level

Thus far, I have outlined a structural levels-of-reality parallel between the products of two different cultural representations, a short story and a scientific journal article. What can be said about the purposes underlying the generation of these levels; how are they related? Scientists create their system of levels as a method to model the workings of a physical system, the reality of which, they assume, underlies their physical perceptions. They believe that each successive representation still contains some essence of that reality. By comparing relationships among quantities within each of those levels, they better acquaint themselves with the rules governing that reality. However, scientists are increasingly recognizing that no representation of reality can be entirely value-free. The scientific method, as it evolved, required the presumption of objectivity, separating the scientist from his object of study. But as science entered the twentieth century, its theories came to reflect the impossibility of achieving objectivity absolutely. The two great physical theories of this century—relativity and quantum mechanics—emphasize the position of the subject (or instrument) making the observations. Relativity theory eliminates an absolute time frame, thereby emphasizing the local time of any observation. Quantum mechanics postulates states of a system that depend on the measurement protocol, thereby implying that one cannot know the state of the system without the perspective implicit in a given experimental apparatus.

Literature necessarily employs consciousness as a tool with which to study consciousness. Recognizing the inherent subjectivity of their work, many writers during this century do not attempt to impose an assumption of objectivity on their work. Thus they do not create levels of reality to model a reality as scientists do. Italo Calvino discusses the subject in his essay "Levels of Reality in Literature":

> Literature does not recognize Reality as such, but only *levels*. Whether there is such a thing as Reality, of which the various levels are only partial aspects, or whether there are only the levels, is something that literature cannot decide. Literature recognizes the *reality of the levels*, and this is a reality (or "Reality") that it knows all the better, perhaps, for not having come to understand it by other cognitive processes. (120)

By juxtaposing and comparing the various levels of a fictive reality, one studies the processes of the mind, of consciousness. The fabrication process is thus of interest, as well as the connections and symmetries between the levels. If we assume the truth of Serres's argument that literature anticipates science, then perhaps it is literature's abandonment of objective hypotheses and its application of subjective techniques that have led to its advanced discoveries of aspects of complex dynamics, which have already been appropriated as suitable models to characterize some kinds of neural activity.

In "Garden," Borges uses an additional level of reality which I like to think represents literature's acknowledgment of inherent subjectivity. In science, the closest analogue to Borges's fourth level is the affirmation of the referee. The scientist's article must be approved by a referee (another scientist in the same field) before it may be published in a journal. Although this referee may not alter the contents of the article, he does subjectively determine whether or not the work meets the consensual standards for "good science." He therefore subjectively mediates between the article and the reader.

In the first paragraph of "Garden," we are informed that the story we are about to read is the edited deposition of Dr. Yu Tsun and that the first two pages are missing. Therefore we have an additional, implicit level of reality between the reader and the first reality I mentioned above: it is the level of the fictive editor who has presented us with this deposition. In the only footnote in the story, which is expressly attributed to the manuscript's editor, we are told that a certain statement by Tsun, accusing Captain Richard Madden of murder, is a "malicious and outlandish statement." Considering that Captain Madden is Tsun's captor, this note leads us to believe that Madden himself is the manuscript's editor and that he intentionally removed the first two pages of the text. Within the deposition, we are given reason to believe that Madden is very concerned with how his actions appear to his superiors:

> Madden was implacable. Rather, to be more accurate, he was obliged to be implacable. An Irishman in the service of England, a man suspected of equivocal feelings if not of actual treachery, how could he fail to welcome and seize upon this extraordinary piece of luck: the discovery, capture and perhaps the deaths of two agents of Imperial Germany? (89)

Madden knew that Tsun's deposition would be read by his superiors, and it is likely that his concern was reflected in how he edited the manuscript. This additional level of reality, which Borges wedges between the story and the reader, represents the insurmountable subjectivity inherent in man's pursuit of self-knowledge. The reader does not have direct access to the story. He is forced to realize that he is reading a subjectively edited version, and like writers and scientists, he must develop methods of coping with this subjectivity.

A Direct Influence: Relativity as Modernism

As is well known, Einstein's special theory of relativity refuted Newton's absolute spatial reference frame. The scientific verification that there is no absolute frame of reference with which one may orient oneself in relation to other reference frames had a direct influence on Borges's work. Relativity theory was generally available to the scientific reader well before "Garden," and considering Borges's propensity for scientific thought, we may safely assume that he was familiar with it. Borges embeds relativity in his "Garden" in two very distinct patterns. In the beginning of the story, Tsun muses that "Things happen only in the present," thereby prompting us to begin thinking about subjective time frames (90) and foreshadowing the more complex idea of an infinite number of local time frames implied by Ts'ui Pên's web of time: "He believed in an infinite series of times" (100). These local time frames relate directly to relativity theory's elimination of Newton's absolute time frame.[6]

The text implicates these apparently disordered time frames as "the cause of the contradictions in [Ts'ui Pên's] novel," recalling the many controversial contradictions, such as the twin paradox, which arose immediately following Einstein's publication of the special theory.[7] The complete resolution of the twin paradox, which I will not elaborate here, lies in the general theory of relativity, which includes not only time frames but accelerations and the structure of spacetime itself. Whereas the contradictions in Ts'ui Pên's novel may be an indirect reference to the general theory, Borges more directly refers to it in the following: "I thought of a maze of mazes, of a sinuous, ever growing maze which would take in both past and future and would somehow involve the stars" (94). This is clearly a reference to spacetime, as conceived in the special theory, and to the general theory because of the involvement of the stars. (It was measurments of stellar positions that provided the first experimental confirmation of the general theory. In addition, in the general theory, the curvature of space results from the gravitational fields associated with the universe's mass, which in Borges's time was believed to be largely deposited in the stars.)[8]

Just as Borges's "Garden" parallels the relativistic garden of subjective reference frames, its caretaker is a simulacrum of relativity's "gardener"—that is, Mr. Albert is clearly modeled after Einstein himself. Beyond an obvious allusion in the name Albert, Borges reinforces the parallel by crediting Mr. Albert with finding the solution to the web of time: "I have worked out the plan from this chaos" (100). Albert, true to his namesake, deciphered the order in Ts'ui Pên's novel and recognized the importance of its subjective time frames. Einstein's theories refute Newton's; Mr. Albert specifically explains to Tsun that his reconstruction of Ts'ui Pên's work also differs from Newton's: "Differing from Newton and Schopenhauer, your ancestor did not think of time as absolute and uniform" (100).

The parallels between Albert and Einstein are further reinforced by secondary and indirect allusions. Consider the story's reference to Goethe. Goethe, a theorist in both science and literature, was like Einstein in that he presented a chromatic theory that opposed Newton's theory of light. However, in this case, Newton's theory prevailed (Burwick; Sepper). Tsun identifies Goethe with Mr. Albert: "I did not speak with him for more than an hour, but during that time, he *was* Goethe" (91). If we have here a reference to Einstein mediated by a reference to Goethe, perhaps the divergence between the outcome of the two scientific theories is as important as the consequences, for it may foreshadow Albert's defeat in time, as well as his victory over time.

Modern physics, specifically relativity with its multitude of local reference frames, was a version of the multiplicity inherent in modernism. Supported by the new scientific theories, this multiplicity eventually led to the fragmentation inherent in postmodernism. The collapse of textual boundaries and the abandonment of enforced unity were imminent when Borges published his "Garden." Striving for closure and unity, the "Garden" carried the seeds of postmodern literature. In an almost Freudian trope, Albert is killed at the end of the story. Dr. Tsun, the terminator whose commitment to his mission is stronger than his humanity, identifies the premeditated act with an exemplary chaotic one: "I swear that his death was instantaneous, as if he had been struck by lightning" (101). Seen as chaotic event, Mr. Albert's assassination seems to prophesy the coming paradigm shift from global to local theorizing, from modernity to postmodernity.

A Paradigm Shift: Global to Local

Modernist scientific paradigms, dominant in 1941 when Borges wrote "Garden," were founded upon global theorizing. Global theories define the behavior for all locations within the system. Only recently have local theories been developed—theories which define behavior rules for isolated local regions within the larger system. The spontaneous appearance of islands of order from a sea of chaos illustrates why local theories are necessary.

Chaos theory has its roots in the mathematical catastrophe theory of René Thom and the physical theory of modern dynamics. Modern dynamics has succeeded classical mechanics as the science of motion for macroscopic physical systems. Unlike its predecessor, modern dynamics does more than just acknowledge the existence of irreversible, nonlinear processes and of randomness; it investigates their behavior in detail and highlights their roles in nature. Although this science is relatively new, it has already been absorbed into the larger interdisciplinary science of chaos. Proponents of the new science of chaos have called it revolutionary and have even claimed that it represents a new paradigm in physics. Whether or not these claims are justified remains to be seen. However, because they are made by reputable scientists, we would expect that such a science would force philosophers of science to reconsider some of their basic conceptualizations, which is indeed the case. The practice of global theorizing was one of the first items on the agenda for reevaluation.

Throughout the history of Western science, theorists automatically formulated their theories to explain phenomena universally. One of the crucial assumptions underlying universal theories is that contiguous parts of the system behave similarly. However, the behavior of nonlinear dynamical systems violates this assumption. As scientists began to study these systems more closely, they observed radically different behavior among nearby parts of the system—i.e., when they started with slightly different initial conditions, very different behavior resulted.[9] Fluid turbulence is the best known and perhaps the most interesting example of such a system.[10] In her paper "Chaos in Contemporary Literature and Science: Local Sites and Global Systems," Hayles discusses the problem of global theorizing and local sites:

> In critical theory, for example, *global* is understood to mean not only cultural systems considered as a whole, but any theory which subsumes particular texts or phenomena into a universal explanation. Marxism, relativity theory, and grammar are global theories in this sense. Similarly, *local* connotes not only a small subsection of a geopolitical area, for example the western coast of Brittany rather than Europe, but also particular textual or cultural sites that resist assimilation into the generalizations of a universal theory.

The newly discovered sensitive dependence on initial conditions, because it defies an important assumption of global theories, forces theorists to develop theories that can account for localized behavior. The turn away from global theories and concentration on local ones could be interpreted as a major shift.[11] Whereas global theorizing is associated with the textual closure and unity often associated with modernism, local theories are analogous to recognizing intertextuality and abandoning a forced textual unity.[12] This aspect of chaos theory is postmodern.

Acknowledging the inapplicability of global theorizing to complex systems is only half the story. The other half is recognizing that, although entirely new regimes of order are being discovered in previously unexplained physical behavior, truly random behavior persists within certain completely deterministic systems. Bifurcation theory is one of the best-understood models of how ordered structures can arise from disorder. This model posits intricately bound local sites of randomness within larger patterns of order, which exemplifies the complementarity of chaos and order. Bifurcation theory is a local theory because it specifies the behavior of the system at local sites but it does not describe the system's behavior away from these regions.

These discoveries are recent and were consequently not available to Borges in 1941 when he planted his "Garden." The science which influenced him was modern, not postmodern. A transitional figure, Borges anticipated bifurcation theory in the nonlinearity of his web of time and thereby added it to the reservoir of postmodern thought. Although I cannot claim direct influence between the "Garden" and postmodern science, it may have circulated through the culture and set up subsidiary currents that did have some influence.

Borges and chaos theorists both use bifurcations to create structural levels of narrative and experimental reality. For the chaos theorist, this is now standard practice, but for Borges it was a new method of structuring apparent randomness. Given that Borges used a bifurcating structure similar to the formalized bifurcation theory of modern chaos theory, bifurcation theory can serve as an interpretive matrix that, in a reflexive turn, can be used further to harvest Borges's "Garden."

Bifurcations: A New Matrix

A bifurcation is a splitting, a decision point where the system must take one path or another. As a simple physical example, consider a leaf floating down a stream. As this leaf comes to a boulder in the middle of the stream, it must obviously go around one way or the other. An essential characteristic of bifurcations in nonlinear physical systems is that, at the splitting, chance determines which path the system will take. We know that the system will go one way or the other, but we cannot know which way it will go; its solution is truly unpredictable. In our example, the leaf has an equal probability of going either way around the obstruction. In complex physical systems, including laboratory experiments (first representations), bifurcations are frequent and, in fact, are desirable in any experiment designed to represent a real nonlinear physical system. Researchers have found that frequent successive bifurcations in the flow of a physical system indicate the system's transition to chaos. Similarly and somewhat symmetrically, the flow may, at some later time, make the transition out of chaos into some ordered state via a succession of reverse bifurcations or convergences.

In the first level of "Garden"—the narrative present—Dr. Tsun's mission is to kill a man named Albert so that he and Albert will get into the newspapers. The papers will in turn be read by his superior in Germany, who will interpret correctly that the Germans should bomb a French town named Albert, where a new British-run artillery park has recently been built. But the nature of his mission leaves Dr. Tsun uncertain as to which Albert to kill, because anyone named Albert would suffice. Pursued by the implacable Captain Richard Madden, Tsun's time is short. The essential bifurcation in this first representation is Tsun's choice of which Albert to terminate. Tsun traverses this bifurcation by randomly choosing an Albert from the telephone directory. Of all the Alberts in England, random chance brings him to the Mr. Albert whose great achievement is the successful understanding and interpretation of a novel written by Ts'ui Pên, the great-grandfather of Dr. Tsun. Here we have a bifurcation in Tsun's life leading, coincidentally, to a convergence with the flow path of Mr. Albert.

In the second level of scientific representation, bifurcations within chaotic systems manifest themselves as multivalued functions in the nonlinear differential equations. (By multivalued, I mean that for a given value of the independent variable, the associated dependent variable may have several different, equally likely values.) The multivaluedness of these nonlinear functions conveniently represents the uncertainty and randomness of a system's behavior at a bifurcation point. Thus the bifurcations, originally present in the physical model, are encoded into the second level of representation with their essential characteristics intact.

In "Garden," the discussion within the first level of representation by Mr. Albert and Dr. Tsun of the latter's ancestor constitutes the second level of representation—the body of information concerning the life, or flow path, of Ts'ui Pên and its bifurcations. Ts'ui Pên's activity during the last thirteen years of his life represented a "mystery" to his heirs (although a "transparent" one to Mr. Albert) because conjecture bifurcated in its interpretation of his stated intentions. Mr. Albert describes this bifurcation and how it could have arisen: "At one time, Ts'ui Pên must have said: 'I am going into seclusion to write a book,' and at another, 'I am retiring to construct a maze.' Everyone assumed these were separate activities" (96). Although Albert subsequently proves that these were in fact the same activity, at the intermediate level of representation they are the two divergent paths of a bifurcation.

At the diachronic demonstration level of chaos theory, the bifurcations appear as forkings in the graphic flow lines in Cartesian coordinates. At each forking, the number of possible flow paths doubles, quickly generating a vast number of possibilities and creating a visual garden of forking paths. Even though all these possibilities exist, only one path can be traversed by a particular system. Once a path is chosen, the others are excluded. However, as mentioned above, it is possible for two flow paths to come together at a convergence point. Then different systems can arrive at the same point via different paths.

Once Albert realized that the novel and the labyrinth were one and the same, that Ts'ui Pên had visualized life as a maze of possibilities, he was able to see the possibilities in Ts'ui Pên's novel as a vast network of bifurcating and converging paths:

> Your ancestor did not think of time as absolute and uniform. He believed in an infinite series of times, in a dizzily growing, ever spreading network of diverging, converging and parallel times. This web of time—the strands of which approach one another, bifurcate, intersect or ignore each other through the centuries—embraces every possibility. (100)

Here, at the third level of representation, we discover that the novel itself—which considers each of the possible flow paths of its characters—is a metaphor for the other levels of representation, visually demonstrating the properties of the bifurcation theory. Ts'ui Pên had written chapter upon chapter following the flow of his characters, and at each bifurcation, instead of choosing one path for his characters to follow, he allowed them all. The chapters, read in a linear fashion by Ts'ui Pên's heirs, seemed to make no sense. All they could find were many chapters of conflicting events: "the hero dies in the third chapter, while in the fourth he is alive . . . " (96). It was Albert's great achievement to transcend these linear expectations and identify the novel's bifurcating, nonlinear structure. Like flow lines in Cartesian coordinates, the novel is a diachronic demonstration of the essential bifurcations in the system it represents. Albert explains all he has found to Dr. Tsun and offers him this, their chance meeting, as an example of bifurcation theory: "Time is forever dividing itself toward innumerable futures and in one of them I am your enemy" (100).

Before allowing his characters to consider alternate realities, Borges specifies that a person, like a physical system, can only flow along one path. In the beginning of the story we find Dr. Tsun saying: "all things happen, happen to one, precisely *now*. Century follows century, and things happen only in the present. There are countless men in the air, on land and at sea, and all that really happens happens to me" (90). That the flow of the characters must choose one path implies that narrative boundaries are upheld. Thus Borges is rooted in modernism. Playing with intertwining narrative levels and nonlinearity, he nevertheless gives his story closure and an explanation. In a postmodern story, the bifurcating narratives and even the story itself need not observe textual boundaries, as for example in Italo Calvino's "The Count of Monte Cristo." Borges managed to process the multiplicities of modernism in a nearly postmodern way, yet he maintained his modernist stance. Let us now examine the methods of Borges to find the modernism in his postmodernism.

Synthesis: The Method of Borges

Clearly, Borges anticipated the two essential characteristics of the bifurcation theory's method of modeling natural systems—i.e., the frequent occurrence of random splittings in a system's dynamic flow and the inexorable nonlinearity of nature. He used a structure which is, as I have shown, remarkably similar to the structure of theoretical modeling currently used in chaos theory. What led Borges to use bifurcation theory as a model? One possible answer is that he recognized it as a Hegelian synthesis of the two polar states of existence, chaos and order. Stanislaw Lem argues that Borges would be interested in finding just such a synthesis:

> Considered from a formal point of view, the creative method of Borges is very simple. It might be called *unitas oppositorum*, the unity of mutually exclusive opposites. What allegedly must be kept separate for all time (that which is considered irreconcilable) is joined before our very eyes, and without distorting logic. The structural content of nearly all of Borges's stories is built up by this elegant and precise unity. Borges calls the one and the same the conflicting notions of the orthodox and the heretic, Judas and Jesus Christ, betrayer and betrayed, the troglodytes and the immortals, chaos and order, the individual and the cosmos, the nobleman and the monster, good and evil, the unique and the repeated, etc. (239)

If Lem is correct about Borges's method, then I would argue that "Garden" is just such an exercise, with chaos and order as the subjects. Perhaps Borges saw, as chaos scientists are beginning to, that bifurcation theory functions logically to join chaos and order.

To understand how order and chaos coexist within bifurcation theory, consider the third level of representation of scientific modeling—graphic flow paths in Cartesian coordinates. Given a set of equations and a digital computer, a scientist may plot the possible states of the system on a two-dimensional coordinate system. In the bifurcation region, the picture looks like a binary tree diagram going from left to right (in the usual convention). One line bifurcates into two and then into four, eight, etc., on out to the limit of the computer's resolution. Theoretically, this successive bifurcation sequence continues until there are infinitely many paths. But it does this in a finite region of space, because the distance to each successive stage decreases with each bifurcation. That this sequence can be plotted by computer implies that its structure is completely ordered. In fact, Mitchell Feigenbaum discovered this principle of order and published it in his landmark paper, "Qualitative Universality for a Class of Nonlinear Transformations." Even though this sequence is completely ordered, chaos (in the form of randomness) still plays a vital role. The system as it flows along the path can only be on one path at a time. Thus when it gets to a bifurcation point, it randomly decides which branch to take. Once it chooses, its path is completely ordered until it reaches the next bifurcation. In this way, chaos is integrally embedded within a completely deterministic system.

Bifurcation is a local theory, which according to my argument classifies it as postmodern science. Does Borges's use of it make his work postmodern? Perhaps if Borges allowed both order and chaos to coexist in his story, as chaos theory does, we might have more certainty where to position Borges in the modernism-postmodernism scheme. The boundary between modernism and postmodernism is, as Borges's canon illustrates, a vast and topologically complex surface. In this story, Borges rejects chaos in favor of order, and postmodernism in favor of modernism. However, some of his other work is definitely postmodern. Classifying Borges exactly through an analysis of his entire body of work would be an impossible task. Nevertheless, in "Garden," the evidence indicates an underlying determinism is at work.

The Denial of Chaos: Determinism

Scientists use bifurcation theory because it best explains the behavior of their physical perceptions. Borges intuited this behavior without complicated scientific investigations. Perhaps he recognized that it could be interpreted as a synthesis of chaos and order, which would perfectly align with his method of *unitas oppositorum*. For Borges the determinist, however, this apparent merging of chaos and order could only be a fictional game, just as his use of the field concept is a game (Hayles, *Cosmic Web* 138). In fact, chaos in "Garden" is never given equal ground. Borges undermines every instance of it that appears in the story.

As Albert explains the bifurcation theory to Tsun, he describes Ts'ui Pên's novel as "chaotic." But he is the one finally to understand the order of its bifurcating chapters. He is the one who discovers that the labyrinth and the novel are one and the same. The labyrinth that Borges uses as a metaphor for life and the universe is a classic symbol of his determinism, for the labyrinth is a completely ordered maze, appearing chaotic only to the uninitiated.[13] The Borgesian universe is an ordered maze, the chaos of which is only an artifact resulting from our not being able to perceive the order. Ts'ui Pên's novel is completely ordered by its author, and the apparent chaos arises only from noncomprehension of its order. Thus at the innermost level of representation, chaos is negated.[14] At the second level, the bifurcation comes from a lack of knowledge about the actions of Ts'ui Pên, not from randomness. Since both of Ts'ui Pên's choices of activity for the last years of his life merged into one, the implied chaos as well as the apparent bifurcation is removed.

In what I originally called the first level of the story—the actions of Tsun and Albert—several possibilities for chaos arise. What I claimed to be the essential bifurcation in this level was resolved by Tsun's use of the telephone directory. On the surface this appears to be a completely random event; however, Borges implies that it could have been avoided. At the very end of the story, a clue to the resolution of this chaotic event comes when Tsun describes his feelings about killing Albert: "He does not know, for no one can, of my infinite penitence and sickness of heart" (101). Tsun is sick at heart because he had to kill this particular Mr. Albert, whom he believes to be a great man for giving Tsun the resolution to the great mystery of his ancestor. But he was forced to kill Albert because he had to elude the Captain: "His hands and voice could, at any moment, knock and beckon at my door" (91). These statements imply that Tsun could have found another Albert to kill, or an entirely different, less murderous method of passing the information if only he had had the time, if he wasn't being pursued by the implacable Captain Madden. Thus the randomness is itself an artifact of the plot, rather than intrinsic to the structure of the represented reality.

Similarly, there is an implied fatality in Tsun's seemingly random selection of Albert—the one person who holds the key to the mystery of his ancestor—and fate belies the random. When Tsun catches a train to Albert's village and narrowly avoids the Captain, he attributes this lucky break to fate, thus ruling out chaos. Albert, discussing his ingenious feat of deciphering the labyrinthian novel, says he has "corrected errors introduced by careless copyists," implying both that such random errors could have been avoided and that the errors are removable. Although it is possible in theory to avoid errors in copying, locating and correcting these errors without seeing the original text seems improbable, thus pointing to a certain inexorability in Albert's reading of the manuscript. Here then, at the most tangible level of the story, Borges manages to undermine the randomness of events which, at first reading, seemed quite chaotic.

Finally, at the outermost level of "Garden," we are told in the introductory paragraph which comes before the beginning of Tsun's deposition that in his *A History of the World War*, Captain Liddell Hart attributes the five-day delay of a British attack on the German line to a torrential rainstorm, a random event. But we are also told that the deposition we are about to read "casts unsuspected light on this event." The implication is that the success of Tsun's mission caused the delay instead of the rain. By this stratagem Borges removes the random cause for the delay and substitutes an ordered one. Indeed, the stated intent of "The Garden of Forking Paths" is to substitute order for randomness.

Lem claims that Borges is a man of another time, that there exists an impenetrable barrier between us and his work:

> It is simple to predict the future of a purely mechanical phenomenon. In its utmost depths, the structural topology of Borges's work acknowledges its relationship with all mechanistic-deterministic kinds of literature, including the mystery novel. The mystery novel always incorporates unequivocally the formula of Laplacean determinism. . . . Even this great master of the logically immaculate paradox cannot "alloy" our world's fate with his own work. He has explicated to us paradises and hells that remain forever closed to man. For we are building newer, richer, and more terrible paradises and hells; but in his books Borges knows nothing about them. (241–42)

With intuitive insight, Borges used bifurcations to achieve an apparent synthesis of order and chaos long before chaos scientists discovered it in their studies. But for him, it was only a phenomenological tool for explaining life's disorder. Underlying his narrative is a belief that in life, as in the labyrinth, chaos is uncomprehended order. Surrounded by the tendency toward global theorizing that characterized his time, Borges was pressured to choose either chaos or order as fundamental. In "Garden" he chose order, and, therefore determinism. In other stories he chose otherwise. Thus, even though some of Borges's techniques and narratives lead into postmodernism, this work carries the signature of determinism and modernism. Borges stands in the cultural stream between modernism and postmodernism, with the currents of physics and literature mixing in his wake.

Notes

1. The temporal philosophies of Bergson and Einstein are considered by Paul Harris in his paper "The Fluidity of Time in the Melting-Plot: Einstein, Bergson,

Faulkner," given at the Conference of the Society for Literature and Science in Albany in October 1988.

2. Also see the essays by Maria L. Assad and Eric Charles White in this volume.

3. In *Physics and the Ultimate Significance of Time*, Griffin claims that a post-modern physics would be defined as re-including the subjective experience excluded by modern science, and he cites David Bohm's "implicate order" project as being the closest thing to a postmodern science (15). I do not disagree with Griffin on this point, but I argue that Bohm's project is mainly philosophy rather than a practice in physics. I regard chaos theory as postmodern because of a rather different criterion, i.e., because it abandons the requirement of a deterministic global theory of nonlinear dynamics in favor of a catalog of local theories. But it is also true that the results of chaos theory, like those of quantum mechanics, are pressing the notion of scientific objectivity out of existence. The second revolution, quantum mechanics, is not discussed in this paper because relativity theory and chaos theory are more appropriate to Borges's work.

4. Dependent variables act as indicators of the system's state; they should never be altered by the experimenter directly. The experimenter alters the independent variables and monitors dependent variables to see changes in the system.

5. Shlomith Rimmon-Kenan discusses the parallels between Albert and Ts'ui Pên in "Doubles and Counterparts."

6. In "Negation of Time," Holly Mikkelson relates Ts'ui Pên's novel to the cosmology of the Hopi Indians, which is remarkably similar to the special theory.

7. The original discussion of the 'twin paradox' occurs in P. Langevin, "L'Evolution de l'Espace et du Temps," *Scienta* 10 (1911): 31ff. A pair of identical twins are separated into two different inertial frames, one in space flight traveling near the speed of light and the other remaining on Earth. Once rejoined, they are found to have paradoxically aged at different rates as determined by special relativity. The solution lies in general relativity's treatment of accelerating reference frames.

8. In more recent cosmological theories, the bulk of the matter in the universe is believed to be hidden in black holes or to exist as so-called "dark matter" (Tucker and Tucker, 1988). In *The Cosmic Web*, N. Katherine Hayles, using mainly an argument from mathematics, discusses Borges's comprehension and application of the field concept. She argues for Borges's fascination with set theory and infinite series (138–67) and discusses the field concept with respect to relativity (45–49). Borges's reference to the general theory implies his comprehension of the field concept and supports Hayles's argument.

9. By "nearby," I mean as close as the adjacent discrete element of the grid resolution.

10. In fluid turbulence, a scale-invariant fractal structure spontaneously arises which mediates the transfer of kinetic energy back and forth between microscopic molecular motion and macroscopic fluid motion. Thus small (microscopic) changes in the system can have very large effects on the whole system.

11. The problem with the language of paradigms is that there is no clean break or transition from one paradigm to another. Although it is true that local theories must be formulated in nonlinear systems, it is also true that global theories are still valid within certain specified limitations. As Hayles indicates, relativity theory is a global theory, but, as yet, its validity has not been challenged by the problem of nonlinearity.

12. Here I have associated the characteristics of closure and textual unity with literary modernism. Although this is the usual assignment, there is still much debate whether it is an appropriate one. Certain scholars argue that boundary-breaking is a more appropriate attribute to define modernism.

13. Carlos Navarro discusses the labyrinth as a metaphor for the universe in "The Endlessness in Borges' Fiction" (402). Also see the discussion of logically deconstructing Ts'ui Pên's labyrinth by Lawrence R. Schehr in "Unreading Borges's Labyrinths" (180).

14. The solution to the chaos of Ts'ui Pên's novel reflects how texts are treated in modernist literary criticism. All things fit and can be explained or otherwise reconciled into a textual unity by those who are sufficiently trained to do so.

Works Cited

Barrenechea, Ana Maria. *Borges the Labyrinth Maker.* New York: New York University Press, 1965.

Bergson, Henri. *Durée et Simultanéité.* Paris: Alcan, 1924.

Borges, Jorge Luis. "The Garden of Forking Paths." *Ficciones.* Ed. Anthony Kerrigan. New York: Grove Press, 1962. 89–101.

Burwick, Frederick. *The Damnation of Newton: Goethe's Color Theory and Romantic Perception.* Berlin: De Gruyter, 1986.

Calvino, Italo. "Levels of Reality in Literature." *The Uses of Literature.* Trans. Patrick Creagh. San Diego: Harcourt Brace Jovanovich, 1986. 101–21.

———. "The Count of Monte Cristo." *t-zero.* Trans. William Weaver. San Diego: Harcourt Brace Jovanovich, 1969. 137–52.

Feigenbaum, M. J. "Qualitative Universality for a Class of Nonlinear Transformations." *Journal of Statistical Physics* 19 (1978): 25–52.

Griffin, David Ray, ed. *Physics and the Ultimate Significance of Time.* Albany: State University of New York Press, 1986.

Hayles, N. Katherine. "Chaos in Contemporary Literature and Science: Local Sites and Global Systems." Twentieth Century Literature Conference. Louisville, 27 February 1988.

———. *The Cosmic Web: Scientific Field Models and Literary Strategies in the Twentieth Century.* Ithaca: Cornell University Press, 1984.

Lem, Stanislaw. *Microworlds.* Trans. Franz Rottensteiner. San Diego: Harcourt Brace Jovanovich, 1984.

Mikkelson, Holly. "Borges and the Negation of Time." *Romance Notes* 17 (1976): 92–98.

Navarro, Carlos. "The Endlessness in Borges' Fiction." *Modern Fiction Studies* 19 (Borges Volume): 395–405.

Rimmon-Kenan, Shlomith. "Doubles and Counterparts: 'The Garden of Forking Paths'." *Critical Inquiry* 6 (1980): 639–47.

Schehr, Lawrence R. "Unreading Borges's Labyrinths." *Studies in Twentieth Century Literature* 10 (1986): 177–89.

Sepper, Dennis L. *Geothe contra Newton*. Cambridge: Cambridge University Press, 1988.

Serres, Michel. *Feux et signaux de brume, Zola*. Paris: Grasset, 1975.

———. *Hermes: Literature, Science, Philosophy*. Ed. Josué V. Harari and David F. Bell. Baltimore: Johns Hopkins University Press, 1982.

Thom, René. *Structural Stability and Morphogenesis*. Trans. D. H. Fowler. Reading: W. A. Benjamin, 1975.

Tucker, Wallace, and Karen Tucker. *The Dark Matter*. New York: William Morrow, 1988.

11

Modeling the Chaosphere: Stanislaw Lem's Alien Communications

Istvan Csicsery-Ronay, Jr.

I. Carousel Reasoning

If there is a single theme informing all of Stanislaw Lem's mature work, it is the problem of "carousel reasoning." To different degrees and in different solutions, in grotesque, playful, realistic, and discursive modes, as explicit dilemma or implicit inscription, Lem's writing returns incessantly to this problem, which represents for him the fundamental paradox of both writing and scientific cognition. "Carousel reasoning," according to Lem's clearest fictional mouthpiece, the mathematician Hogarth of *His Master's Voice*, consists of carrying reflection within a logical system to the point where premises and conclusions exchange places. As the rational carousel goes round, the reasoner discovers that her premises depend upon her conclusions, that her thinking consists of combining and recombining elements and operations until proving and assuming are indistiguishable.

For Lem, as for Hogarth, "carousel reasoning" is an inherent flaw in all philosophical world models, since "they do not contain appeals to some decisive factor in favor of a given proposal" (Csicsery-Ronay 250). Philosophers, Hogarth claims, strive to define the position of human intelligence in the order of things by "equating [themselves] with the norm of the species" (*His Master's Voice* 30); unlike scientists, their thinking does not "come up against some hard focal point of facts that sobers and corrects it" (29). Without such a focal point, reason has no horizon toward which to orient itself, and no world of difference against which to test itself.

Scientific rationality escapes the carousel as long as it has "the facts." But as Hogarth knows very well, no facts exist truly outside the matrix of science's rationalistic assumptions. When scientists encounter completely unfamiliar or unstructurable phenomena—anomalies, "aliens," "chaoses"—they, too, become philosophers on the carousel. Lem, the fantasist of methodology, specializes in inventing such flummoxing anomalies. His fantastic scientists can break down these rational enigmas and combine their parts according to previously effective norms of science; but the resulting knowledge is in bad faith, since the enigmas call into question not the details but the ground of scientific reason itself. The "carousel," moreover, need not be confined to any one discipline or ethic. Given the great variety of theoretical models available from the history of the different sciences, the search for the right combination of hypotheses and norms leads to the relativization and devaluation of all. Hence the recurring motif of the informational labyrinth through which Lem's protagonists wander: the Library in *Solaris*, the maze of messages in *Memoirs Found in a Bathtub*, the carousel of hypotheses in *His Master's Voice*, culminating in the anthologies of imaginary textual fragments in *A Perfect Vacuum* and *Imaginary Magnitude*. In the face of truly alien realities, there is little left for science to do but affirm their existence.

II. The Chaosphere

Katherine Hayles has described the "dialectic" between textual closure and openness in Lem's writing as an incessant movement from one extreme to the other, in which the logical development of either principle leads to its own undermining.[1] "Whether the text begins with order or chance, referentiality or self-referentiality, by the end both polarities have been so enfolded into each other by the operation of the dialectic that they are inextricable" (Hayles 297). This mutual enfolding of opposites is a version of carousel reasoning. The term "dialectic" is Lem's own ("Chance and Order" 88); yet it is saturated with irony. For dialectical interplay in Lem's work is implosive. Rather than extending the text, the dialectic sets up diverse significations that fuse with one another. Like Hegel's dialectic of the Spirit of Reason, Lem's also has a "spirit" propelling it toward a moment of absolute self-consciousness. Lem's spirit, however, is parodic,[2] driving thought with inexorable and binding logic toward self-enclosure in a sealed bubble of solipsism. Finally, it is articulated only by echoes, reflections, and replications of itself.

Science, the reference and concern of Lem's textual play, is involved in the same mutual enfolding of polarities that Hayles describes. Fundamental categorial differences that enable scientific epistemology to work—space, time, mathematical relations—dissolve when the horizon implodes. In a universe where origin becomes telos, containment becomes emptiness, identity becomes difference, the spatial universe becomes a temporal moment, and information becomes mass and energy, Lem's protagonists are haunted by a parodic consummation of history. Instead of the systematic unification of the ontological oppositions, Lem's scientists experience self-conscious isolation within the self-elaborating confines of systematic rationalization. Once absolute differences have been dissolved and enclosed within rationality, there is no way to set up a logical system—a matrix, a language, a unified science—which can be tested against anything distinct from itself. The system has become so self-inclusive that it no longer has an outside; it has neither ground nor relation. It is nowhere, at no time, compared to nothing. A singularity, in which *ratio* steps into the the chaos of absolute nonrelation. The chaosphere.

III. Fantastic Science

Lem's fantastic science allows for three possible stances toward this singularization. One option is represented by the constructors. These technologists of artificial intelligence[3] treat the enclosing chaos as a void into which they project models of human consciousness. They are "creators" of new realities, in that their self-projecting constructs also imitate freedom. The constructors' creations, from the "personoids" of "Non Serviam" and the homunculi of "Doctor Diagoras" to the meta-AIs of "Lymphater's Formula" and *Golem XIV*, are creatures of universes that are "fully rational, but they are not fully rational inhabitants of them" (see for example "Non Serviam"). Although the promise of self-organization, and indeed transcendence, exists in these creations (cf. Hayles 300–301), this self-projective modeling inevitably leads to an extension of the circular chaosphere and not a liberation.

Another school of Lem's fantastic scientists pursue randomness. There is Inspector Gregory's dogged search for a criminal perpetrator in *The Investigation*; ex-astronaut Adams's analogous quest in *Chain of Chance*; the indissoluble uncertainty in *His Master's Voice* about whether the message is an intentional object or an unknown natural process. Since randomness can never be brought into relation, *ratio*, it cannot open up a hyperrational system. Except perhaps for *The Investigation* (and satirically in "De Impossibilitate Vitae et De Impossibilitate Cognoscendi" of *A Perfect Vacuum*), the chance-novels tend to resolve the problem randomness poses for plot construction by dissolving it or explaining it away.

Lem discusses the privileged role of randomness has for contemporary fiction in his "Metafantasia." He argues that the last set of norms against which writers can try to perform their fictive transgressions is the all-embracing causal explanation of conventional science.

> If we posit that the task of literature is not to ever give a definitive explanation of what it presents, and is therefore to affirm the autonomy of certain enigmas rather than to enter into explanations, then the most enigmatic of possible enigmas is a purely random series. ("Metafantasia" 65)

The affirmation of literature's enigmatic qualities appears in such hypermodern experiments as chance-poetry, cut-ups, and the random sequencing of some of Robbe-Grillet's works. But the enigma of chance, intended to liberate the work by undermining an overdefining cultural context, is itself undermined by the defining enclosure of the text. The reader knows that the chance is a product of the writer's design and hence an intentional procedure, not the "real thing." Instead of opening the field of signification to chance, such works merely *simulate* chaos, representing it as a usable disorder at the service of an expansive determining order.

There are thus two forms of modeling—constructing something out of oneself, which leads to the infinite labyrinths of self-referentiality, and simulating chance, which paradoxically neutralizes randomness's capacity to open the solipsistic enclosures of Lem's self-referential texts. Lem entertains one other form of fantastic modeling, balanced between these two. It is a mode that simultaneously valorizes the unexpected and forges a relation that will make these outposts of the inexplicable relevant for human concerns. This mode is the encounter of human reason with alien minds that are fundamentally different, and yet which share with human consciousness the ability to make models. The quest for extraterrestrial contact informing much of Lem's best writing, from *Solaris* and *His Master's Voice* to *Summa Technologiae*, is not so much a search for a ground for reason, or for an absolute Other, as for a difference in relation to which human consciousness can position itself. The alien represents the only chance of the human species for a relation not subsumed within the chaosphere.

Modeling in Lem's fiction can be a passive reading of nature or an active construction that commands nature to behave a certain way. Both of these orientations result in anthropomorphic self-projection, since both end in a universe defined in human terms. A third mode also exists: *appealing* to nature. This form of modeling represents science as a Heisenbergian activity, asking nature to make specific responses to specific questions.[4] In Lem's novels of alien communications, scientist-protagonists search the universe for signs of design and intention other than their own, and they are usually strict in their empirical standards. Their goal is the construction of communication, not an oceanic fusing with science-fiction demiurges who require the self-annihilation of the questing intellect.[5] Since humanity in its autoevolutionary phase is, for Lem, a self-constructing species condemned to making its own decisions about its fate, it cannot expect to communicate with a subject incapable of constructing itself. Human beings can only hope to extract a response from something that also desires a response. Lem's questers are out to gain greater purpose through freely entered dialogue. To maintain autonomy, the communication must be constructed; it must be dialogical. This can only happen through models of communication that mediate between one self-conscious species and another.

What kind of match will there be between terrestrial and extraterrestrial models? Even if mind is a normal phenomenon in the universe, it may not be uniform. In Lem's fiction there is only one form of naturally evolved mind that can have a lasting interest for us: an analogue of our own, that is, mortal self-consciousness, having in common with us awareness of its incompleteness, ignorance of its direction, and knowledge of the tools of communication. The mind with which we can communicate must be an imaginative intellect with the power to have goals (cf. Minsky 126–27), and to make models of them. It must have the capacity to imagine the universe as a totality in which intelligence has a part, and the certainty that its image of the universe is inevitably incomplete—that it can, at best, only model the chaosphere. Only such beings can offer the promise of modeling which may lead to alien contacts, or at least to the knowledge that such contact might have been possible, once, for someone, even if not for us.

In the pages that follow, I will sketch how three of Lem's most important fictions of extraterrestrial communications show his stances toward escape from the chaosphere: the romantic contact of *Solaris*, the indeterminacy of *His Master's Voice*, and the demonic recursion of Lem's latest novel, *Fiasco*.

IV. *Solaris*: The Human Model

In Lem's first major work, *Solaris* (1960), the chaosphere is represented by Solaristics, the branch of earthly science that evolved through humankind's encounter with the gigantic sentient colloidal ocean of the planet Solaris. The planet is known to be capable of incredible self-regulation, governing its macroprocesses by controlling its orbit around two suns, and also its microprocesses by the manipulation of neutrino-fields to create phantasmic simulacra of human beings. After the discovery of Solaris, the desire to understand the ocean became for a time the greatest quest in science. But when the novel's protagonist, psychosolarist Kris Kelvin, arrives at the Solaris Research Station, Solaristics is a badly degenerating research program. After a hundred years of study, the Solaris Project has produced only stalemate and paradox.

The planet has resisted scientific categorization so much that each scientist, and each discipline, are caught in escalating complexities, ultimately forcing them to step out of scientific rationality altogether. First came the competition of very general hypotheses. The biologists defined the ocean-planet as a gigantic "prebiological" quasi-cell; the astrophysicists as an extraordinarily evolved organic structure; the planetologists proposed that it was a "parabiological" plasmic mechanism; some even argued that it was merely a very unusual geological formation. The evolutionary view entered with the hypothesis that it was a "homeostatic ocean" which had evolved into total adaptive control of its cosmic environment in a single bound, bypassing the phases of cellular differentiation (*Solaris* 23–25).

Some things have been determined with precision: the planet controls its orbital periodicity directly, and discrepancies of time-measurement are discovered even along the same meridian. But very little mathematical certainty is possible, since the planet often changed the measuring devices applied to it, and the human scientists no longer know exactly what it was their readings were registering. Solaris acts as a macrocosmic uncertainty generator.

In the "golden age" of Solaristics, bold theorists and heroic explorers willing to risk their lives established that the ocean is alive, in some sense. But because the planet did not respond to the Solarists' probings, the work increasingly declined into taxonomy—an excruciatingly ironic taxonomy since everything about the planet was unprecedented in human science, and all relevant categories had to be invented from scratch, without comparisons.

It was in this phase that the Solarists descended to the surface of the planet to study its "inventions," the so-called symmetriads, mimoids, and asymmetriads. These are gigantic, intricate structures that emerge for short periods from the ocean's surface. Neither their purpose nor their principles of construction are intelligible. The scientists cannot determine whether they are adaptive, exploratory, or aesthetic—but each commentator betrays unconscious perceptual biases in his descriptions of them. The strongest proponents of the planet's intelligence were the explorers who approached and entered these awesome constructions. But the same things that led to artistic awe led also to scientific despair. A bitter split developed between those who wished to withdraw from the project, believing the planet was somehow willfully refusing to communicate, and those for whom contact with Solaris became the be-all and end-all. Lem brilliantly ties this to a psychoanalysis of science. The demoralized Solarists are like insulted suitors, and the contact-men are only made more ardent by the lack of response. Ultimately, the planet will understand this better than the men themselves, when it sends embodiments of their unconscious erotic fantasies and guilts to "visit" them.

Frustration at their inability to understand the planet gradually leads the Solarists to make increasingly psychological hypotheses. The planet's silence is viewed by some as a sign of autism, by others as a sign of an "ocean yogi." Ultimately, the Solarists are compelled toward models of intentional behavior taken from terrestrial religions. Observers plausibly depict the scientists' obsession with communicating with the ocean as narcissistic projection or religious mania, the desire for union with the Godhead. For other scientists, the uncategorizable translates into indifference, or even active hostility. The scientific gain from the study of Solaris is nil. The sciences are ironically unified by their universal failure to interpret the Alien, ultimately collapsing into one another. The myriad models used by the generations of Solarists prove to be anthropomorphic analogies with no demonstrable relevance to the sentient ocean. At the moment of complete stalemate (the actual beginning of the novel's action), the planet appears to have defeated human science altogether by establishing unpassable limits. Knowledge appears to have become a library of self-contained cross-references; the only thing leading out of the system is the human desire to make contact with what is not itself.

A breakthrough comes with the "Visitors," who appear to be materialized forms of the scientists' own unconscious thoughts (142). To be more precise, the breakthrough is the Visitor-Rheya, a simulacrum of Kelvin's dead young wife, for whose suicide he has carried feelings of guilt for several years. Rheya differs from the other scientists' Visitors primarily because she is clearly a model of the human original. She is like the authentic Rheya in several ways: indistinguishable in appearance, she attempts to emulate the original's suicide and inspires feelings of love and guilt in Kelvin. She is also, like a model, different than the original in important ways. She has superhuman strength and no memory of her previous life. Physically, she is a *formal* copy taken from Kelvin's memory. Her dress, for example, has no buttons or zippers (implying perhaps that Kelvin was not the most attentive lover of the original woman).

The Visitor-Rheya appears to be a model created by the ocean, modeling Kelvin's memory of the authentic Rheya. But it is a property of the Visitors that the longer they remain with their hosts, the more autonomous they become. The simulacral Rheya gradually becomes almost as human as Kelvin—as aware of her ignorance of her origins, as capable of cunning judgment, as capable of self-sacrifice. She appears to evolve into a self-regulating Solarian model of a human being. Physically, she is different; she is composed of neutrinos organized in a neutrino field apparently under the planet's control. But the structure of her behavioral autonomy is isomorphic with human behavior.

Rheya's modeling appears to mediate between Solaris and Kelvin. Her form is imparted by Kelvin through his unconscious memories; her material, the neutrino field, apparently by Solaris. Her modeling-function also appears to be determined by the ocean, at least in the beginning. But though she may be an observation device, she is most likely a self-programming one (Philmus 185). Her autonomy—which may be functionally necessary in order to evoke Kelvin's love and her own sacrifice—appears to be necessary for the Solarian ocean to make experimental changes in Kelvin's unconscious mind (or whatever it is that keeps him, like other humans, from getting through to the ocean). For the ocean, she may be a model of Kelvin's emotions and the psychic behavior of individual (male) human beings. For Kelvin, she is a model of the ocean's power to model and hence to understand. She is the most complete example of the Solarian models of the human image that the ocean had begun, in a process the scientist Messenger had called "Operation Man," to create, beginning with the grotesque gigantic baby drawn from the memories of Berton, an early Solarist who crashed into the planet (*Solaris* 97). The fact that the Visitor Rheya does not return in a new version after her annihilation, as all other versions did, seems to indicate that her purpose was successfully achieved, and now she can be discarded. One might argue that several powerful distracting disanalogies in the model—such as the growing love-affair between her and Kelvin, and possible future misunderstandings between two fully autonomous beings—threaten to develop, which might severely interfere with the planet's project of cognition.

Whatever the case may be, Rheya's existence and behavior imply the planet's prodigious model-making sentience. Through the anthropomorphic model, the ocean-consciousness appears to have established a form of contact with the alien world of human beings. Thus the Solarian ocean appears to be capable of higher-level modeling (the imitation of intelligent organic structures) than human beings. It is capable of the highest conceivable form of model, a creature capable of acting with increasing freedom, and whose illuminating and theory-fertilizing function can only be realized by a response from her human co-creators.

The problem with such a model is that its "fit" cannot be judged by a putatively autonomous, conscious human being without some independent standard to which both creatures might be compared, a metamodel. It is precisely the freedom of human consciousness to recognize and act on something other than anthropotropism that Lem's novel questions. The Visitor-Rheya is a human model, a simulacrum, who reproduces the uncertainty of human cognition, the uncertainty of humanity's place in the universe. Indeed, she redoubles this uncertainty. This redoubling might mean one of two things. Either the further displacement of meaning, in the shape of yet another unanswered question embodied in the universe; or, on the contrary, the creation of a being with which there can be affective contact, of a *medium* of communication. The human image is then either a form of chaos, an echo instead of a reply, or the agent of order, the messenger, the reply to a question at the heart of being human.

Everything in *Solaris* hinges on how we decide this question. But it cannot be decided. The unified chaos of the novel does not appear in the action; it is not an object of storytelling. It appears, rather, in the wedge between two mutually incompatible but equally compelling readings: a straightforward romance of superrational contact with the alien, and an ironic, self-deconstructing satire of it.[6] However one reads the novel, it is undeniable that contact in terms of the novel can only be made through the mediation of models. In *Solaris*, human science was fortunate enough to happen on a being willing and able to initiate the kind of modeling that can get to the heart of intelligent beings through mutually constructed models of communication.

V. *His Master's Voice*: The Pure Model

With *His Master's Voice* (1968) the theme of alien communications again comes to the center of Lem's fiction. As in *Solaris*, terrestrial science is faced with a mysterious cosmic phenomenon it does not have the conceptual tools to comprehend. An enormously long and regularly repeated emission of neutrinos that has apparently been streaming from space for millennia is discovered by chance. A team of scientists is quickly assembled by the Pentagon to study the emission in a program explicitly reminiscent of the Manhattan Project. As with Solaristics, the study of the "letter from the stars" begins with a wide range of general hypotheses which are gradually narrowed down, and whole sciences and disciplines are discarded in the process. Unlike Solaris, the impenetrable phenomenon in *His Master's Voice* is most likely not a naturally evolved structure but an informational-artifact, perhaps a message sent by intentional beings.

One can read *Solaris*'s conclusion to say that the Solaris problem is ultimately resolved by benevolent telepathy; Rheya's modeling-mediation was apparently made possible by the ocean's penetration of the human Solarists' minds. The sentient ocean worked in a domain where autonomy is a given. Since earthly science does not work with self-transforming models, communication with an alien whose entire being seems to express itself in creating such objects was largely out of human hands from the start. Human science deals either with already coded information, or it translates ("interprets") phenomena into coded information, which is one of the functions of a scientific model.

In this sense, the "letter from the stars" is the opposite of the sentient ocean. It cannot be changed, it cannot respond, but as a severely regular text it might at least provide the information necessary for understanding its structure. In fact, it proves too complete even for this. Where the Solarian ocean was too fluid to be understood by human scientists, the "letter from the stars" is too self-integrated; while the ocean granted Kelvin a modeled context in the form of a personal love, the "letter" lacks all context.

Even so, the novel's narrator, Hogarth, a curmudgeonly Nobel laureate in mathematics, is able to discover two unambiguous pieces of information about the neutrino stream. First, the "message" is closed, it has the structure of an object. Second, it is not binary, i.e., the code's form is finer than the interpretive capacities of the terrestrial receiving instruments. The first discovery means that the message can be viewed as a whole, but also that it does not point to a context necessary for its interpretation. It may be hermetically self-reflexive, capable of being understood only by those who already know what it means. The second discovery means that the message cannot be accurately interpreted even if the "code" as it is received is "cracked," since the instrumental limitations of earthly science prevent the whole text from arriving. Neither of these discoveries discourages the project's physicists from extracting what they consider usable information from the letter and treating it as if it were inherent in the message.

The ocean had intervened in the Solaris project in response to the human Solarists' earlier attempts to intervene in its processes. (The apparent immediate catalyst to the Visitations was an emission of X rays from the Solaris Station to the surface of the planet.) *Solaris* thus depicts a dialogue of models. But the letter of *His Master's Voice* can work in one direction only; furthermore, if it truly had Senders, they certainly would have been dead for millions of years. Neither the code nor its Senders can respond. And since the letter is closed, the potential analogies to be used in models of it must also be completely closed. The only such analogies available to the "His Master's Voice Project" are objects defined by the Schwartzschild radius, like black holes, or perhaps the universe itself—in other words, an object whose only outward-directed information is the absence of information.

This stalemate is dislodged when it is discovered that the neutrino-stream is "biophilic," in two senses. First, it slightly strengthens the survival chances of certain macromolecules essential for life in systems already tending toward biological self-organization. Hence, since the Earth was evidently in the path of the emission for millions of years, the code may have influenced the emergence of life. Later, another aspect of its biophilia is discovered. The project's physicists manage to synthesize from their putative decodings a substance capable of a "tele-explosive," or TX-effect. In the end, Hogarth determines that the TX-effect is limited in range. Moreover, beyond a certain short distance it has an inverse effect. Thus the substance cannot be used a weapon, since it would annihilate its wielders and not its target. Hogarth speculates that this limitation is a matter of an internal precaution deliberately inscribed throughout the whole message by the Senders. The message therefore represents for him a pure communication, in which technical and ethical purity coincide: it is incapable of being misunderstood and incapable of being used for harm.

In both *Solaris* and *His Master's Voice*, the creator-emitter is inaccessible, primarily because its powers of organization and energy-emission are unimaginably greater than humanity's. Their models, however, are not completely alien. Rheya is a person of sorts (although she can be viewed as a kind of self-elaborating readout of certain preprogrammed instructions, in this she is not unlike human beings vis-à-vis their genetic "programs"). The neutrino emission is a piece of organized information (although it still might be a natural phenomenon rather than a letter). There the similarity ends. A person can be engaged, influenced, loved. As Kelvin falls deeper and deeper into his quixotic love for the Visitor-Rheya, the lovers gradually transform each other, and he appears to make purely intuitive contact with her "source." Although the "personality" of the ocean cannot be inferred from its person-creature, there are reasons for believing that the ocean has the power to make precisely this sort of inference. The greater Kelvin's needful love for Rheya, the greater the mysterious apparent contact between the ocean and Kelvin.

The "letter," by contrast, cannot be engaged or influenced. The message's closure forces the project scientists to discard more and more hypotheses, yet they are given almost no new information about how they might approach the text. As the message resists all the methods and hypotheses in the armory of human science, the image of Senders becomes proportionally remote and finally empty. It is never decided whether the stream is an artifact or a natural cosmic phenomenon. At the end of the novel, several models for the message's function are proposed: that it is the order-regenerating legacy bequeathed by the last civilization of a dying universe, that it is the imperishable impersonal information core of a universe that has passed into antimatter, that it is a purely unintentional cosmic excretum. These hypotheses can go nowhere, because the "letter" cannot be approached sufficiently for hypotheses to be verified or falsified. Because this hyperorganized stream of information cannot be deciphered, it is a model of the Cosmic Secret. Unlike the silence of the universe, which like chaos can never be studied since it implies nothing, the neutrino message may mean a great deal. Indeed, it may mean *everything*, the very existence of the universe.

Hogarth wants to believe that the stream truly is a message. That would mean that the Senders are sufficiently like humanity to be limited by mortality, a situation that requires them to use instrumental mediation. And precisely because they are mortal and scientific like us, the message shows that they are perfect by comparison. A culture whose moral perfection matches its technological perfection, capable of devoting their whole civilizational energy to broadcasting such a biophilic—and perhaps even cosmophilic—letter, means that the message is not for us. Still, given humanity's genius for destruction, it is comforting to know that such beings might exist.

In a sense, Lem's "letter" is a model of language as if seen from outside. Language, after all, is the fundamental modeler, the paradigmatic case of modeling. All the models of science must draw from or refer to the larger linguistic framework (including mathematics), or they would never fit into the larger pattern of scientific theory, let alone help to establish it. In *His Master's Voice* Lem "estranges" this self-evident and apparently trivial point, which ceases to be trivial whenever someone tries to imagine how to communicate with extraterrestrials. For the "letter" exemplifies the point at which information and noise become indistinguishable, just as the chance evolution of a material object and an intentional artifact become indistinguishable. The "letter" models an absence that unifies all human language, context, and reference into what is "outside" it. Paradoxically, human modeling is sufficiently parallel to the neutrino stream's order for us to be aware of our exclusion. Thus Hogarth's hope duplicates Kafka's: "There is hope, but not for us."

The Soviet critic Kagarlitsky called *Solaris* one of the most romantic works of science fiction (37). Contact is achieved (*if* it is achieved) through love, and the creation of a woman is used as an instrument of dialectical sublation. Moreover, the novel itself, considered as a model encompassing a story about modeling, similarly requires identification with the characters' struggles for contact in order to make sense of the whole. The novel can be simultaneously earnest and satirical, for it assumes the reader's interest in the alien. The novel's self-reflexive dimension appears once we identify the problem of reading the novel with the problem of understanding the Alien. If Kelvin can love Rheya and experience strange dreams apparently issuing from the planet, the reader can feel that the novel models the problem of knowing another subject directly.

His Master's Voice is, from this perspective, a more sophisticated and essentially postmodern work than *Solaris*, for it assumes that the problem of knowledge is the problem of language. At every turn in the novel, the reader is faced not merely with interpreting the mode in which the events of the text occur, but also with deciding whether the text even allows the possibility of interpretation. The Solarian ocean does not use language, or at least no language recognizable as such by the Solarists. By contrast, all that appears of the Senders in *His Master's Voice* is their language. Just as the text of the letter cannot be deciphered (or even determined to be intentional information) because it is so precisely defined that it reduces redundancy to zero, so the meaning of the story's events remains completely undecideable. In the end, one cannot tell whether *His Master's Voice* is a tale about the yearning for graceful cosmic power that can control the ambiguities of information, or about the impossibility of ever distinguishing chance from causal order.

VI. *Fiasco*: The Infernal Model

Lem did not return to the theme of extraterrestrial communication for twenty years. During that time, his fiction revolved almost exclusively around the theme of human self-projection; carousel reasoning replicated the human image in a technosphere that encountered no resistance from the material universe. With *Fiasco* (1987), Lem went back to the subject matter of his most successful fiction, returning as well to the ethical tone of his early novels. One can read *Fiasco* as a step that closes a grand circle in Lem's project of depicting the prison of carousel reasoning. The reprise of the naive ethical mode of Eastern European scientific fantasy occurs in such a tragic and mournful key, it would be unimaginable in any youthful work. *Fiasco* completes the circle of tales of alien contact, supplying the missing piece: contact leads the human explorers to destroy what they came to meet, and to destroy the ideals that led them to desire contact in the first place. Where *Solaris* conjured a romantic vision of miraculous contact with a superhuman being, and *His Master's Voice* the epistemological hopelessness of contact with superior civilizations, *Fiasco* depicts a version of chaos unmitigated by higher beings, a chaos in which utopian ideals and technology lead inexorably toward their own disintegration and war against the Other.

The novel's main action concerns an expedition of human scientists who set out to make actual physical contact with a distant stellar civilization that appears to be within the "window of contact," the historical interval during which a planetary civilization is no longer pretechnological but not yet technologically suicidal or isolationist. The expedition itself represents the acme of human civilization. The mother ship is a gigantic hydrogen-flow-stream vessel; the crew is capable of "sidereal engineering"; their science is able to transform a black hole into an onion of time-space "skins," through which the mother ship can navigate several centuries in a few days. Yet, like the ideal Greek figures evoked by the novel's relentless allusions to classical myths of descent into the Underworld, the crew of the mother ship *Eurydice* and the landing vessel *Hermes* are drawn into an inexorable and terrible struggle with an inverted image of themselves.

As in all of Lem's fiction, the intellectual action moves from extremely broad hypotheses to ever narrower and more specific models, as the encounter with the mysterious phenomena intensifies. While the *Hermes* journeys toward its destination, its crew passes the time discussing various theories about the distribution of intelligent life in the universe, the probable kinds of intelligence, the plausible courses of civilizational development, and the possibilities for communication. As soon as they arrive in the solar system of the target planet, Quinta, the vague hypotheses are quickly sharpened by the practical need to make models of the inscrutable, but clearly dangerous, phenomena they observe. For everywhere, the *Hermes*'s crew finds signs of war, extending to the outermost regions of the Quintan solar system.

This war appears to be conducted exclusively by self-regulating killer satellites, synthetic viruses, and related automata. The human explorers abandon their original plan of appearing directly to the Quintans; instead, they conceal themselves in order to study the civilization and its satellite-armory better. The weapons they observe are clearly cybernetic artifacts, not protean beings or uncrackable codes, yet even so they are impenetrable. The human emissaries are neither passively tolerated as in *Solaris*, nor simply in the path of a message as in *His Master's Voice*. Threatened by aggression, they must make their decisions quickly, under pressure. With the help of a pompous supercomputer named DEUS, the crew of the *Hermes* tries to make sense of the technology of the aggressive automata and of many unintelligible phenomena they observe on the planet itself. From this they hope to build up an image of the Quintans' motives, if not the rationale of the whole Quintan civilization. They must do this through abstract modeling—for the Quintans refuse to answer their attempts at communication, except through bizarre and apparently self-destructive acts of high-technological violence.

Gradually, DEUS and most of the crew agree that Quintan civilization occupies a special phase of technological evolution, in which it has transformed its whole solar system into a "cosmic war sphere." The aggression appears to have usurped the civilization's normal goals and functions, so that the war continues automatically. The entry of human beings endowed with sidereal engineering threatens, at the least, to inspire the Quintans to destroy themselves by appropriating that technology, or, at worst, to extend the boundaries of the war-sphere further into the universe. A representative of the Vatican accompanying the crew urges them to withdraw, leaving the Quintans to their own path lest, despite their good intentions, the human messengers should incite physical and ethical catastrophe. The astronauts believe they have no choice but to engage the Quintans in a form of escalating confrontation until the Others submit to civilizing communication with the representatives of Earth. The heroic honor of Earth's emissaries cannot allow either attacks or questions to go unanswered—the code of exploratory desire takes precedence over humble renunciation.

At first, the Quintans do not respond at all except by violence. Later, when they do respond, they cannot be trusted. The human crew is consequently entirely at the mercy of its models of the situation. These models allow them to make rational moves toward their goal. The Quintans appear to be similar to human beings in their technology. Indeed for the reader, their runaway Star Wars technology is far more familiar than the transgalactic utopian magnanimity of the human emissaries. Nor is the Quintans' behavior without terrestrial analogues. Even so, modeling leads the *Hermes*'s crew not to rational compromise but to participation in a war game, and the desire for contact and communication becomes a catastrophic exercise in end-game theory. Where in the earlier novels the modeling-work of superior beings led to autonomy and grace (the "genesis" of Rheya and "The Word" of the Master's Voice), in *Fiasco*'s mindless conflict of technological civilizations, the modeling is demonic. When the novel's hero finally arrives on the surface of the planet (possible only after the humans threaten to destroy the planet by burning away its atmosphere and searing through the surface to the core) he discovers he is caught in a modeling-trap:

> He found himself in a situation whose structure was typical of the algebra of conflicts. A player made a model of his opponent, a model that included the *opponent's* model of the situation, then responded to that with a model of a model of a model, and so on, ad infinitum. In such a game there were no longer any clear, reliable facts. Very tricky, he thought—devilish. Better than instruments here would be an exorcist. (315)

The models do in fact lead to answers, but these look unnervingly like reflections of the modelers. Even DEUS's models can only be based on human analogues inscribed in its original program. Consequently, the models begin to draw the same human beings who undertook the project of contact with the greatest imaginable altruism and benevolence into a process in which they increasingly resemble the Quintans. Because of their superior technology, they become meta-Quintans. Rather than saving the alien civilization, they become agents of its destruction—and perhaps of its self-destruction.

The models of the Quintans' autonomous war-sphere can only describe abstract relations, through mathematical systems like game theory, decision theory, and information theory. The domain of motives, values, and sufferings remains unintelligible in these systems. Neither the Quintans nor the human beings can create self-sacrificing "autonoms" or messages that cannot be misprised. In *Fiasco*, the model for the dialogue of models is war. Lacking the direct power of the Solaris ocean or the ideal tact that Hogarth attributes to the Senders, the one thing that the Quintans and the Earthmen have in common is their aggressive self-consciousness, which is thoroughly and ironically reflected in their technospheres. The models, rather than mediating between alien beings, create the illusion of understanding and the delusory promise of a reply. It ultimately proves easier for human beings to dock in a black hole than to establish communication with another cosmic culture.

In the sad and harrowing conclusion of the novel, its hero, Mark Tempe, whose whole life has been dedicated to actually *seeing* the Quintans, arrives on the planet as the emissary of Earth. He cannot be sure that anything he sees is authentically Quintan and not an ironic, inverted Quintan image of *human* models. When he begins a desperate search to find what a true Quintan looks like, Tempe loses his sense of time and forgets to send the crucial flare to alert his shipmates that he is still alive. At the very moment that he does make *physical* contact with one of the Quintans, unexpectedly so unlike the humans with whom they have engaged in a war of hostile models, the hovering spaceship unleashes its solar laser, destroying its emissary along with the civilization with which it had come thousands of light-years to communicate.

With *Fiasco*, Lem depicts a vision of infernal modeling in which the models and their technological analogues prevent true encounter. In the absence of significantly different kinds of communication, the increasingly autonomous models draw the human emissaries into representing the alien Quintans' behavior in terms all too familiar in Earth's history: the familiarity of war, impatient honor, and the self-destruction of intelligence.

Notes

1. Much of my argument is indebted to Hayles's essay, "Space for Writing: Stanislaw Lem and the Dialectic 'That Guides My Pen.' "

2. Cf. Jarzebski's remarks on Lem's mockery of Hegelian rationalism in "Stanislaw Lem's 'Star Diaries' " (370).

3. "Personetics" is the science of constructing artificial sentient beings via cybernetics (cf. "Non Serviam"); "fantomology" is the study of artificial realities "that are in no way distinguishable from normal reality by the intelligent beings that live in them, but which nonetheless obey rules deviating from that normal reality" (*Summa Technologiae* 4:171); "imitology" is the study of the construction of informational imitations of natural phenomena (4:162).

4. See Heisenberg's *Philosophic Problems* (73).

5. This is the burden of Lem's critique of Stapledon's Star Maker ("On Stapledon's Star Maker," 4).

6. The present reading of Solaris may appear rather "positive," especially compared with my own considerably more skeptical reading in an earlier essay ("The Book Is the Alien"), on which much of the present discussion draws. I do not believe I am contradicting myself, however. The reading of Solaris as an ironic metafictional and metascientific parable of undecideability requires that readers build up a realistic story and its negation simultaneously. The effect of undecideability could never happen if the realistic tale were not extremely care-

fully developed, and indeed the most attractive alternative. Solaris's particular virtue in Lem's corpus is the power of the positive reading to resist the metalinguistic undermining—a power usually reserved for the Pirx tales, now including *Fiasco*.

Works Cited

Csicsery-Ronay, Istvan, Jr. "The Book is the Alien: On Certain and Uncertain Readings of Lem's Solaris." *Science Fiction Studies* 35 (1985): 6–21.

———. "Twenty-Two Answers and Two Postscripts: An Interview with Stanislaw Lem." Trans. Marek Lugowski. *Science Fiction Studies* 40 (1986): 242–60.

Hayles, N. Katherine. "Space for Writing: Stanislaw Lem and the Dialectic that 'Guides My Pen'." *Science Fiction Studies* 40 (1986): 292–312.

Heisenberg, Werner. *The Philosophic Problems of Nuclear Science*. New York: Pantheon, 1952.

Jarzebski, Jerzy. "Stanislaw Lem's 'Star Diaries'." *Science Fiction Studies* 40 (1986): 361–73.

Kagarlitski, Julius. "Realism and Fantasy." In *Science Fiction: The Other Side of Realism*. Ed. Thomas Clareson. Bowling Green: Bowling Green University Popular Press, 1971. 29–52.

Lem, Stanislaw. *A Perfect Vacuum*. Trans. Michael Kandel. New York: Harcourt Brace Jovanovich, 1983.

———. *Chain of Chance*. Trans. Louise Iribane. New York: Harcourt Brace Jovanovich, 1984.

———. "Chance and Order." Trans. Franz Rottensteiner. *The New Yorker* 30 January 1984.

———. "De Impossibilitate Vitae et De Impossibilitate Cognoscendi." *A Perfect Vacuum* 141–66.

———. "Doctor Diagoras." *Memoirs of a Space Traveller. Further Reminiscences of Ijon Tichy*. Trans. Joel Stern and Maria Swiecicka-Ziemianek. New York: Harcourt Brace Jovanovich, 1983.

———. *Fiasco*. Trans. Michael Kandel. New York: Harcourt Brace Jovanovich, 1987.

———. *Golem XIV*. Trans. Marc E. Heine. *Imaginary Magnitude* 97–248.

———. *His Master's Voice*. Trans. Michael Kandel. New York: Harcourt Brace Jovanovich, 1983.

———. *Imaginary Magnitude*. New York: Harcourt Brace Jovanovich, 1986.

———. "Lymphater's Formula" ("Formula Lymphatera"). *Ksiega robotow*. Warsaw: Iskry, 1961.

———. *Memoirs Found in a Bathtub*. Trans. Michael Kandel and Christine Rose. New York: Harcourt Brace Jovanovich, 1986.

———. "Metafantasia: The Possibilities of Science Fiction." Trans. Etelka de Laczay & Istvan Csicsery-Ronay, Jr. In *Microworlds*. Ed. Franz Rottensteiner. New York: Harcourt Brace Jovanovich, 1984.

———. "Non Serviam." *A Perfect Vacuum* 167–96.

———. "On Stapledon's Star Maker." Trans. Istvan Csicsery-Ronay, Jr. *Science Fiction Studies* 41 (1987): 1–8.

———. "Professor A. Donda." *Maska.* Krakow: Wydawnictwo Literackie, 1976.

———. *Solaris.* Trans. Joanna Kilmartin and Steve Cox. New York: Berkely, 1971.

———. *Summa Technologiae.* Budapest. 1976.

———. *The Investigation.* Trans. Adele Milch. New York: Harcourt Brace Jovanovich, 1986.

———. "The New Cosmogony." *A Perfect Vacuum* 197–229.

Minsky, Marvin. "Why Intelligent Aliens Will Be Intelligible." *Extraterrestrials: Science and Alien Intelligence.* Ed. Edward Regis, Jr. Cambridge: Cambridge University Press, 1985. 117–28.

Philmus, Robert. "The Cybernetic Paradigms of Stanislaw Lem." *Hard Science Fiction.* Ed. George Slusser and Eric Rabkin. Carbondale: University of Southern Illinois Press, 1986. 177–213.

Stapledon, Olaf. *Last and First Men.* New York: Dover, 1968.

12

Negentropy, Noise, and Emancipatory Thought

Eric Charles White

In *Order Out of Chaos*, their account of the philosophic implications of recent research in nonequilibrium thermodynamics, Ilya Prigogine and Isabelle Stengers argue that contemporary physical speculation is moving toward a conception of the universe "that might be called 'historical'—that is, capable of development and innovation" (252). This new view of nature involves a shift from theories insisting on the deterministic character of natural process to an emphasis on "statistical, stochastic description." Nonequilibrium thermodynamics has revealed that the evolution of a physical system is only partially destined to follow a particular course of development. Under certain "far-from-equilibrium" conditions, random "fluctuations" have the potential to propel a system toward a "bifurcation point" at which the direction of change becomes unpredictable. As Ervin Laszlo points out in *Evolution: The Grand Synthesis*, a physical system far from thermal and chemical equilibrium may act indeterminately: "By an apparently random choice one among its possibly numerous fluctuations is amplified, then the fluctuation spreads with great rapidity," radically altering the existing system (35). The point here is that microscopic random fluctuations—purely chance occurrences—can bring about macroscopic transformation. Nature changes its shape in a moment of truly protean metamorphosis.

Physical systems are not, then, constrained to follow a single evolutionary path. Instead, Laszlo says, they have "bundles of trajectories. . . . Given identical initial conditions, different sequences of events unfold" (20). Cosmic and biological evolution have not followed a single line of development laid down at the dawn of time. The future does not derive from the past in strict linear fashion. The deterministic unfolding of the universe has repeatedly been punctuated by catastrophic bifurcations, by chance swervings in unforeseeable directions. Nature, in more venerable parlance, is ruled by Fortune.

This insistence on the aleatory character of natural process entails an important corollary. Although change can destroy a system, at the critical moment of transformation matter may spontaneously organize itself into a more complex structure. That is, at a stochastic bifurcation point in far-from-equilibrium conditions, the famous second law of thermodynamics (according to which entropy never decreases) is consistent with *local* decreases in entropy. Rather than toward disorder, the local direction of change may be toward a more highly differentiated state. A new steady state may come into being that, in Laszlo's words, is "more complex than the structure in the previous steady state" (36). For instance, beyond a certain threshold, laminar flow inevitably becomes turbulent. The onset of turbulence constitutes an increase in complexity. According to Prigogine and Stengers:

> The multiple time and space scales involved in turbulence correspond to the coherent behavior of millions and millions of molecules. Viewed in this way, the transition from laminar flow to turbulence is a process of organization. Part of the energy of the system, which in laminar flow was in the thermal motion of the molecules, is being transferred to macroscopic organized motion. (141)

"Order" comes out of "chaos," then, through "active matter's" capacity for "self-organization." Contradicting the general drift toward entropic degradation, a pocket of negentropy spontaneously appears.

The role ascribed to stochastic self-organization in this vision of natural history is an emancipatory one. As the stochastic leap toward the unprecedented liberates nature from determinism, so the emergence of order out of chaos overcomes entropic degradation. Nature is thus both "free" and "progressive." Indeed, unpredictability and complexity exist in a reciprocal relationship: the more complicated the cosmos becomes, producing in succession physical, chemical and biological levels of organization, the more open is it as well to further evolutionary innovation. Laszlo can therefore assert that evolution "is always possibility and never destiny" (20). Natural history is not fated to pursue the same dull round until the dissipation of energy and disintegration of matter are complete. Instead, there is the ever present possbility of a break with the past, a radical departure from the chain of causality, enabling a new beginning.

Michel Serres has found an important precursor to this view of nature in the physical theory of ancient Epicurean philosophy. In *The Birth of Physics in the Text of Lucretius*, Serres argues that the Epicurean physics of Lucretius posits chaos as the originary condition of the universe. Lucretius images this state of disorder as the eternal fall of atoms through space, an image that approximates a model of laminar flow. At uncertain times and indefinite places, the universal fall of the atoms is interrrupted by what Lucretius calls the *clinamen*, "the smallest conceivable condition for the first formation of turbulence" (*La naissance* 13). The *clinamen* can be thought of as a stochastic swerving or random fluctuation whose subsequent amplification, as the atoms begin now to collide with one another, ensues in the emergence of a vortex or whirlpool—precisely, turbulence beginning in a previously laminar flow—a spiralling motion that heralds the beginning of the world. Reality does not, then, exist necessarily but has come into being by chance. Only because of the accidental swerving of the *clinamen* is there "something rather than nothing. . . . That which exists is improbable" (*La naissance* 33).

To understand this process of stochastic self-organization from chaos we require, Serres argues, a new physics, a science that would be "an organon of miracles, a discourse of the miraculous," a "theory and practice of circumstances" akin to the *bricolage* of savage thought that will point the way to a cultural renaissance, to a new vision of cultural practice. Such a science would be a science of Venus, goddess born from the ocean, traditional emblem of flux and dissolution: "We exist by the grace of Aphrodite. . . . The great Pan is reborn" (*La naissance* 98, 82, 123, 33). By the grace of Aphrodite, by the agency of the *clinamen*, we emerge from chaos as "negentropic islands on or in the entropic sea . . . pockets of local order in rising entropy" (*Hermes* 75). The physical science at issue here would thus posit "a plurality of worlds and . . . temporary existence." In other words, reality "does not have an origin, but is always in the process of being born" as random fluctuations provoke the emergence of ever new pockets of negentropy. Aphrodite is an improvisational artist who turns continually to chaos, disorder, the "other side" of causal determinism as the source of novelty and renewal. And we ourselves must become improvisational artists, *bricoleurs* who live "close to the clinamen, where Nature is born," close to the fertile chaos from which form is continually emergent. Serres calls this stance toward experience the "wisdom of the Garden," of the local, the circumstantial, the occasional, a wisdom shared with Lucretius by Montaigne, whose essays embody a spirit of improvisationality that Serres would emulate in his own writing (*La naissance* 229, 169, 226, 236).

But if we would return to the Garden, to the domain of Venus, a small cultivated locality surrounded and impinged upon constantly by an unmasterable wildness, we must forsake another deity: Mars, the war god, who proposes a science and an ethics of "totalization, force, mastery, and empire" (*La naissance* 236). The respective orientations of Venus and Mars are absolutely antithetical to one another, diametrically opposed scientifically, aesthetically, and politically. Thus, where Venus proposes a "science of caresses" valorizing stochastic processes, Mars insists on a "science of death" according to which reality can be reduced to deterministic trajectories. Again, Venus would promote multiple perspectives on a world in flux, multiple tales of cosmic evolution, while Mars demands representational closure, a definitive image of reality, a single master narrative commanding the entire sweep of natural history. Finally, Venus's desire for nonagonistic cooperation in the social realm can never coexist with a martial politics of domination and totalizing control. *The Birth of Physics in the Text of Lucretius* therefore contains a lengthy denunciation of the god of war.

Following Lucretius, for whom the twin evils of human society—ambition and greed—issue from a fear of death, Serres condemns the regime of Mars as a futile attempt to evade the ultimate omnipotence of entropy. Mars cannot accept the fact that since "everything flows . . . nothing is of invincible solidity." His science is driven by a hatred of fluidity, impermanence, and uncertainty: there is a "horror of the world at the heart of the theoretical." Mars's hatred of aleatory change provokes an attempt to prolong his reign indefinitely and thus achieve a condition of invulnerable, eternal mastery over reality. But by a tragic irony, we worshipers of Mars deliver ourselves over to death at the very moment we seek to evade our mortality, our inability to remain the same amidst the buffetings and shocks of the swarming chaos that surrounds us. Only as an "open system," open to perturbations from without that may provoke a miraculous transformation, can life flourish. To seek finality, autonomy, perfect mastery is to become a "monster . . . in itself and for itself, autistic and dead" (*La naissance* 12, 162, 176). The monstrous regime of Mars, the "manic-depressive syndrome" (*La naissance* 81) of those who quest frantically but futilely after total control over themselves and their environment, depends on violence: violence against the self, when it is denied access to renewal in chaotic spontaneity; violence against others, when they are subjected to martial discipline and regimentation; violence against nature itself, which becomes merely the raw material for the augmentation of power. Mars thus initiates a "thanatocratic" world order in which "there is nothing to be learned, to be discovered, to be invented" because totalizing mastery has rendered diversity homogeneous: "nothing new under the sun of identity" that reduces difference to sameness (*Hermes* 100).

Dispelling martial delusions of grandeur depends, then, on acceptance of the ineradicable mortality of every form and structure, of every organism, which must henceforth be understood as a momentary eddy, a transitory whirlpool in an ever-flowing stream. Increasingly desperate attempts to channel and contain this onrushing flood must ultimately come to nought "for they are all just a little brownian motion on the surface, superficial disturbances hiding the incurable erosion of . . . the world" (*Hermes* 120). But recognizing the impermanence of reality also reveals the secret of its creation. The entropic degradation of every object is merely a continuation of the process by which it was constituted by the swerving *clinamen*: "Nature declines and that is its act of birth" (*La naissance* 45). Endless passing away and coming into being presuppose one another. *Only metamorphosis truly exists.* As one eddy disappears, another forms downstream. Acceptance of mutability thus restores us to Venus, to her "creative science of change and circumstance" which "breaks the chain of violence, interrupts the reign of the same" as we learn to identify not with any particular image of fulfillment but with the process of stochastic self-organization itself (*Hermes* 99, 100).

In *The Parasite*, Serres considers the prospects for an Epicurean cultural practice that would remain always open to the unforeseen spontaneity of the *clinamen*'s swervings. It turns out that the model for stochastic self-organization from chaos applies not only to physical and biological systems but describes equally well the production of meaning from noise. From a martial perspective successful communication between two interlocutors depends on the exclusion of a third person, "the prosopopeia of noise" who threatens constantly to disrupt the transmission of messages (*Hermes* 67). Since the optimum performance of any system depends upon communicative transparency, noise must be eliminated. Martial communicators, who believe only in "simple, rough, causal relations" and for whom "disorder always destroys order," thus build their "homogeneous, cruel systems upon the horror of disorder and noise." The elimination of noise is meant to perfect the functioning of the system and enable the worshipers of Mars henceforth to reside in "the good, the just, the true, the natural, the normal" as complacent dogmatists, forever impervious to the specter of change (*The Parasite* 14, 68).

As the dogmatic character of martial social life suggests, the exclusion of noise amounts to an exclusion of genuine information. Information, understood in Gregory Bateson's phrase as the "difference that makes a difference," is excluded in favor of information-free, wholly redundant messages. The system endlessly reiterates, endlessly ratifies itself. But such a system, however self-coherent or optimally efficient, is nevertheless doomed to entropic degradation. Like any closed system, it can only run down. The achievement of redundancy—when everything that needs to be said has already been said—is analogous to entropic homogeneity when matter-energy settles into terminal equilibrium. In cultural systems, then, just as in physical systems, noise or chaos amounts to a force for renewal. Serres thus imagines a "parasite"—precisely, static in a communication channel—who intervenes to interrupt normal communications. By perturbing the routine exchange of messages, the parasite can provoke the production of novelty. The parasite's introduction of confusion into a logically closed system enables the generation of alternative logics. Like a "simple fluctuation, a chance event, a circumstance," noise too can produce a new system of meaning (*The Parasite* 18).

We would thus have a conception of the production of meaning as a process of transformation involving the integration of noise. Though noise may destroy one system, this destruction permits the emergence of another, potentially more complex system in its place. As in nonequilibrium thermodynamics, we cross a bifurcation point at which the system leaps to a new steady state of augmented significance. The parasite, an abusive guest at the table of communication, renews a conversation which would otherwise grow stale by posing a topic that has nothing whatsoever to do with what has been said before. Just as physical structures appear by miraculous transformation rather than at the end of a chain of linear causation, so meaning "appears locally, here, there, yesterday, tomorrow. . . . Local and plural, it is aleatory and stochastic" (*La naissance* 181). The emergence of genuine information is thus the antithesis of the closed system that eternally repeats a finite set of messages. True meaning is unprecedented, prophetic, the "unheard-of."

For Serres, communicational "harmony," understood as the consensus achieved between interlocutors who understand each other perfectly, is only "an antechamber to death." Cultural vitality depends on "parasitic dissonance," the redemptive non sequitur (*The Parasite* 126). As order comes out of chaos, so sense requires nonsense. Meaning emerges not as predictable derivative but as stochastic departure from tradition, as *invention*.

In this sense, Serres shares the hope that William Paulson attaches to literature in *The Noise of Culture: Literary Texts in a World of Information*. According to Paulson, literature has become a marginal activity in a society in which instrumental, purely communicative modes of discourse are dominant. But the postmodern legitimacy of literature derives precisely from its marginality: literature will henceforth be a "source of differences" whose purpose is to "enrich thought and inspire new moves in the language games of society." For Paulson, literature creates meaning precisely by placing meaning in jeopardy. Literature is a "noisy channel" that "assumes its noise as a constitutive factor of itself." Literary discourse is thus distinct from instrumental discursive modes that seek communicative transparency. Literature functions as the "noise" of culture. By perturbing existing systems of meaning, it enables the invention of new ideas, and ultimately, new domains of knowledge. The moral of the story is a "politics of criticism": literary studies provide a space in which resistance can be mounted against the totalizing power-effects of technocratic modernity.

Such a theory of textual production can be extrapolated from Thomas Pynchon's *The Crying of Lot 49*. Pynchon's heroine, Oedipa Maas, resists the entropic degradation of the narcissistically self-contained social system of "San Narciso" by producing the hitherto hidden meaning of history: the possibly sinister, possibly redemptive "Tristero"/"Trystero" conspiracy whose revelation might so destabilize her society as to propel it toward a bifurcation point, a moment of unprecedented transformation. Taking the noise and "W.A.S.T.E." of her increasingly claustrophobic milieu as point of departure, Maas "keeps it bouncing," or reverses the slide toward terminal stillness, by engaging in a deliberate invention of meaning: "*Shall I project a world?*" (134, 159).

This daughter of Venus would thus undo the "endless, convoluted incest" of San Narciso, a martial society that disciplines, manipulates and casually inflicts violence on its members the better to produce and consume redundant messages and useless goods; a thoroughly wasteful society that wastes everyone in it, transforming its inhabitants into alienated, isolated, loveless "inamorati," failed suicides who desperately crave community. *The Crying of Lot 49* is a "parable of power," of technocratic manipulation and totalizing control. In America, "with the chance once so good for diversity," Oedipa discovers an all-encompassing system dedicated to optimum performance that pursues this end by reducing every member of society to the status of "generic," interchangeable bits in the great social cybernetic network. She therefore seeks to bring about a "miracle"—"another world's intrusion into this one"—which would renew her society, transforming the "printed circuit" of San Narciso into Pentecostal glossolalia. The noisy babbling of one speaker would then provide an opportunity for the next noisy speaker to invent meaning stochastically, on the spur of the moment, without reiterating what has always already been said (Pynchon 5, 83, 35, 136, 104, 88, 13).

The respective critiques and alternatives to life-denying totalization proposed by Pynchon and Serres are obviously reminiscent of one another. They also converge with the concerns of other contemporary thinkers. For instance, Pynchon's image of San Narciso as a printed circuit and Serres's model of "dialogue" as the exclusion of a "noisy" third person can appropriately be compared to Jean-François Lyotard's indictment of late capitalism in *The Postmodern Condition* as determined by an overriding criterion of "performativity," the optimization of productive force.[1] Similarly, Oedipa Maas's deliberate projection of the meaning of history might be understood as an example of Lyotardian "paralogy" in which paradoxical moves actually reconstitute the language games of society. Again, Serres's excoriation of the reign of Mars recalls Deleuze and Guattari's polemic in their *Anti-Oedipus* against paranoid forms of psychic and social organization. Serres's ethic of the Garden and celebration of stochastically emergent, strictly local meaning would then correspond to Deleuze and Guattari's schizocultural alternative: the unstructured mobility of schizophrenic desiring intensities.[2] All of these thinkers—Serres, Pynchon, Lyotard, Deleuze, and Prigogine as well—attribute an emancipatory potential to an unmasterable, unimaginable "outside": precisely, "chaos" or "noise" as source of vitality and renewal.

Emancipation is not, however, a feat that can be accomplished once and for all. Every effort to resist totalizing power constitutes a new domain within which power will again seek to maximize its control. Recapitulating the career of her namesake, Oedipa Maas may herself be the criminal she seeks. Oedipa sets out on her quest determined to undo the ever increasing organization of a closed system dedicated to optimizing its own efficiency. The irony is that in undoing one system, she lays the foundations for another one, equally cruel and homogeneous. The rendering uniform of all dimensions of social life, or society as a perfectly organized cybernetic network of Muzak, Tupperware, freeways, shopping malls, and tract homes, can proceed anew on the basis of the Tristero, whose acronym—"W.A.S.T.E.," or "We Await Silent Tristero's Empire"—does not bode well for a fresh start. Oedipa may have become a producer rather than a passive consumer of meanings, may have overcome the drift toward "generic" sameness, but from whatever interpretation she attributes to reality, another totalizing system can emerge. Is the *Tryst*ero a communications network by means of which genuine information can be transmitted, "the direct epileptic Word, the cry that might abolish the night?" Or is the *Trist*ero just another term for the nightmare of history? Pynchon does not credit without qualification the possibility of an alternative to the totalizing mobilization of productive force in late capitalism which maintains global control through the programmatic incitement and normalization of desire. Oedipa's paralogical intervention may merely install an essentially similar world order (127, 87).

This problematic recurs in Pynchon's next novel, *Gravity's Rainbow*. Pynchon proposes in this later work that the act of cognition itself is inextricably implicated in a futile, delusive quest after mastery. According to Katherine Hayles in *The Cosmic Web*, one cannot "speak from within a field," or inhabit a Garden and thus forego the temptation of imperialist expansion, without "betraying it to the linear processes of articulation" that seek relentlessly to enlarge the scope of intelligibility. There is no Edenic language; every form of discourse, even the most apocalyptically paralogical, will in the end seek to "consolidate control by extending the image of human consciousness to all creation" (188, 172).

Relevant here is Oedipa's encounter with the Nefastis Machine. In this delirious version of a perpetual motion device, information entropy, understood as a measure of uncertainty and hence of potential meaning, appears able to overcome thermodynamic entropy, the tendency of any closed system to decline toward thermodynamic equilibrium. The metaphorical import of the Nefastis Machine would then entail the view that by taking advantage of the opportunity presented by noise and uncertainty to produce unprecedented significance, one might renew a decaying social system. But Pynchon does not endorse so optimistic a reading. The Demon in the Nefastis Machine requires the collaboration of a telepathic sensitive in order to complete its task successfully: "The sensitive must receive that staggering set of energies [transmitted by the Demon], and feed back something like the same quantity of information. To keep it cycling." Information here furthers the routine functioning of the machine. As Oedipa learns, on the "secular level all we can see is one piston, hopefully moving. One little movement, against all that massive complex of information, destroyed over and over with each power stroke" (77). The production of meaning thus contributes to the prolongation of power.

Pynchon's qualification of Oedipa's invention of the meaning of history entails a second, even more disheartening possibility. In producing the Trystero conspiracy, Oedipa can never be certain that she is not merely "paranoid," an atomized bit, isolated and closed in on herself, already a victim of the system's drive for total control. Her attempt to decipher the meaning of history may have been foreseen and recuperated in advance by a system that encourages an incessant circulation of redundant messages. And if redundant, Oedipa's message is incapable of perturbing the system's functioning: it is not really "noisy." As she watches the world around her "becoming generic," Oedipa is advised to "cherish" her fantasy, "for when you lose it you go over by that much to the others. You begin to cease to be." But if her fantasy begins and ends in herself, if she has failed to tap into a force for renewal outside the closed system of San Narciso, if the putative "noise" she attempts to integrate into a new system of meaning is not truly informative but only redundant, then her decipherment of the meaning of history is futile. Indeed, the possibility that the Trystero "is all a hoax" perpetrated by the trickster Pierce Inverarity implies that Oedipa's pursuit of the Tristero has from the beginning been a trick meant to integrate her more fully into the martial regime of San Narciso. If the "legacy of America" is only Inverarity's "need to possess, to alter the land, to bring new skylines, personal antagonisms, growth rates into being," then his once promising slogan—"Keep it bouncing . . . that's the secret, keep it bouncing"—amounts to a curse. The production of meaning fails to reveal "truth's numinous beauty." Instead, it ratifies a "power spectrum" (104, 103, 126, 134, 136).

Oedipa is never able to determine whether Pierce Inverarity has really sent a message containing genuine information, whether indeed there is "a revelation in progress around her," or contrariwise, whether she is trapped, Rapunzel-like, in a high tower, "embroidering a kind of tapestry . . . seeking hopelessly to fill the void." She can never be certain the "hieroglyphic sense of concealed meaning" that she hopes will transform the world is not a solipsistic projection (28, 10, 13). Like the proto-punk band "The Paranoids," perhaps Oedipa too inhabits a private reality. Such would serve the interests of an advanced technocratic order that depends precisely on social atomization and accordingly welcomes the proliferation of private meanings. As Serres puts it in *The Parasite*, to the "cruelty of systems with one norm" must be added the "implacable power of systems with several norms": "Tolerance is part of the panoply of intolerance" (68).

Does this nymph who resides at a motel appropriately called "Echo Courts" simply return the name of Narcissus? Does she in the end reiterate the system's routine functioning? As Tony Tanner points out in his book on Pynchon, Oedipa needs to "discover which information really works against entropy as opposed to the kind of non-information ('newsless' letters) that effectively accelerates it" (68). But this discovery is permanently deferred. Oedipa never ascertains whether the Trystero system is genuinely informative, whether it will make a difference or merely repeat the same dull round, the same repetitive, cyclical movement suggested by the corporate name "Yoyodyne" (Pynchon 14).

Oedipa's dilemma can thus be summarized: if she has not discovered a way to constitute an alternative social arrangement, then the hitherto hidden "plot" she uncovers as executrix of Inverarity's will reiterates the existing system. But if she has succeeded in positing an alternative, this alternative is only relatively emancipatory. *Tryst*ero is *Trist*ero: the prospect of community fails to avert the nightmare of history.

This dilemma can also be stated with reference to Laszlo's *Evolution: The Grand Synthesis.* According to Laszlo, the "processes of evolution create systems on multiple hierarchical levels." Going up the hierarchy of structural complexity in nature, passing from particles to atoms to chemical molecules to organisms to ecologies and finally to human social systems, we encounter increasing indeterminacy, increasing freeplay. Thus, catastrophic transformation is more frequent in cultural than in natural history. But this hierarchy, which seems to promise freedom, "is not only a structural but is also a *control* hierarchy." Higher levels control the behavior of levels lower in the hierarchy, forcing lower levels "into a pattern of collective behavior." The emergence of a higher-level system can in fact be understood as a "*simplification* of system function" (35, 25). Each new hierarchical level reduces and constrains the teeming variety of lower-level systems.

Moreover, when a higher level comes into existence, controlling what lies below it, the system then becomes more complex by means of what amounts to redundant amplification of the founding premises of the new level. This process of redundant amplification—in effect, a drive toward ever more comprehensive control—now appears as a principal feature of evolution. That is, at each new level in the evolutionary hierarchy, the process continues, producing, as it were, ever more elaborate variations on a single theme until, inevitably, it overreaches itself, leading to a phase of chaotic fluctuation and the onset of renewed evolutionary innovation. In Laszlo's words, "once a new hierarchical level has emerged, systems on the new level tend to become progressively more complex," an increase in complexity that renders them increasingly vulnerable to catastrophic destabilization and the appearance of a new hierarchical level. The destabilization of far-from-equilibrium systems pushes them "up the ladder of the evolutionary hierarchy . . . complexifying system structures on each organizational level and simplifying them again on the next level" (25, 46). Our endorsement of the stochastic emergence of the unprecedented—a promise of freedom—must therefore be a limited one. First of all, the unprecedented emerges at the top of a hierarchy whose lower elements it now controls. Second, the evolutionary innovation thus accomplished is then endlessly reiterated as the new system elaborates a range of possibilities all of which are implicitly contained in the original innovation.

Such a recognition troubles Michel Serres in *The Parasite*. Serres finds himself increasingly unable to maintain a distinction between the respective sciences of Mars and Venus. The term "parasite" is ambiguous: on the one hand, it is equivalent to Lucretius's *clinamen*, whose stochastic swerving produces novelty. The parasite is defined as "a differential operator of change" that enables the stochastic emergence of islands of negentropy. It renews decaying systems and is therefore on the side of life. Thanatocratic systems accordingly seek to expel parasites. For instance, by chasing every parasite the Cartesian ego was able to constitute modern rationality, technology, mastery, and possession, "the proliferating multiplication of a certain type of sameness" (*The Parasite* 196, 180). The parasite is inimical, then, to the ethos of performativity that determines a society like that of San Narciso.

But a different portrait of the parasite emerges from the standpoint of biology and sociology. Here Serres argues that every parasite dreams of presiding over the system it brings into being. The parasite is now on the side of control, absolute power: precisely, death. Serres tries at several points to theorize a way out of this dilemma. He thus proposes three models of communication. The first he associates with the name of Leibniz. In this system, individuals relate to one another "by the intermediate of God. As the unique mediator, he is all-knowing and all-powerful." This first system amounts to a martial regime from which all heterogeneity has been banished. One power, one criterion determines the totality: "the local moves toward the global and the plural toward the singular." A second model of communication is associated with the name of Hermes. Here we have a regime of strictly local chieftaincies: Hermes, god of the crossroads, is a parasite who "has placed himself in the most profitable positions, at the intersection of relations." For Serres, this system is also suspect because it is agonistic: a welter of parasites struggle for control. Finally, a third system is "the invention of the Paraclete, on the Pentecost." In this system, ever-multiplying ideolects replace a coercive common language. Relations among speakers can therefore "be considered to be many-many and the network that describes them is decentered." But Serres concedes the utopian character of this Pentecostal glossolalia: such "a graph has never been seen" (*The Parasite* 43–44). The best that can be hoped, apparently, is the agon of the many among themselves—the reign neither of Venus nor of Mars but of Hermes, trickster, liar, and thief who plays constantly for temporary advantage.

Elsewhere in *The Parasite*, Serres proposes as an alternative to the parasite what he terms the "producer:" an "archangel because he bears information, news, novelty, and because he is necessarily at the head of the line in relation to the parasitic chain." Again, this distinction collapses; there never has been any "producer," only parasites parasiting one another. The "producer" is always already a parasite occupying an entirely relative position on the parasitic chain. Hence the dramatic and surprising renunciation with which Serres concludes this book: since the book of the parasite is "irreparably a book of Evil," Serres proposes now to find *ataraxia* or tranquility by identifying not with Venus, daughter of the ocean, but with the ocean itself, image of absolute dissolution: "Quiet, serene, no anxiety. The high seas" (*The Parasite* 167, 253).[3]

The emergence of order out of chaos and meaning from noise is irresolvably ambiguous. As soon as order does come into being, there ensues a process of progressive elaboration as the new system strives toward maximum scope and power. Emancipatory innovations inaugurate disciplinary norms. What was paralogical now becomes performative, as Lyotard perhaps intends to suggest when he notes that postmodernist artistic innovations inevitably succumb to a modernist nostalgia for wholeness.[4] The local becomes the basis for a new global. From the moment Venus intervenes to perturb and so renew a decaying system, she becomes her antithesis. A martial regime of deterministic trajectories can never permanently be overcome by stochastically emergent turbulence. Life is forever out of balance: the singular and specific quests relentlessly after totality. What was once information, through endless reiteration, becomes merely redundant orthodoxy. The parasite is not Paraclete, and Oedipa *is* Oedipus.

We are thus caught in a double bind. Oedipa Maas "had heard all about excluded middles; they were bad shit, to be avoided;" but at the end of *The Crying of Lot 49* Oedipa finds herself in just such a state of tragic suspension (136). Serres arrives at a similar impasse: "Life works; life is work, energy, power, information. It is impossible to translate this description into an ethical discourse" (*The Parasite* 88). The emergence of meaning from noise is, so to speak, "beyond good and evil." Although there is the possibility of resistance to totalizing power, every act of resistance itself inaugurates yet another closed system committed to the total control of every object that falls within its domain. Serres's apparent abandonment of the distinction between Mars and Venus in favor of the ambiguously charged term "parasite," in this sense recalls Foucault's definition of "power" in *The History of Sexuality* as simultaneously repressive and constitutive (92ff.). Serres and Pynchon would also have to concur with Derrida's insistence throughout his career that even the most skeptical iconoclasm remains vulnerable to metaphysical illusions. A Sisyphean perspective, then, on cultural practice: the activity of making sense by way of recourse to chaos, noise, and chance circumstance is an interminable task because power, like desire, is protean and omnipresent.

Notes

1. According to Lyotard, contemporary technocratic society is "terroristic" because the system's efficient functioning depends on the manipulated or coerced "adaptation of individual aspirations to its own ends." As in the generic world of San Narciso, the "criterion of performativity" reduces individual agents to interchangeable bits (63).

2. Deleuze and Guattari contrast the "paranoiac, reactionary, and fascisizing pole" of libidinal investment with the "schizoid revolutionary pole" as follows:

> The paranoiac and the schizoid investments are like two opposite poles of unconscious libidinal investment, one of which subordinates desiring-production to the formation of sovereignty and to the gregarious aggregate that results from it [i.e., present-day disciplinary, normalizing society], while the other brings about the inverse subordination, overthrows the established power, and subjects the gregarious aggregate to the molecular multiplicities of the productions of desire. (366, 376)

3. In the next chapter in this volume, Maria Assad assesses Serres's more recent effort in *Genèse* (1982) to think beyond this impasse by returning to Venus, but Venus now reimagined as "Venus turbulente" whose always just-emergent form remains inseparable, indeed, indistinguishable finally from eternally generative flux.

4. Lyotard, "Answering the Question: What is Postmodernism?" *The Postmodern Condition* 71–82.

Works Cited

Deleuze, Gilles, and Félix Guattari. *Anti-Oedipus: Capitalism and Schizophrenia.* Trans. Robert Hurley, Mark Seem, and Helen R. Lane. New York: Viking, 1977.

Foucault, Michel. *The History of Sexuality*. Vol. I: *An Introduction*. Trans. Robert Hurley. New York: Random House, 1978.

Hayles, N. Katherine. *The Cosmic Web: Scientific Field Models and Literary Strategies in the Twentieth Century*. Ithaca: Cornell University Press, 1984.

Laszlo, Ervin. *Evolution: The Grand Synthesis*. Boston: Shambhala, 1987.

Lyotard, Jean-François. *The Postmodern Condition: A Report on Knowledge*. Trans. Geoff Bennington and Brian Massumi. Minneapolis: University of Minnesota Press, 1984.

Paulson, William. *The Noise of Culture: Literary Texts in a World of Information*. Ithaca: Cornell University Press, 1988.

Prigogine, Ilya, and Isabelle Stengers. *Order Out of Chaos: Man's New Dialogue with Nature*. New York: Bantam, 1984.
Pynchon, Thomas. *The Crying of Lot 49*. New York: Bantam, 1966.
Serres, Michel. *Hermes: Literature, Science, Philosophy*. Ed. Josué V. Harari and David F. Bell. Baltimore: Johns Hopkins University Press, 1982.
———. *La naissance de la physique dans le texte de Lucrèce: Fleuves et turbulences*. Paris: Minuit, 1977.
———. *The Parasite*. Trans. Lawrence R. Schehr. Baltimore: Johns Hopkins University Press, 1982.
Tanner, Tony. *Thomas Pynchon*. London: Methuen, 1982.

13

Michel Serres: In Search of a Tropography

Maria L. Assad

On the back cover of *Genèse* Serres tells of an original disagreement between himself and those to whom he had entrusted a first reading of his new work. He, the author, had entitled it "Noise"[1] which in old French means "noise [bruit] and furor, the tumultuousness [le tumulte] of things and rivalrous dissension among human beings. 'Noise' indicates [désigne] chaos." In contrast, his readers insisted that he call his essay "Genèse," because it "tries to listen to the fragile formation of things and messages coming from this brouhaha" like Venus emerging from the turbulent seas. The fact that they won this debate cannot hide, however, a significant shift in Serrean discourse which his readers were not ready to accept at that point. Exiled from its titular position, Serres sneaks "noise" in through the back door, so to speak, and defends it as a *word* that has vectorial power, directing our attention *to* chaos. It does not state chaos, is not analogous with chaos. Serres's only, but effective, defense for his choice of title is rhetorical; by indicating and pointing to something, "noise" is a trope.[2] It circumscribes a "tropological space"[3] where chaos brings forth order, though neither author nor reader knows the precise point of juncture. It is a black box examined by the reader as to its output, "genèse": the birth, formation and generation of things and messages that give order to our lives.

All the while the author remains fascinated with the black box itself: "This book attempts to describe *as closely as possible* what in nature or our culture is chaotic and multiple. It is a book on beginnings" (my emphasis). It seems that the author is pleading with his authoritative reader: Yes, do read my book as a genesis, if you must, for I do so also. But never forget that this is not a primary, absolute, divinely ordained Genesis by the Word that was in the beginning. Rather it should be understood "in its most humble and abounding [foisonnant]" signification of multiple geneses, "small generations, numerous becomings, abounding possibilities, and disappearances." Understood as a plurality of creative moments that surge forward, cross, intersect, fall back, link and relink, "genèse," in spite of the insistent reader, becomes itself a trope, designating, not a unitary point of beginning, but multiple possibilities by which various epistemological discourses generate themselves. Not only does Serres thus sum up what *Genèse* had already "attempted" to describe, he also and, in a manner of speaking, backhandedly establishes an intimate relationship between "genesis" and "noise" in their commonly shared tropical nature.[4]

This discussion of what, after all, is a printed afterthought, points to a bifurcation in Serres's entire work that puts reader and author at a fork in the road. The reader avidly follows his thoughts when she reads in them the genesis of our episteme as a cosmogony (literally, the birth of order), out of the abyss that is chaos. Serres himself reads the same genesis but also hears noise, turbulence, that tumultuous chaos that is the nurturing, primal plasma of stochastic moments of invention. Where the thoughtful reader witnesses the birth of Venus out of the waters of chaos, Serres sees "Vénus turbulente" (*Genèse*, back cover), that strange contradiction of terms indicating a moment where ordered singularity and chaotic background noise cannot be separated but are paradoxically the same and other.[5]

The postscript is not only an affirmation of the figurative nature of the two most powerful terms in *Genèse*, "noise" and "genèse," thereby underscoring the need to pursue multiple directions of understanding for each and any of its many discursive images; it is also, and equally urgently, a redirection for the reading of Serres's earlier works. For *Hermes*, I-V, and *The Parasite* are focused on the birth of order out of disorder, on the many epistemological discourses that describe this cosmogony in their unique, distinctive, often rival and exclusionary ways. The dominant allegorical figures in these texts are Hermes and the parasite; both privilege the attention given to the appearance of order out of disorder, Hermes covering chaos with his wily commerce until order is established, the parasite playing the third man who creates instability, turns things inside out or renders them unintelligible and whose exclusion draws a neat, definitive line between stability and turbulence. Hermes and the parasite have become paradigms for the intimate relation between order and disorder, Hermes favored for his scientific qualities, the parasite for his talent to evoke its historical processes. Both Hermes and the parasite are perfect figural spaces in which the sciences, philosophy, mythology, literature, and the visual arts are shown to produce unique discourses on that relationship.

Pulling together into one logical system these most varied allegories of our episteme, Hermes becomes, in the slightly modified words of Harari and Bell, "the reconnector of many explanatory systems" (xxxii). The parasite fulfils the same function but by different methods, or in a different figural space, that of power, words, economics, and human relations. He is a "humanistic" model, complementary to Hermes' mytho-scientific one. "By virtue of its power to perturb [any system], the parasite attests from within order the primacy of disorder; it produces by way of disorder a more complex order" (Harari and Bell xxvii). Characteristic of almost all of Serres's writing until *Genèse* is the demonstration of a fundamental epistemological law, namely the suppression of chaos and its knowledge, as the condition of possibility for any organizing principle to function historically. But chaos does not disappear. Lying just beneath the surface, a thin mantle of patchwork systems that can tear at any moment or wear away, chaos reappears at random moments and is always explained away as contingency, unintelligent multiples, mixtures—understood to be impure, improper or irrational, depending on the discourse—all variabilities that need to be tamed and reharnessed into unitary systems or binary oppositions. They are bubbles of disorder that produce a reaction which makes a vigilant system only the stronger for it.

Boiling up from turbulent chaos, these "bubbles" are the object of Serres's investigation in *Genèse*. Functioning as openings ("manholes," the author calls them) that bring chaos to the surface, they are connectors, analogous to Hermes and the parasite, between chaos and order. But when Serres attempts to define the object of this text as the "multiple as such [le multiple tel quel]" (18), he does indeed pursue the same quest as in *Hermes*, I-V, or *The Parasite*, but from a different perspective, if not to say an opposite angle. Instead of conceiving chaos as the negation of order (dis-order reflecting this conceptual urge lexicologically), Serres goes backward ("en amont") in this text and climbs down the manholes to find the generative power of chaos. It is a natural consequence of his previous investigations:[6] if the "bubbles" are so creative and inventive in their interruptive occurrences within systems, how much more creative and fertile must be chaos, the foundation and emitter of these bubbles. *Genèse* is Serres's invitation to his reader to follow him as he opens the manholes, peers into the darkness, and listens to the cacophony of the turbulence below.

The expression "the multiple as such" is one of the rare occasions in Serres's many texts when he uses a key term in its literal, proper sense. Its "nominal definition" (de Man 57) promises to determine a "stable system of signification" (de Man 2) for an object phenomenologically analyzed by a discourse the limits of which are apparently clearly defined in the introduction entitled "The Object of this Book."[7] What makes for a disquieting reading, however, are nine pages replete with figurative approximations culminating in a sentence of utter indeterminacy: "It [the multiple] is perhaps somewhat viscous [Il est peut-être un peu visceux]" (*Genèse* 19). By his Greek tradition, Hermes represents the multiple; by his function as the third man, the parasite breaches any binary system and exposes it to the multiple, the variable. But neither Hermes nor the parasite is ever portrayed in Serres's previous writings in as formless and undecidable a fashion as this "multiple" which is viscous, but then only somewhat and without any certainty. The "multiple" of *Genèse* widens the tropological space of Serrean definitions, until it becomes "perhaps, somewhat" limitless, or close to it.[8] In its nonprogrammatic implication—three expressions of indeterminacy (perhaps, somewhat, viscous) undermining the certainty of being—the statement on the multiple is "programmatic" for all models Serres employs in *Genèse* and the writings that follow. Previously, Hermes and the parasite offered allegorical explications for the emergence of ordered systems out of the multiple, but at a price Serres considers to grow exponentially intolerable. For the demonstration depends on a successful cover-up of disorder. On the other hand, the implication inherent in "the multiple is perhaps somewhat viscous" does not explicate or demonstrate anything; on the contrary, the act of defining is its undoing, the demonstration is in fact an obfuscation, the ex-plication closes in on itself, becomes an implication, what Serres describes elsewhere as the "baker's logic."[9] The implication of the "multiple" in *Genèse* does not cover up disorder, it points towards the chaotic, gives it the largest, most shapeless space possible, while just barely holding on to a shred of ordered form, namely the sentence that states the implication, so that both author and reader can share a common plateau of understanding from which together both can venture on the journey into chaos.

What is at stake in *Genèse* is a shift in Serres's writings, not a change of purpose. Usually tilting toward the ultimate aim of explaining the birth of order, the delicate balance between chaos and system is reversed in *Genèse*. Now chaos looms larger, while order remains the foothold that anchors us while we descend into the turbulence. Nothing has really changed in Serres's discourse. Its "object" is still the black box where chaos changes to order. But the emphasis shifts to the observation of a region that edges closer to chaos. This becomes visible in the tropes that now begin to dominate his works. Indeed, the difference in Serrean writing before and after *Genèse* is purely tropical.[10] The arguments in earlier works are carried by models that hide the viscous, perhaps, somewhat, but are nonetheless constructed on the same principle as his later works.

Chaos has always been a participator in Serres's thought constructs. The very fact that the early models' main endeavor is to hide chaos posits its overpowering presence.

The model that galvanizes all discursive elements in *Genèse* is "la belle noiseuse," lexically an adjectival noun that may perhaps, somewhat, approximately, mean: she who makes noise, creates furor and turbulence; she who stirs up trouble, creates smoke screens, makes the ashes and the dust swirl (204) until all contours—including her own—all forms and order have broken down, the foundations dissolved, the abyss laid open. At the same time the creature is beauty itself, thereby aesthetically validating what appears to be a string of negative connotations. The source for this model is Balzac's short story *Le chef-d'oeuvre inconnu* in which two painters, Poussin and Porbus, are allowed the singular privilege of viewing a secret work by their greatly admired master, Frenhofer, a painting of a woman by the name of Catherine Lescault. What their eyes behold, however, is a canvas covered with an incomprehensible jumble of colors and lines. Endlessly attempting to perfect his masterpiece, the artist has destroyed all recognizable forms and outlines; an excess of perfected form has annihilated perfection and created chaos. In vain do Porbus and Poussin search for some visual meaning. They are confronted with nonmeaning, except for one corner of the canvas, forgotten in the artist's increasingly feverish work, where they recognize the exquisite outlines of a woman's foot, a trace of what would have been.

This unknown masterpiece is "baptized 'La Belle Noiseuse'" (*Genèse* 31),[11] a decisive Serrean gesture that the reader needs to appreciate fully in order to read the rest of *Genèse* under the equally significant title of "Noise." The text creates a double homology that cuts across several levels of representation: the reader (of the back cover) mesmerized by Venus rising from the turbulent waters is to the author, venturing into abyssal furor, what the two painters in their eager search for recognizable forms on Frenhofer's canvas are to the latter's ecstasy before the chaos of colors and destroyed forms, what the "chef-d'oeuvre inconnu" is to "la belle noiseuse," what "Genèse" is to "Noise." The delicate foot of Catherine Lescault/Venus is the passage that holds together the separate, antinomic elements of the homology. By resurrecting Catherine Lescault's surname, Serres does not make the unknown masterpiece "known," give it form or render it intelligible. Rather "noiseuse" indicates "noise" which indicates chaos. "La belle noiseuse" is thus a trope that frees the masterpiece from the search for forms, from the "connu"—meaning or concept—which conventionally determines the "inconnu."

"La belle noiseuse" frees chaos from any antithetical determination in relation to order. Not quite but "somewhat" like "noise," it is a tropical space in which order and chaos meet, become one: "Vénus turbulente," says the back cover. "'La Belle Noiseuse' . . . is the black box that encompasses [comprend], implies, envelops, that is, entombs all profiles, appearances, all representations, finally the work itself" (41). Serres insists on the name "beautiful trouble-/noise-maker" because it "is not a painting, it is the 'noise' (turbulence) of beauty, the naked 'multiple,' the many waters from which arises or does not arise, as the case may be, beautiful Aphrodite" (40). Hermes and the parasite projected exactly the same tropical space in Serres's discourse, but the desire to see Venus rising sets logical limits for these tropes. "We always see Venus without the ocean, or the ocean without Venus, we never see the 'physical' surge forth, 'anadyomène,' from the metaphysical" (40).

The back-cover discussion on "genèse" and "noise" is therefore more than an enlightened and enlightening postscript to a difficult and at times opaque discussion in *Genèse*. It may be read as a word of caution for any reading of *Genèse* that overshadows "noise." For the genesis of Venus is a singular event, albeit an extraordinary one, as we learn aesthetically from Botticelli and others. But "as soon as a phenomenon manifests itself, it leaves primal noise; as soon as something appears, surges forth and blossoms [point], it reveals itself by hiding [voilant] the 'noise.' Therefore the primal noise is not part of phenomenology" (33). We see Venus, clearly and eagerly, and welcome her, because she represents the beauty that is inherent in order and logic as they are perceived in our cosmology. This explains why we do not see "La Belle Noiseuse" and why we call it in our blindness—together with Balzac—the unknown masterpiece, because what it "represents" cannot be represented.

In spite of its title, *Genèse* is Serres's attempt to unveil what has always been hidden in its starkness and nudity, what even Hermes and the parasite could not achieve: Venus *and* the chaotic waters, Venus *in* the ocean, the ocean clinging to Venus, chaos at the core of order and order within chaos. Therefore Serres must call her "la belle noiseuse," a creature that is neither Venus nor the primal noise/turbulence/waters. She is Venus to whom viscidly cling the waters as she emerges and merges, in a thousand births, in multiple birthings, possibilities, and disappearances. She is equally the turbulent abyss that harbors limitless multiples and from which boils up, now and then, the exquisite form of one of those possibilities. She is the beautiful manifestation of order that is in chaos and chaos that is beautiful in its innumerable possibilities of phenomenal manifestations. She is Frenhofer's canvas covered with chaotic colors that "may or may not, as the case may be," coalesce into a beautiful form.[12]

"La belle noiseuse" becomes Serres's dominant trope for his subsequent texts, because she multiplies—chaotically—the possibilities he already had discovered in Hermes and the parasite, whose very marketability, however, as theoretical models in the arena of scientific debate as well as in the humanities, all too often reduces them to binary, conceptual dimensions. "La belle noiseuse" cannot rely on a mythophilosophical background for the kind of conceptual grid the many-talented Hermes operated by. She has no "sitos" for referential refuge as does the parasite, the "para-sitos." She cannot be marketed and will never be a theoretical model, because beautiful chaos cannot be reduced to unitary concepts and binary systems. *Genèse* and all post-*Genèse* works could be linked under the common heading "Noise," I, II, III, etc., in a kind of parallel development to the series of *Hermes* texts. But mindful of his readers' advice, Serres chooses titles that merely allude to the "multiple" (foundations, statues, the senses, Hermaphrodite) and skirt the chaotic, but may always be recuperated by the reader through conceptual references. But never again do they have the theoretical power of a Hermes or the parasite; they can no longer relate order and disorder in an antithetical system; none—to my knowledge—have become models in a kind of tropological grid similar to the one Hermes and the parasite created. Not because they are weak, but because their texts reveal that they function within the chaotic tropographic space of "la belle noiseuse" and therefore cannot operate as models for systems that explain the input and output of a given black box. Rather, they are tropes for the black box itself, not explaining but "perhaps, somewhat" implying "la belle noiseuse."[13]

Moreover, *Genèse* effectively deconstructs the models that previously provided a rigorous grid for Serres's thought system and paradoxically thus reveals their latent tropographic power that led Serres on his path to "la belle noiseuse" in the first place. Given that Frenhofer's masterpiece is a space of parasites ("noise" and noise) that change the portrait of Catherine Lescault into "La Belle Noiseuse," Serres shows that any parasite is always parasitically abused by a stronger or louder parasite until the whole parasitic system crumbles under its own devouring tendencies (125). "La belle noiseuse" is the trope par excellence, the "model" that deconstructs all models that sustained Serres's previous investigations of epistemological systems. Thus we see a circuitous meandering through a graphic, not a logical space, that appears both closed and open-ended, and is implied in *Détachement* as a "new knowledge [un nouveau savoir]" (175) free of binary constraints: "I try to free myself from the hell of dualistic thinking" (*Genèse* 210) which always leads to the tyranny of the "One." In the attempt to discourse on chaos, Serres rejects above all the dualism order/disorder, because it is based on the unitary concept of order and the exclusion of all other possibilities, bifurcations, multiplicities. Instead "we should definitely keep the term 'noise' [primal noise, furor, tumult], the only positive word to express a state which we indicate exclusively with negative terms, e. g. disorder" (43).[14]

Serres proposes the introduction into philosophy of the "concept of chaos" (*Genèse* 161–62). But to do so means to abandon the primacy of traditional philosophical methods:

> To attempt to think and to produce supposes taking risks, supposes living precisely in a flux that remains outside the classification process of encyclopedias. . . . All our classified reasoning, all our codices, habits, and methods induce us to speak [of chaos] as an outsider or by negation: outside the law and nonmeaningful. But I speak a positive chaos [Mais je dis le chaos positif]. (161–62)

Therefore, what we call his dominant trope changes the negative term "unknown masterpiece" to a positive one, "la belle noiseuse." "Positive" must be understood, however, purely as a strategic term and not as a conceptual one, since "noiseuse" indicates chaos and thus falls outside the binary opposition of affirmation/negation.

In fact, the full extent of the chaotic power that shapes Serres's tropes in and after *Genèse* becomes obvious when one considers the excellent anthology of some of Serres's pre-*Genèse* writings, *Hermes: Literature, Science, Philosophy*. In their introduction, the editors can still state that "Serres's theoretical program is encyclopedic" in its display of interferences and interreferences (xxix). Of course Harari and Bell are guided by Serres himself who, in *Hermes*, IV, e.g., speaks of the "theory of noise" (Harari and Bell 76), and in *Hermes*, I, uses "noise" as a theoretical model: "Following scientific tradition, let us call '*noise*' the set of interference phenomena that become obstacles to communication" (Harari and Bell 66). But in a radical subversion, portraying allegorically the deconstructive gesture expressed theoretically by Vincent Leitch in the dictum "Repeat and undermine" (177), "la belle noiseuse" explodes the encyclopedia so that its referential dimensions become unlimited multiples, that is, chaotic. "Noise" has become primal noise ("bruit de fond") and is "perhaps the depth of being" (*Genèse* 32). The "noise" that becomes a phenomenon is Venus rising out of the ocean: Hermes and the parasite are theoretical models for the multiple expressions of her birth that constitute our episteme. But "la belle noiseuse" is "Vénus turbulente," chaotic fullness of phenomena, chaos (turbulence) and order (Venus) intertwined and "modeled" by Balzac a full century before chaos theory became a serious object for modern systematic research.

The effect of the "chaotic model," as we very cautiously propose to call "la belle noiseuse," is particularly poignant in the last chapter of *Genèse*, in which Serres uses the allegorical story of the tower of Babel to demonstrate the shift that, alone, will guide us into the new knowledge of "chaos positif." Babel is the figure for the many systems we build. "We even conceive of a general theory of systems, like a universal metasystem reaching the sky. Let us call this great effort a constructivist model" (199). We always think of ruined Babel in terms of failure and insist on rebuilding little Babels to claw our way up out of the ruins of non-meaning, non-sense, chaos. But "Babel is not a failure; only when the tower lies in ruins do we begin to comprehend that one must understand without concepts" (200). Not in the magnificent tower touching the heavens, but in the chaotic heap of stones and in the "strange clamor of confused tongues" (200) do we recognize "la belle noiseuse." "Babel is a non-integrable multiplicity" (200); Babel constructed, or the mere theory of Babel completed, is always a unitary construct based on an "axiom of closure, whereas there is never anything but multiplicity" (204), that is, innumerable possibilities, variabilities, and noise that offers multiple verbal forms. Instead of regretting the loss of the unitary construct—the concept as the basis for all our knowledge—we must understand that "the tower of Babel reverses itself [collapses], and so does the meaning of its text" (203).[15] In this way, we begin to read the story of Babel in a non-Genesis fashion: "Non-completion does not mean ruinous residue or failure, but is the primary status of all things" (203). Neither positive nor negative, this view of chaos is offered within a text dominated by "la belle noiseuse" and deals in particular with noise, din, the confusion of language. Not the tower as the paradigm of a systematic, unique, unilingual solution, but the chaotic ruins and the cacophony of incomprehensible sounds are the "ordinary conditions" of our Lebenswelt (203). Not Frenhofer's wonderful paintings displayed in his studio, but the confusion of strange outlines and colors on a hidden masterpiece is "the formless fountain of all forms" (39). Balzac represses this knowledge behind the "unknown"; Serres resurrects it by "reversing the text," baptizing it "La Belle Noiseuse."

It remains, nonetheless, surprising at first glance that *Genèse*, a text on positive chaos, should end on the apocalyptic model of the tower of Babel, coupled with an injunction to the philosopher to climb down from the height of his logical constructs and descend into the netherworld of formless shadows: "At least once in his life . . . he must refuse to sit on the rock [that caps the abyss] and avoid discourse in the manner of unified systems that are deaf to the primal noise of the multiple" (204).

Thus at the end of *Genèse*, the quest for the multiple acquires a moral aspect which Serres amplifies in *Rome, le livre des fondations* by illustrating the multiple in a concrete historical setting ("faire voir dans le concret [les] multiplicités" [back cover]). Rome is his historical model for a rationally ordered system or a perfectly unified concept, and *Rome, le livre des fondations* is the gradual descent from the height of the Roman tower of Babel into the confusion of its multiple foundations. Serres's "continuous and free-wheeling" (16n) reading of the first book of Livy's *History of Rome* is a marvelously creative archaeological dig that uncovers layer upon layer of multiplicities—be they "Roman street mobs in uproar, Roman legions swarming over the countryside, peasants busy during the harvest seasons, cattle grazing along the river, enemy cavalry charging, forces everywhere, clamor, or acclamations" (*Rome* back cover). Visibly outlined in the table of contents, these multiplicities indicate chaotic states or situations, until the journey into chaos reaches the last chapter (the last stage of the reversal of the tower), entitled "Multiplicity at Peace."

The thoughtful reader quickly becomes aware of the terror-filled tension that characterizes this text. Serres himself tells of a trembling that gripped him at the thought of applying rigorous, scientific argument to an investigation of the unstable fluctuations of time. The text becomes the site where the theoretical models of Hermes and the parasite do battle with "la belle noiseuse." The question Serres repeats over and over again is this: How does Rome, the epitome of order and logic, and the paradigm of organized unity, evolve out of multiplicities that teem with tumultuous variabilities? The answer is terrifying: at each historical layer of chaotic conditions, the steps taken to organize the entity called Rome are accomplished by excluding the third man, that is, the "middle," by eliminating all multiple possibilities except one which becomes the rational "solution." Rome is founded in chaotic multiplicities, but built on violence, murder, death, and rivalry. Each time Serres arrives at this conclusion, he feverishly digs deeper to an earlier founding, only to find the same deadly struggle repeated. In its cathartic outpouring, *Rome* records the bitter fruit of the birth of order out of chaos. The beauty of Venus can also become the tyranny of Rome.[16] Although latent in the *Hermes* series, and certainly in *La naissance de la physique* and *The Parasite*, this insight is here arrived at, not through theoretical models, but in the wake of reflections on the "chaotic model" of "la belle noiseuse." The shift in tropes, away from a tropology and towards a tropography[17] in which "figures of speech" allow us to enjoy almost limitless variabilities, has implications for the moral aspect of the struggle between chaos and order. In *Rome*, Serres takes issue with the looming possibilities of endless violence inherent in the triumph of order which makes us forget to "reverse the text now and then" and to descend into the ocean of multiples, in order to refill our "fountain" of possibilities.

> Our theories lack this fluent field, this moving vastness, and most of all the multiple that freely stirs and swirls there. It is always captured. The multiple is captured. It is captured by the "one," that is, one concept or one synthesis. It is captured by an individual or individuals, in which case it is power or representation, tragedy or polity. Livy's stories tell the capture of the multiple by the One. . . . That capture is also our history. (238)

What *Genèse* bids the philosopher do, *Rome* shows to be everyone's responsibility: the journey into chaos, perilous as it may be, must be undertaken in order to regain the lost wisdom, the richness of the multiple "fragile formations" of our episteme out of the viscous ocean of chaos, so that we free ourselves from the tyrannical stranglehold of exclusionary systems of thought and from the paucity of binary solutions. Yet, Serres discovers in Livy's text one episode that recounts a multiplicity at peace, on a day when no competitive actions are taking place. That day, "no hero was proclaimed, no concept nor unity created, there was pure multiplicity" (*Rome* 281). On the day of rest—respite from violence and terror—the "multiple as such" gushes forth, "like a geyser through a fissure. . . . [It is] the primal matter of history, its conditions, its reality" (281) before it becomes the prehistory and the history of Rome. "La belle noiseuse" thus reappears in her beauty at the end of *Rome*.

Détachement echoes some of the projections of *Rome*, but without the "trembling" that marks the latter. It is a text of resignation, of nostalgia for our innocence lost, and is perhaps the most personal of Serres's works to date. Subtitled "Apologue," it is a mytho-allegorical tale of the philosopher's errantry among the multiples, his plunge into the "noise." Vacillating between hope and despair, it recounts the bittersweet process of detachment from the rival discourses of both the sciences and the humanities, because *Rome* had laid bare the murderous tendencies of the all-pervasive parasite. In the primal turmoil of multiples which is "perhaps the depth of being," Serres searches for that "prodigious knowledge that must once have existed and the traces of which we have lost" (*Détachement* 109). But it is still there and is a "knowledge without death"

(110). Since he cannot rationally arrive at it, Serres implies its existence: "There exists a knowledge outside our knowledge, closed off by our very science and killed off by our very language" (111).[18]

The final and perhaps most eloquent allegory in this text is the story of Diogenes, the barrel-dweller. To illustrate the detachment from our violent episteme, he sheds, one by one, the objects that link him to society and its rivalries, so that he may find "the depth of being," the "new knowledge." We begin to understand "that this mud-encrusted cynic destroys all parasites" (168). When he finally sheds his coat and embraces, naked, the snowman sculptured by some children, his errantry has brought him to the multiple at peace; Diogenes has left all concepts behind; "Diogenes is at peace" (125). What Serres does not say, but what is implied, is the possibility of the snowman slowly melting under the fervent embrace of Diogenes. Covered by the icily viscous waters that bathe his naked body, Diogenes, like "Vénus turbulente," emerges from this text as "la belle noiseuse." But again, Serres sees the multiple threatened by Rome, the One, this time in the allegorical figure of the Great Alexander whose imposing shadow—another trope for any unified concept—falls upon Diogenes. Sadly, Serres concludes that the philosopher's request that the king remove his shadow traps Diogenes once again in the vicious cycle of combat and rivalry:

> Alexander [the concept] reigns over all, including his opponents. His power is so great that none remains who can object. To contradict the king is to belong to the king, to oppose power is to enter into the logic of the powerful. (142)

The sadness at Diogenes' failure[19] to remain "at peace" in the multiple is almost tangible in the last pages of *Détachement*. The very last lines conjure a renewed leave-taking for yet another journey into chaos:

> Leaving, setting sail, disappearing, wandering about, descending slowly into the earth, taking off into the air. . . . Pity on this world! May the new knowledge come! (177)

If one fully comprehends the chaotic implied in Diogenes, his naked body touching the immediately physical, tasting water from his bare hands, seeing the sun ("face au soleil"), certainly smelling the dirt he lives in, it is easy to understand why Serres would choose the sensory as a tropographic space in *Les cinq sens*. Science and language having failed him, he now chooses "that other form of listening which is attuned to 'le donné,' the data of our senses as they map out, each in its specific way, our 'contingencies' with the physical world of which we are a part" (Kavanagh 938).[20] Two points about this text must be emphasized:

1. The descent into chaos is not a journey into a distant and absurd

space; our theoretical constructs have succeeded in creating this illusion. The new knowledge is indeed outside our conceptual world. But *Les cinq sens* illustrates through a series of "chaotic models" ("mélange," circumstances, fluidity, turbulence, the multiple, and especially Ulysses' circumnavigations)[21] within the tropography limitlessly "marked" by the senses, that chaos is the innermost of all human experiences. The descent into chaos is a descent into ourselves as "subjects" who open themselves through the body's windows which are our senses, to the immediate world around us, to "a reality which is always local, multiple, circumstantial and 'mélangé' " (Kavanagh 939). The outside becomes the inside, but this chaotic process also deconstructs the dualism outside/inside itself.

2. Deconstructive processes have, of course, been well established in modern thought.[22] But what sets Serres's "chaotic logic," also named "la logique des circumstances," apart from deconstructive theories, are precisely the absence of theory and the fullness of a fluid blending ("mélange") of subject and object until we truly "understand without concepts." This is the reason why his latest works are rather hastily labeled poetic as opposed to the earlier writings of a supposedly more scientific-philosophical character. It is difficult to pin any label on a work that "speaks positive chaos." Characteristically, a good part of an interview given by Serres in 1986 is devoted to the question of valid distinctions between poet and philosopher (James 789–90). His statement that "logic is inseparable from style" (James 793) alludes to a fateful division in our episteme originating in Aristotle's writings but is also a seminal pronouncement on his own work. For what the *Le Monde* discussion of *Les cinq sens* laments as an endless repetition of metaphors and parables (see note 13), is also a space where tropes do not reveal their proper meaning but multiply (adverbially speaking) inscribe their bifurcations. Instead of a tropology, we prefer therefore to speak of a tropography when defining Serres's discourse on "noise." In other words, we are dealing with a poetics for which Serres, however, rejects any unitary definition. It is a discourse in which traditional logic is only one of many variables, which renders it strange to systemic thought and frightening because of its chaotic traits.

In *Les cinq sens*, "circumstances" is the trope that equals "la belle noiseuse" in richness of Serrean implications and "inventions." The section that carries this trope as its title (308–37) illustrates, in particular, to what extent the "object of this work," the multiple, is mirrored in the act of writing itself: the story of Ulysses' *circum*navigations on the open sea, interrupted now and then by stops (*stances*) providing new experiences, becomes an allegory for "noise," but also for its own tropical nature. For Serres does not reject the "stance," the singular position taken, the unifying locus attained, in short, rational unity, but understands it as one "locality" among the many variables that compose the multifaceted richness, the circum-stances, the Serrean "mélange" of the world we live in and are a part of. Serres argues that Ulysses' "randonnée" is preferable to linear logic as a method of experiencing the world. The "Ulyssean method" is a "route that is oblique, tortuous and complicated" (289), but one which leads to, and is the chaotic encounter with, the multiple: "Circumstances express a multiplicity that is irreducible to unity" (323). Surging from this text, too, is "La Belle Noiseuse," the masterpiece that cannot be reduced to observable forms, units of color or conceptual norms of representation. Here she is generalized into the "logic of circumstances," her exquisite foot being a stance, one of the many stops in the course of Ulysses' vagaries.

This section is equally remarkable for its ability to tie together in a few pages the whole of Serres's enterprise. Subheadings, such as "Logique," "Grammaire," "Statique," "Mécanique céleste," "Thermodynamique," "Zoologie," and "Amour" (on literature), remind the reader of the astounding coherence of his many "livres très pluralistes" (James 789) since *Hermes, I*, while at the same time offering passages that demonstrate the extraordinary rhetorical shift from individual allegorical topics to an allegorical variability that reveals the circum-stantial tropography of our episteme as a whole.

The companion pieces of *Les cinq sens* are *Statues* and, to a lesser degree, *Hermaphrodite*. In *Les cinq sens* Serres's Penelope is the static, the statue around which swirl the many bifurcations of his zigzagging "randonnée" through the sensory. It is a text in which the *subject* is in search of the object in the multiple; Diogenes in search of reality. It is composed of a series of allegorical reflections on the "circum" within the circumstances that have become the tools for Serres's fluid logic. *Statues*, on the other hand, is a text in which the *object* opens up to the subject, the snowman inundating Diogenes with fluid variabilities; it is composed of a series of allegorical reflections on the "stance" within the logic of circumstances. The exquisite foot of "la belle noiseuse" reappears here as "statue." And again, the table of contents is revealing as to the author's approach to positive chaos: the text is a scriptural tower of Babel that *reverses itself* in a movement towards chaos, from the first chapter on a historically recent "statue" to the last where time collapses into dateless mythologies. Here Serres is on a quest for the many statues that "litter" our epistemological landscape. As he slowly descends the statue—

marvelously outlined on the back cover by a trope which can only be said/read to "indicate" Serres's own work!—he arrives again at the chaotic, at the multiple. Now the tropes are statues in the widest sense of the word, including mummies and the wife of Lot, "la belle noiseuse" calcified into a statue of salt.

But the most important aspect of *Statues* is the discovery that the statue/stance/static is not static at all but is itself a tropographic space in which the multiple may be inscribed. Since subject and object are now intertwined, introduced in the twin figure of Diogenes and the snowman, and tropologically defined in the circumstances of *Les cinq sens*, Serres can suppose chaos not only in an objective way, as scientists do when proposing a chaos theory or Serres himself with the trope "noise," but in a way that passes through the subject: now, chaos is indicated by the trope "death" (which he had treated as an "object of his investigations" for the first time in *Hermaphrodite*). The statue is the passage through which we reach death; not violence, murder, and bloody furor, but death that is chaos and the multiple, out of which is born the union of subject and object. Now one understands that Catherine Lescault's beautiful foot, surrounded by chaotic formlessness, is indeed a "stance," but also the trope of a birth: a first experience of subject, object, and death in a primal, chaotic embrace. Like Lot's wife who stands, human cadaver and stone all in one, on the threshold between death (behind her) and life (in front of her) (325), "the statue is a black box: open it and you are staring death in the face" (328). Our history and our sciences are born when we turn, like Lot and his children, away from death (327–28). Serres sees our entire episteme marked by a fatal forgetfulnees. We must look back now and then and remember that our history comes from death (327), our science comes from death (328), that all our knowledge comes from death. The many statues that are part of man's cultural world are reminders (the bubbles rising from the chaotic netherworld) of death as the chaotic birthing place of our conceptual world. It is therefore not surprising that Serres would privilege the mummification rituals of ancient Egypt in this particular text.

In *Hermaphrodite*, based on Balzac's short story "Sarrasine," the investigation of the statue harboring death centers on the mixed bodies ("corps mêlés") of Hermes and Aphrodite. The impact of this slender volume lies less in the local implications of the androgynous than in the relationship that it creates between the trope "death," the early theoretical model of Hermes, and the chaotic model of "la belle noiseuse" under the name of Aphrodite. This relationship is expressed through a trope that is superbly circumstantial, in the Serrean sense of the term: Hermaphrodite, "mixed bodies" whose joint stability is itself undermined by the revelation that its origin is not order, but chaos implied in death. This gives rise in *Statues* to the Serrean implication that death is not a local metaphor but a global condition of the statue, the stable, the stance.

Hermes, the "reconnector" for multiple explanatory systems, has become part of a chaotic model. This suggests, of course, that Serres's work has come full circle. But to say so is to presume that his writing is in fact complete and, more importantly, that it has achieved the status of a philosophical system. These pages have attempted to show otherwise. Serres finds his Penelope in the immediate object, the world which he now "understands without concepts." But the embrace of subject and object, first illustrated by the figure of Diogenes and the snowman, becomes the statue "as such": Hermes revisited. This is the joint that simultaneously ties the circle and breaks it open. The historical descent in *Statues*, down this tropological tower of Babel, brings Serres again to the chaotic, but he now understands it in a different way. "La belle noiseuse" appears here in a phrase that unites the subjective, the objective, and death: "The masterpiece originates in death [Le chef-d'oeuvre vient de la mort]" (105). Hermes revisited is chaotic Hermes, is Hermes originating in death, is Hermes "mêle," is Hermaphrodite. The theoretical model in *Hermes* and *The Parasite*, so eminently marketable, has become a "corps mêlé" which, by its very Serrean "definition," rends any system. Serres's return to Hermes is a bold response to the *Le Monde* reviewer's plaintive cry that he come to the point: by returning to Hermes the author comes indeed to the point, but the "point" turns out to be the limitless space of tropes in which the philosopher must continue his discursive Ulyssean journey. "Genèse" is indeed "noise." Hermes, the borderstone lining the linear paths to order, has become the winged-footed one, the only one of the god's many characteristics Serres recalls at the end of *Les cinq sens*. The god who creates order out of chaos and carves stances out of circumstances, has taken flight in order to follow his own multidirectional odyssey. Serres's logic of circumstances has indeed become inseparable from the tropography of his discourse.

Notes

1. The French noun "noise" is different from the English both in pronunciation (nwaz) and multiple meaning. In the following pages, the French noun will be identified with quotation marks. No marks will be used when the English *noise* is indicated.

2. This definition of the trope is borrowed from Paul de Man's essay "Pascal's Allegory of Persuasion" which examines Pascal's discourse on the relationship between logic and rhetoric in his "Réflexions sur la géométrie en général. De l'esprit géométrique et de l'Art de persuader." In the course of his inquiry, de Man follows Pascal's argument that "not all men have the same idea of the essence of things which [are] impossible or useless to define . . . (such as, for example, time). It is not the nature of these things which I declare to be known by all, but simply the *relationship between the name and the thing*, so that on hearing the expression *time*, all turn (or direct) the mind toward the same entity" (de Man 6). De Man then explicates: "Here the word does not function as a sign or a name, . . . but as a vector, a directional motion that is manifest only as a turn, since the target toward which it turns remains unknown. In other words, the sign has become a trope, a substitutive relationship that has to posit a meaning whose existence cannot be verified, but that confers upon the sign an unavoidable signifying function. The indeterminacy of this function is carried by the figural expression 'turn (or direct) the mind,' a figure that cannot be accounted for in phenomenal terms. The nature of the relationship between figure (or trope) and mind can only be described by a figure. . . . As such, it acquires a signifying function that it controls neither in its existence nor in its direction" (7).

Breaking with traditional definitions of the figurative as a *deviation* from the norm or an *opposite* to the literal, de Man's understanding of Pascal's trope is instead based on a *substitutive relationship*, a *turning toward* something that remains, however, undefined. (See also Gérard Genette's introduction to Pierre Fontanier's *Les Figures du discours*, and Paul Ricoeur's commentary on Fontanier in *La métaphore vive* [63–86].) In the absence of the "something," the trope gains thus enormously in importance and becomes interesting for our reading of Michel Serres.

3. Michel Foucault uses this expression when talking about the "rhetorical dimension of words" in *The Order of Things* (114). The recognition of this dimension in Serres's discourse is crucial for a global understanding of his work.

4. All preceding quotations are my translations of the text on the back cover of *Genèse*. In what follows, all quotations from Serres's texts are my translations, except those from the anthology edited by J. Harari and D. Bell.

5. What Serres expresses here in the allegorically nonrepresentable figure of "Vénus turbulente" is theoretically explored by Paul Ricoeur in a "tropological approach" which would permit the bringing together ("associer") of the Same and the Other in the Similar. "Better yet: in the Analog, it being a resemblance between relationships rather than between simple terms" (*Temps et récit* III: 219). This approach does, in fact, help Ricoeur reduce certain difficulties he encounters trying to solve the aporetic of time. With the appearance of *Genèse*, Serres's discourse increasingly favors tropological approaches with a similar degree of success.

6. In this context, *La naissance de la physique dans le texte de Lucrèce* is a pivotal text. What is considered by many to be Serres's reassessment of the

models of modern science through a series of reflections on Lucretius's didactic poem *De rerum natura*, is also and specifically an expression of his fascination with a historical exclusion of gigantic proportions: the relegation of a seminal scientific discourse, namely Lucretius's text, to the literary-mythological realm, because it posits its origins on stochastic events that cannot easily be systematized. This miscue at the dawn of Western episteme informed all our scientific and philosophical knowledge within the *double limits* of a mechanics of solids and dialectic reasoning, and splintered its totality by exclusionary gestures that produce strife, competition, and poisonous relations among the various fields of expertise.

7. Among Serrean book and chapter titles, this is indeed a rare chapter heading for its composition of "literal" terms. Most are highly figurative and many require a considerable dose of mental agility on the part of the reader to savor their multiple connotations.

8. The disappearance of the "stable system of signification" in the viscous abyss that the introduction turns out to be, is not the only cause for the reader's disquiet. For what he perceives to be tropological models of the "object of this book" becomes simultaneously a subversive text on the trope itself. Not only are the "multiple" and "noise" tropes, but the multiple is, in turn, implied to be the nature of the trope as such. Therefore, the "object of this book" is also an inquiry into the problematic of the trope. Extrapolating from this text, Serres's writing as a whole may be regarded not only as a systemic discourse on the relationship between chaos and order, but also as an allegorical narrative on language "as such" becoming the language of chaos. Serres's use of tropes is therefore not purely heuristic. In his texts they become "chaotic" models of themselves. In this context, the article on Pascal (see note 2) has interesting parallels, when de Man deconstructs the "allegory that is a sequential narration" (12) so that it becomes an allegory that "pretends to order sequentially, in a narrative, what is actually the destruction of all sequence" (23).

Serres's love-hate relationship with language which plays a major (tropical!) role in texts written after *Genèse*, acquires an entirely new perspective, if one keeps in mind what also "bubbles" up in *Genèse*: the aporetic of the trope, or the aporetic that is the trope. It is this reader's hypothesis, rather "implied" than examined for now, that the fascinating interplay between the Greek and French languages, part of the exposition of so many conferences Serres has given and of texts he has written, is an integral element of Serrean tropo-logy, the logic of tropes.

9. *Rome* (87). The baker's "strange" logic of the pliable or the folding-in is also described as the "logique des sacs" in contrast to the "logic of boxes," a figure for classical logic (180). However strange it may be, it has historical roots in what Eric Charles White describes as "a spirit of improvisationality" to be found, among others, in Montaigne and, of course, Lucretius ("Negentropy," chapter 12 this volume).

10. But saying "pure difference of tropes" is, of course, ineluctably implying a multiple uproar or "noise" of differences when applied to Serres's works.

11. In earlier versions of "Le chef-d'oeuvre inconnu," Frenhofer calls his beloved "ma belle noiseuse," but Balzac represses her surname in the definitive version, because its connotations are inconsistent with the artist's chaste passion. A morally "disorderly" name is suppressed in order to suppress a logical disorder within the narrative discourse. Serres's "baptism" reverses this ordering process.

12. Balzac's story of the "unknown masterpiece" is a fine example for one of Serres's major topics: literature projecting scientific discoveries or theories, before the scientific community per se has recognized and incorporated them into its discourse. Frenhofer's chaotic canvas is a "complete" depiction of chaos theory such as it is being formulated by scientists of our era. (For the nonscientist, a very cursory description of the present-day state of scientific research into chaotic "phenomena" may be found in "The Mathematics of Mayhem." He can also consult Gleick, *Chaos*.) The story states explicitly that the artist expended great effort and time on his masterpiece, paying attention to the smallest detail and fearing that it might be lacking in even finer ones. He is describing, of course, an example of fractal geometry. His two disciples, on the other hand, see nothing but "a kind of formless fog" (Balzac 436, my trans.), because their vision is Euclidean. Similarly, the exquisite nude foot in one corner of the canvas is Balzac's literary expression for what today are called "strange attractors, erratic patterns [that] have become emblems of chaos theory. They represent the unsuspected order in chaos" ("The Mathematics" 89).

Another example is illustrated in Thomas P. Weissert's essay "Representation and Bifurcation: Borges's Garden of Chaos Dynamics" which examines key concepts of modern dynamics and finds them fictionally elaborated almost half a century ago in one of J. L. Borges's "Ficciones."

13. The difference between explication and implication is crucial in Serrean discourse. Ignoring it leads Thomas Ferenci, e.g., to conclude his otherwise very perceptive review of *Les cinq sens* with the observation that "*Les cinq sens* is above all a series of allegories and parables that function in a metaphoric sense. . . . Metaphors without a doubt stimulate the imagination, but they indicate, at best, the directions to be explored. Michel Serres has already opened these avenues. One would like to see the author hasten his step a little" (p. 30, my trans.). The reviewer has a clear sense of Serres's discourse, yet misses the point. In this, he resembles the readers who chose "Genèse" over "Noise."

14. The passion for rationalist thought has caused the French to forget the noun "noise," while the English language has retained at least one denotation, noise (*Genèse*, 31–32).

15. It is difficult to convey in English the analogy between the tower falling and the text turning its meaning upside down. In the original "la tour de Babel se renverse, et le sens de son texte"; physical collapse and reversal of meaning become one through the verb "se renverser."

16. Rome is the eminently successful historical concretization of the mythological function Mars fulfills in opposition to Venus in Lucretius's poem, the historical significance of which Serres had already recognized in *La naissance de la physique*. Eric White in this volume acknowledges the leap from myth to

history when he writes in his discussion of this work: "We worshippers of Mars deliver ourselves over to death at the very moment we seek to evade our mortality" ("Negentropy," chapter 12 this volume).

17. Our preference for a tropography over a tropology signals the same distinction that Serres makes between an iconography and a mere "scenographie." The difference is that of "multiplicity of possible profiles and the integration of all possible horizons," while a scenography represents a definite number of phenomena (*Genèse* 41). In *Rome*, this difference becomes even more pronounced in the figure of the writing of the "boustrophedon," a graphic that "does not make choices" but "transforms binary alternatives into a continuum" (25–27).

18. Already recognizing the contours of this "outside" knowledge in Serres's earlier writings, Eric White stresses its "emancipatory potential . . . as a source of vitality and renewal" ("Negentropy," chapter 12 this volume).

19. Eric White expresses a similar "disheartening possibility" when he proposes that Pynchon's parasitic heroine Oedipa, instead of revitalizing a closed system, may "already [be] a victim of the system's drive for total control" ("Negentropy," chapter 12 this volume).

20. The shift in Serres's strategy is dramatic. In *The Parasite* he still held out hope for finding revitalizing "inventions" in literature which Eric White paraphrases as a "space of resistance" against modern technological domination.

21. To express Ulysses' wanderings, Serres very aptly uses the term "randonnée" which means "a rambling long walk" and which shares the same etymological root as the English "random." As a "random walk," Ulysses' "randonnée" is thus a well chosen "chaotic" model.

22. The deconstruction of the dualism inside/outside is particularly effective in Jacques Derrida's *De la grammatologie* (42–108), where traditional linguistics is the target. An example of a more recent deconstructive reading is Barbara Johnson's *A World of Difference*. The author states that her "understanding of what is most radical in deconstruction is precisely that it questions [the] basic logic of binary opposition, but not in a simple, binary, antagonistic way" (12). Resembling Serres's discourse, Johnson's text becomes even more Serrean when she formulates what "deconstruction shows, [namely] that there is *something else involved* [that] ceaselessly escapes the mastery of understanding and the logic of binary opposition by exhibiting some 'other' logic one can neither totally comprehend nor exclude" (13).

Works Cited

Balzac, Honoré de. *La Comédie Humaine*. Vol. 10. *Etudes philosophiques* Paris: Gallimard, 1979. 413–38.

De Man, Paul. "Pascal's Allegory of Persuasion." In *Allegory and Representation*. Ed. Stephen J. Greenblatt. Baltimore: Johns Hopkins University Press, 1981. 1–25.

Derrida, Jacques. *De la grammatologie*. Paris: Minuit, 1967.

Ferenzi, Thomas. "Une philosophie du chahut." Review of *Les cinq sens*, by Michel Serres. *Le Monde* 15 November 1985: 27–30.

Foucault, Michel. *The Order of Things*. New York: Vintage Books, 1973.

Genette, Gérard. Introduction. *Les Figures du discours*, by Pierre Fontanier. 1830. Paris: Flammarion, 1968. 5–17.

Gleick, James. *Chaos: Making a New Science*. New York: Viking, 1987.

Harari, Josué V., and David F. Bell, eds. *Hermes: Literature, Science, Philosophy*, by Michel Serres. Baltimore: Johns Hopkins University Press, 1982.

James, Geneviève. "Entretien avec Michel Serres." *The French Review* 60 (1987): 788–96.

Johnson, Barbara. *A World of Difference*. Baltimore: Johns Hopkins University Press, 1987.

Kavanagh, Thomas M. Review of *Les cinq sens*, by Michel Serres. *MLN* 101, no.4 (1986): 937–41.

Leitch, Vincent B. *Deconstructive Criticism*. New York: Columbia University Press, 1983.

"The Mathematics of Mayhem." *The Economist* 8 September 1984: 87–89.

Ricoeur, Paul. *La métaphore vive*. Paris: Seuil, 1975.

———. *Temps et récit*. Vol.III. Paris: Seuil, 1985.

Serres, Michel. *Les cinq sens*. Paris: Grasset, 1985.

———. *Détachement*. Paris: Flammarion, 1983.

———. *Genèse*. Paris: Grasset, 1982.

———. *Hermaphrodite*. Paris: Bourin, 1987.

———. *Hermes* I, II, III, IV, V. Paris: Minuit, 1969–80.

———. *La naissance de la physique dans le texte de Lucrèce: Fleuves et turbulences*. Paris: Minuit, 1977.

———. *The Parasite*. Trans. Lawrence R. Schehr. Baltimore: Johns Hopkins University Press, 1982. Trans. of *Le Parasite*. Paris: Grasset, 1980.

———. *Rome, le livre des fondations*. Paris: Grasset, 1983.

———. *Statues*. Paris: Flammarion, 1987.

Weissert, Thomas P. "Representation and Bifurcation: Borges's Garden of Chaos Dynamics." Chapter 10 this volume.

White, Eric Charles. "Negentropy, Noise, and Emancipatory Thought." Chapter 12 this volume.

Contributors

Maria L. Assad, an assistant professor in French at the State University College at Buffalo, has published articles on Racine, Mallarmé, Artaud, and Barthes. She is the author of *La fiction et la mort dans l'oeuvre de Stéphane Mallarmé* (1987) and translator for Raymund Schwager's *Must There Be Scapegoats?* (1987).

Istvan Csicsery-Ronay, Jr., is associate professor of English and World Literature at DePauw University.

Sheila Emerson, assistant professor at Tufts University, has published articles on Byron, Ruskin, and other nineteenth-century writers. She is at work on a book entitled *Ruskin: The Genesis of Invention.*

N. Katherine Hayles, Carpenter Professor of English at the University of Iowa, holds advanced degrees in chemistry and English. She is the author of *The Cosmic Web: Scientific Field Models and Literary Strategies in the Twentieth Century* (1984) and *Chaos Bound: Orderly Disorder in Contemporary Literature and Science* (1990).

Linda K. Hughes is associate professor of English at Texas Christian University in Fort Worth, Texas. She is the author of *The Manyfacèd Glass: Tennyson's Dramatic Monologues* (1987) and, with Michael Lund, coauthor of *The Victorian Serial* (forthcoming). Her articles on Victorian literature include "Turbulence in the 'Golden Stream': Chaos Theory and the Study of Periodicals," *Victorian Periodicals Review* (1989).

Kenneth J. Knoespel is associate professor and Acting Head of the School of Literature, Communication, and Culture at Georgia Institute of Technology. In addition to his articles on medieval and Renaissance science and hermeneutics, he has published a monograph on late medieval Ovidian commentary, *Narcissus and the Invention of Personal History*, and his book *Newton and the Failure of Messianic Science* (with Robert Markley) is forthcoming.

Michael Lund, professor of English at Longwood College in Virginia, is the author of *Reading Thackeray* (1988) and numerous articles on Victorian fiction.

Robert Markley is associate professor of English at the University of Washington, editor of *The Eighteenth Century: Theory and Interpretation* and series editor of *The Series in Science and Culture* for the Oklahoma Project for Discourse and Theory. He is the author of *Two-Edg'd Weapons: Style and Ideology in the Comedies of Etherege, Wycherley, and Congreve* (1988), *Fallen Languages: Representation, Science, and Belief in England, 1660–1740* (forthcoming), and coauthor (with Kenneth Knoespel) of *Newton and the Failure of Messianic Science* (forthcoming).

Adalaide Morris is professor of English at the University of Iowa. She is the author of *Wallace Stevens: Imagination and Faith*, a co-editor of *Extended Outlooks: The Iowa Review Collection of Contemporary Women Writers*, and the editor of the H.D. centenary issue of *The Iowa Review*. In addition to her work on H.D., she has also published essays on Emily Dickinson, Adrienne Rich, and the contemporary American canon.

William Paulson, associate professor of French at the University of Michigan, Ann Arbor, is the author of *The Noise of Culture: Literary Texts in a World of Information* (1988). His study of Flaubert's *Sentimental Education* will be published in the Twayne's Masterwork series this year.

David Porush is associate professor of literature and science at Rensselaer Polytechnic Institute, where he has been teaching since 1981. He is the author of *The Soft Machine: Cybernetic Fiction* (1985) and *Rope Dances* (1979), a collection of short stories.

Peter Stoicheff, an associate professor of English at the University of Saskatchewan, Canada, is the author of articles on modern American poetry and fiction, and on closure in the modern long poem.

Thomas Weissert is a student of natural philosophy and physics at the University of Colorado. He is currently engaged in an interdisciplinary study of the physics, philosophy, and history of modern dynamics and chaos theory. With his involvement in the Society of Literature and Science, he has discovered people whose minds have recovered from the two-culture delusion.

Eric Charles White teaches twentieth-century literature and critical theory in the English Department of the University of California at Santa Barbara. He is the author of *Kaironomia: On the Will-to-Invent* (1987).

Index